Sustainable Development and Management of the Shallow Subsurface

The Geological Society of London
Books Editorial Committee

Chief Editor
Bob Pankhurst (UK)

Society Books Editors
John Gregory (UK)
Jim Griffiths (UK)
John Howe (UK)
Howard Johnson (UK)
Rick Law (USA)
Phil Leat (UK)
Nick Robins (UK)
Randell Stephenson (UK)

Society Books Advisors
Eric Buffetaut (France)
Jonathan Craig (Italy)
Tom McCann (Germany)
Mario Parise (Italy)
Satish-Kumar (Japan)
Gonzalo Veiga (Argentina)
Maarten de Wit (South Africa)

Geological Society books refereeing procedures

The Society makes every effort to ensure that the scientific and production quality of its books matches that of its journals. Since 1997, all book proposals have been refereed by specialist reviewers as well as by the Society's Books Editorial Committee. If the referees identify weaknesses in the proposal, these must be addressed before the proposal is accepted.

Once the book is accepted, the Society Book Editors ensure that the volume editors follow strict guidelines on refereeing and quality control.

More information about submitting a proposal and producing a book for the Society can be found on its web site: www.geolsoc.org.uk.

It is recommended that reference to all or part of this book should be made as follows:

De Mulder, E. F. J., Hack, H. R. G. K. & Van Ree, C. C. D. F. 2012. *Sustainable Development and Management of the Shallow Subsurface*. Geological Society, London.

Sustainable Development and Management of the Shallow Subsurface

E. F. J. DE MULDER
Earth Science Matters Foundation, The Netherlands

H. R. G. K. HACK
University of Twente, Faculty of Geo-Information Science and Earth Observation, The Netherlands

and

C. C. D. F. VAN REE
Deltares, Department of Scenarios and Policy Analysis, The Netherlands

2012
Published by
The Geological Society
London

THE GEOLOGICAL SOCIETY

The Geological Society of London (GSL) was founded in 1807. It is the oldest national geological society in the world and the largest in Europe. It was incorporated under Royal Charter in 1825 and is Registered Charity 210161.

The Society is the UK national learned and professional society for geology with a worldwide Fellowship (FGS) of over 10 000. The Society has the power to confer Chartered status on suitably qualified Fellows, and about 2000 of the Fellowship carry the title (CGeol). Chartered Geologists may also obtain the equivalent European title, European Geologist (EurGeol). One fifth of the Society's fellowship resides outside the UK. To find out more about the Society, log on to www.geolsoc.org.uk.

The Geological Society Publishing House (Bath, UK) produces the Society's international journals and books, and acts as European distributor for selected publications of the American Association of Petroleum Geologists (AAPG), the Indonesian Petroleum Association (IPA), the Geological Society of America (GSA), the Society for Sedimentary Geology (SEPM) and the Geologists' Association (GA). Joint marketing agreements ensure that GSL Fellows may purchase these societies' publications at a discount. The Society's online bookshop (accessible from www.geolsoc.org.uk) offers secure book purchasing with your credit or debit card.

To find out about joining the Society and benefitting from substantial discounts on publications of GSL and other societies worldwide, consult www.geolsoc.org.uk, or contact the Fellowship Department at: The Geological Society, Burlington House, Piccadilly, London W1J 0BG: Tel. + 44 (0)20 7434 9944; Fax + 44 (0)20 7439 8975; E-mail: enquiries@geolsoc.org.uk.

For information about the Society's meetings, consult *Events* on www.geolsoc.org.uk. To find out more about the Society's Corporate Affiliates Scheme, write to enquiries@geolsoc.org.uk.

Published by The Geological Society from:
The Geological Society Publishing House, Unit 7, Brassmill Enterprise Centre, Brassmill Lane, Bath BA1 3JN, UK

The Lyell Collection: www.lyellcollection.org
Online bookshop: www.geolsoc.org.uk/bookshop
Orders: Tel. + 44 (0)1225 445046, Fax + 44 (0)1225 442836

The publishers make no representation, express or implied, with regard to the accuracy of the information contained in this book and cannot accept any legal responsibility for any errors or omissions that may be made.

© The Geological Society of London 2012. No reproduction, copy or transmission of all or part of this publication may be made without the prior written permission of the publisher. In the UK, users may clear copying permissions and make payment to The Copyright Licensing Agency Ltd, Saffron House, 6–10 Kirby Street, London EC1N 8TS UK, and in the USA to the Copyright Clearance Center, 222 Rosewood Drive, Danvers, MA 01923, USA. Other countries may have a local reproduction rights agency for such payments. Full information on the Society's permissions policy can be found at: www.geolsoc.org.uk/permissions

Publishing Disclaimer

General disclaimer
No responsibility or liability is assumed by the Geological Society of London or any copyright owner for any injury or damage to persons or property as a consequence of the reading, use or interpretation of its published content. Whilst every effort is made to ensure accuracy, the Geological Society of London, the authors, Editors and copyright owners cannot be held responsible for published errors. The views or opinions expressed do not necessarily reflect views of the Geological Society of London or copyright owners. Inclusion of any advertising material does not constitute a guarantee or endorsement of any products or services or the claims made by any manufacturer.

Safe working practices
The Geological Society of London endeavours to ensure that all published images and descriptions of working practices contribute to the objectives of the individual papers in which they appear. The inclusion of such content should not be taken as an endorsement of the practices depicted or described by The Geological Society of London, the authors, Editors, copyright owners or any other person or body.

British Library Cataloguing in Publication Data

A catalogue record for this book is available from the British Library.
ISBN 978-1-86239-343-1

Distributors

For details of international agents and distributors see:
www.geolsoc.org.uk/agentsdistributors

Typeset by Techset Composition Ltd, Salisbury, UK
Printed by CPI Antony Rowe, Chippenham, UK

This publication is dedicated to and in honour of **Dr. Wissam Al-Hashimi**, one of the contributors to the First International Symposium on Sustainable Development and Management of the Subsurface, who was kidnapped and brutally murdered in Baghdad, Iraq, August 2005. Dr. Al-Hashimi served the Geosciences in many capacities, one of them as a Vice-President of the International Union of Geological Sciences, from 1996–2002.

Contents

Preface ... xii
List of Abbreviations ... xiii
Acknowledgements ... xv

Chapter 1 Introduction ... 1
1.1. Perception and reality ... 1
1.2. Definitions ... 1
1.3. Options for the subsurface ... 1
1.4. The book ... 1
1.5. Earth scientists and the International Year of Planet Earth ... 2

Chapter 2 Historic use ... 3
2.1. Extraction ... 5
    2.1.1. Earth materials ... 5
    2.1.2. Renewable resources ... 5
    2.1.3. Non-renewable resources ... 6
        2.1.3.1. Base metals (copper, lead, zinc and tin) ... 6
        2.1.3.2. Iron ... 10
        2.1.3.3. Precious metals ... 11
        2.1.3.4. Rare metals ... 12
        2.1.3.5. Construction materials ... 12
        2.1.3.6. Chemical materials ... 12
        2.1.3.7. Fuel minerals ... 12
    2.1.4. Impact of mining ... 14
2.2. Subsurface space and infrastructure ... 15
    2.2.1. Human settlement ... 15
    2.2.2. Storage ... 16
    2.2.3. Stable construction ... 17
    2.2.4. Space for infrastructure ... 20
2.3. Concluding remarks ... 21

Chapter 3 Current use ... 23
3.1. Extraction and storage ... 23
    3.1.1. Renewable resources ... 23
    3.1.2. Non-renewable resources ... 24
    3.1.3. Environmental concern and reclamation ... 29
    3.1.4. Storage ... 32
3.2. Subsurface space and infrastructure ... 34
    3.2.1. Tunnelling ... 35
    3.2.2. Underground space ... 39
3.3. Other functions ... 40
    3.3.1. Bearing capacity ... 40
    3.3.2. Life support ... 41
    3.3.3. The subsurface as an archive ... 41
3.4. Concluding remarks ... 41

Chapter 4 Technical challenges and assets ... 43
4.1. Surface v. underground engineering ... 43
    4.1.1. Mining v. civil engineering ... 43
4.2. Geology and ground materials ... 44
    4.2.1. Rock–soil cycle ... 46
    4.2.2. Cementation and interlocking ... 48
    4.2.3. Igneous rocks and unstable minerals ... 48
    4.2.4. Mass ... 49
4.3. Ground properties ... 49
    4.3.1. Strength, deformation, and failure ... 49
    4.3.2. Intact ground strength ... 50
    4.3.3. Deformation, stress and strain of soil and rock ... 51
    4.3.4. Time effects ... 51
    4.3.5. Discontinuities ... 51
    4.3.6. Groundmass and discontinuities ... 51
    4.3.7. (Ground) fluids and gases ... 54
    4.3.8. Influence of water pressure ... 57
    4.3.9. Weathering ... 58
    4.3.10. Crusts and other 'strong' layers ... 60
    4.3.11. Erosion ... 61
    4.3.12. Swelling ... 62
    4.3.13. Heat flow and heat insulation ... 62
    4.3.14. Biological characteristics ... 62
4.4. Excavation and support ... 62
    4.4.1. Stress around underground excavation ... 62
    4.4.2. Failure modes of an excavation ... 63
    4.4.3. Need for support ... 63
    4.4.4. Stand-up time, time effects, and life times ... 64
    4.4.5. Underground excavation under surface structures ... 65
    4.4.6. Excavation and support techniques ... 65
    4.4.7. Small underground infrastructure, dredging and boreholes ... 66
    4.4.8. Water ... 67
    4.4.9. Dewatering ... 68
    4.4.10. Ground improvement: freezing, grouting, compaction, pre-loading, heating, chemical and biological treatment ... 68
    4.4.11. Water pressure ... 69
    4.4.12. Aggressive or polluted water ... 70
    4.4.13. Impact of excavation method on groundmass ... 70
    4.4.14. Economic viability of excavation methods ... 71
4.5. Examples of excavation in various groundmasses ... 71
    4.5.1. Rocks with high intact rock strength ... 71
    4.5.2. Flexible and rigid support ... 72
    4.5.3. Loose materials and very weak ground ... 72
    4.5.4. Permafrost and permanent snow and ice ... 74
    4.5.5. Mining, existing underground structures, and karst caves ... 75
    4.5.6. Strong and weak elements in groundmass ... 78
    4.5.7. Underground excavations in swelling or squeezing material ... 80

4.5.8.	Excavation in volcanic areas	80
4.5.9.	Excavation in man-made materials	80
4.5.10.	Cut and cover	84
4.5.11.	Interaction between foundations and subsurface	84

4.6. Designing underground structures — 84
- 4.6.1. Classification systems — 84
  - 4.6.1.1. Bieniawski's Rock Mass Rating (RMR) — 86
  - 4.6.1.2. Q-system — 86
  - 4.6.1.3. Mining Rock Mass Rating (MRMR) — 87
  - 4.6.1.4. Geological Strength Index (GSI) — 88
- 4.6.2. Excavation damage — 88
- 4.6.3. Monitoring — 91
- 4.6.4. New Austrian Tunnelling Method (NATM) — 93
- 4.6.5. Analysis of Controlled Deformations (ADECO) methodology — 93
- 4.6.6. Numerical modelling — 93

4.7. Operational risks and geohazards — 95
- 4.7.1. Routine — 95
- 4.7.2. Unexpected ground conditions, data limitation and use of expert knowledge — 96
- 4.7.3. Risk management — 97
- 4.7.4. Geohazards — 97
  - 4.7.4.1. Earthquakes and volcanoes — 97
  - 4.7.4.2. Gas, explosions and spontaneous combustion — 98
  - 4.7.4.3. Biohazards — 98
  - 4.7.4.4. Water, snow and ice — 98

4.8. Concluding remarks — 99

Chapter 5   Information about the subsurface — 101

5.1. Site investigation — 101
- 5.1.1. Site investigation standards and codes — 101
- 5.1.2. Maps with surface features only — 101
- 5.1.3. Subsurface maps — 103
- 5.1.4. Remote sensing — 103
- 5.1.5. Field campaigns — 104
- 5.1.6. Pits, trenches, and inspection holes — 105
- 5.1.7. Boreholes — 105
- 5.1.8. *In-situ* strength and deformation tests — 106
- 5.1.9. Laboratory and simple field tests — 107
- 5.1.10. *In-situ* stress tests — 108
- 5.1.11. Fluid pressure, permeability and flow — 108
- 5.1.12. Geophysics – non-destructive sampling — 110
  - 5.1.12.1. Seismic methods — 111
  - 5.1.12.2. Electro-magnetic methods — 112
  - 5.1.12.3. Geo-electrical methods — 113
  - 5.1.12.4. Magnetic, micro-gravity, gamma-ray, and neutron & gamma–gamma density — 113
  - 5.1.12.5. Geophysical borehole logging — 114
- 5.1.13. Site investigation in areas with a historical record (archaeological sites) or caves — 115
- 5.1.14. Forensic site investigation — 115

5.2. Geological and geotechnical models 116
    5.2.1. Model dimensions 116
    5.2.2. Data formats 118
    5.2.3. Data file formats 119
    5.2.4. Scale, detail, resolution and data density 119
    5.2.5. Temporal data 119
    5.2.6. Computer aided modelling 120
5.3. Data processing and information 125
    5.3.1. Data harmonization and standardization 125
    5.3.2. Data and the internet 125
        5.3.2.1. Geography Markup Language (GML) 125
        5.3.2.2. CityGML 125
    5.3.3. Data sources 125
        5.3.3.1. Small infrastructure 125
        5.3.3.2. Military 126
        5.3.3.3. Topography 126
        5.3.3.4. Public works 126
        5.3.3.5. Archaeology 126
        5.3.3.6. Geology 127
        5.3.3.7. Geotechnology 127
    5.3.4. Geo-databases 127
        5.3.4.1. Information management for civil engineering 128
5.4. Models and uncertainty 128
    5.4.1. Geological–geotechnical subsurface models 128
    5.4.2. Model uncertainty 129
    5.4.3. Time 131
5.5. Visualization 132
5.6. Concluding remarks 132

Chapter 6 Legal aspects, policy and management 133
6.1. Legal aspects 133
    6.1.1. Landownership and spatial planning 134
        6.1.1.1. (Full) ownership 136
        6.1.1.2. Condominium law and cooperative model (multiple ownership for multiple use) 136
        6.1.1.3. Lease and easement, building and planting rights 137
        6.1.1.4. Spatial planning 137
    6.1.2. Natural resources, subsurface construction 138
        6.1.2.1. Mining law, extraction of construction materials 138
        6.1.2.2. Groundwater 144
        6.1.2.3. Subsurface structures and infrastructure 146
    6.1.3. The environment 148
        6.1.3.1. (Subsurface) waste disposal and storage 149
        6.1.3.2. Remediation and reclamation 150
        6.1.3.3. Risks and liabilities 151
    6.1.4. Evaluation 152
6.2. Policy 154
    6.2.1. General policies 154
    6.2.2. Mineral resources 155

6.2.3. (Ground) water policy	157
6.2.4. Transport, infrastructure and planning	157
6.3. Management	159
6.3.1. Sustainable development	160
6.3.2. Management tools	161
6.3.2.1. Land registration systems	161
6.3.2.2. Permitting	161
6.3.2.3. Data-collection	162
6.3.3. Mineral resources	162
6.3.4. Groundwater resources	162
6.3.5. Subsurface structure and infrastructure	163
6.3.6. Comparison at regional and national levels	163
6.4. Concluding remarks	165

Chapter 7 Future use   169

7.1. Future developments and trends	169
7.1.1. Population	169
7.1.2. Urbanization	169
7.1.3. Quality of life	169
7.1.4. Environmental awareness	170
7.1.5. Technological development	170
7.1.6. Future developments	170
7.1.7. Uncertainties concerning underground development	171
7.2. Extraction and storage	171
7.2.1. Mining	171
7.2.2. Energy	172
7.2.3. (Ground) water	172
7.2.4. Storage	172
7.3. Infrastructure and public space	172
7.4. Future subsurface management	173
7.4.1. Geological quality	173
7.4.2. Combined use and planning	173
7.4.3. Urban planning	175
7.4.4. Future underground constructions	175
7.4.5. Non-construction future use of the subsurface	175
7.5. Closing remarks	175
7.5.1. City of the future	178

References	179
Index	201

Preface

This book is a compilation of topics and issues the authors think are required for a Sustainable Development and Management of the Subsurface. It attempts to arrive at a logic structure for dealing with the wide variety of aspects concerning the use of underground space, both time wise and geographically. The book aims to address engineers, urban developers, lawyers, policy makers, insurance professionals, geoscientists, and all others involved in subsurface construction providing an overview of not only technical aspects but also legal, governmental and policy making issues. It is a reference for the general public interested in the Earth science aspects of future cities and its citizens.

It also contributes to the ambitions of the International Year of Planet Earth (2007–2009) aiming at increasing awareness of the public and politicians to a safer, healthier and wealthier society on this planet, and more in particular to the Megacities theme of the International Year.

E. F. J. de Mulder, H. R. G. K. Hack & C. C. D. F. van Ree
March 2012

List of Abbreviations

AGS	Association of Geotechnical and Geoenvironmental Specialists
ASTM	American Society for Testing and Materials
BGR	Bundesanstalt für Geowissenschaften und Rohstoffe
BLM	Bureau of Land Management
BS	British Standard
BTS	British Tunnelling Society
CAD	Computer Aided Design
CCS	Carbon dioxide Capture and Storage
CEN	Comité Européen de Normalisation, European Committee for Standardisation
COB	Centrum Ondergronds Bouwen
DIN	Deutsches Institut für Normung e.V.
EC	European Commission
EEA	European Environment Agency
ESA	European Space Agency
GIS	Geographical Information System
IAEA	International Atomic Energy Agency
IAEG	International Association for Engineering Geology and the Environment
ICMM	International Council for Minerals and Metals
IMF	International Monetary Fund
INTI	Instituto Nacional de Tecnología Industrial
IPCC	Intergovernmental Panel on Climate Change
IPPC	Integrated Pollution Prevention and Control
InSAR	Interferometric Synthetic Aperture Radar
ISO	International Organization for Standardization
ISRM	International Society for Rock Mechanics
ISSMGE	International Society for Soil Mechanics and Geotechnical Engineering
ITA	International Tunnelling Association
ITIG	International Tunnelling Insurance Group
IUGS	International Union of Geological Sciences
IWMI	International Water Management Institute
JMA	Japan Meteorological Agency
LIDAR	Light Detection And Ranging
NAM	Nederlandse Aardolie Maatschappij B.V.
NASA	National Aeronautics and Space Administration
NEN	NEderlandse Norm
OECD	Organization for Economic Cooperation and Development
OGC	Open Geospatial Consortium
RPD	Rijks Planologische Dienst
SEG	Society of Exploration Geophysicists
TCB	Technische Commissie Bodembescherming
TCFE	Technical Council on Forensic Engineering
TBM	Tunnel Boring Machine
UN	United Nations
UNCED	United Nations Conference on Environment Development

LIST OF ABBREVIATIONS

UN ESCAP	United Nations Economic and Social Commission for Asia and the Pacific
UNESCO	United Nations Educational, Scientific and Cultural Organization
UNEP	United Nations Environmental Program
USGS	United States Geological Survey
VROM	Volkshuisvesting Ruimtelijke Ordening en Milieubeheer (Dutch Ministry of Housing, Spatial Planning and Environment)
W3C	World Wide Web Consortium

Acknowledgements

The authors wish to acknowledge all those who have contributed to the First International Symposium on Sustainable Development and Management of the Subsurface, 3–5 November 2003 in Utrecht (The Netherlands). We greatly appreciate the contribution of the Session chairs and keynote speakers: J. Bosch, J. Besner & G. Galipeau, J. M. Hebly, B. Marker, H. von Meijenfeldt, P. Nathanail, G. Nardi, H. Overbeek, H. Speelman, T. Torp and R. Waterman. Many thanks go to all authors, workshop leaders, and other contributors including H. Admiraal, S. Apostolaki, G. Arends, C. Bremmer, J. Buma, P. Dorsman, P. Duffaut, P. van der Gaag, H. Gehrels, J. Herbsleb, W. Al-Hashimi, J. A. Hudson, J. T. Hudson, S. Karstens, R. van der Krogt, V. Kutepov, L. Maring, A. Mavrikos, G. van Meurs, K. Millar, O. Mironov, V. Navarro Torres, D. Ngan-Tillard, V. Osipov, A. den Outer, D. Pereboom, L. Platteeuw, H. Ploeger, R. Rijkers, O. van Sandick, O. Schuiling, B. Sman, J. Streng, R. Stuurman, F. Taselaar, V. Tolmachev, K. Turner, B. Wassing, H. Weerts and W. Zigterman. Special thanks are due to the Session's Secretaries: S. van der Lugt and H. Hooghart.

J. G. A. M. Arnoldus is acknowledged for reviewing Chapter 6. The authors are very grateful to those who allowed re-production of illustrations to the text, and to J. G. A. M. Arnoldus, S. Carelsen, E. M. Charbon, W. Tegtmeier, and W. Verwaal for the photos they provided for this book.

The authors also wish to acknowledge Professor J. S. Griffiths and the other reviewers from the Geological Society for their efforts, encouragement and constructive remarks in producing this book.

The financial support by the Delft Cluster research programme, The Netherlands Ministry of Housing, Environment and Physical Planning (currently Ministry of Infrastructure and Environment), TNO, Deltares unit Geoengineering (formerly GeoDelft), and the Faculty of Geo-Information Science and Earth Observation (ITC) of the University of Twente are greatly acknowledged.

Chapter 1 Introduction

1.1. Perception and reality

Humankind has always been more interested in its visible surroundings, as landscape, air and water, rather than in the subsurface. For most, the subsurface is a mysterious perhaps even threatening realm often perceived as the domain of worms, dirt, darkness, danger, evil, decay and death. The public, including many intellectuals and policy-makers, has generally no idea about the subsurface, let alone the processes that govern it and how these may determine their lives. This lack of awareness points to a major knowledge gap and has hampered major development of the underground so far.

In reality, the subsurface always played a key role in fulfilling the needs of societies and in sustaining ecosystems. However, the public has been mostly unaware of that. Planet Earth offers much more to humankind than just its surface. As affordable and sustainable options for an expanding (and increasingly urban) human population are running short, the subsurface may provide a last and challenging resort to future generations. With this book, the authors aim to bridge that gap in knowledge and to shed some light on the many options for a sustainable use of the subsurface in the past, at present and in the future.

1.2. Definitions

In general the 'subsurface' is everything between the Earth's surface and the very centre of the planet, 6370 km below the surface. However, humans have never physically penetrated the planet deeper than about 12 km (Section 7.5) and it is not likely that this will become much more in the next few decades. Moreover, the scope of this book is limited to the 'shallow subsurface'. In practice, this consists of the first 250 m underground, with a special focus on the first 100 m below the Earth's surface.

The Earth is a dynamic system with interactions between the geosphere, the hydrosphere, the biosphere (including humankind), the atmosphere, and outer space. Such interactions are often quite complex and sometimes occur over (very) long timescales. Many of these interactions are not yet fully understood, challenging reliable predictions for future behaviour of System Earth, including the subsurface. In the past half century, however, the understanding of how the Earth and the processes work have improved dramatically.

'Sustainable development' is a key issue in this publication. Many definitions exist for this term, each reflecting a different political or philosophical perspective. The original definition is by the Brundtland Commission (1987): 'Development that meets the needs of the present without compromising the ability of future generations to meet their own needs'. That definition has been adopted in 1992 by the Earth Summit in Rio de Janeiro in its Agenda 21 Programme (UNEP 1992) and is used in the context of the subsurface. In line with most authors on this topic (Shields 1998), 'sustainable development' is addressed along three pillars or perspectives: economic, environmental and social in this book (Section 6.3.1).

1.3. Options for the subsurface

Since 1975, the world population rose 70% and increased from 4 to 7 billion in 2011. In the same time span, urbanization rose from 37.9–50.6%. By 2050, the UN (United Nations) expects 9.15 billion people to inhabit this planet, most of them in cities. Billions more people should thus be accommodated in existing and new cities generating unprecedented urban expansion. They also need resources, which normally are and will be extracted from the subsurface. Improved standards of living put yet another pressure on the use of the Earth as they demand for more personal space. Higher demand for urban space is reflected in the already rocketing prices of urban land. Increasing pressures on urban land will make underground development a more realistic option for future generations than generally perceived today.

The subsurface provides excellent opportunities for a sustainable development as, in principle, more than sufficient space is available underground and constant temperatures can generate very significant savings on energy bills. Subsurface development is and will not be restricted to urban areas only, but will become a more viable option in rural areas too. Sustainable use of the subsurface also implies a smart use of Earth processes to the benefit of human societies and ecosystems.

1.4. The book

This book aims to inform civil engineers, scientists and decision makers dealing with the urban underground and its wider surroundings. Information from this book may prepare them for situations when Earth processes act otherwise than expected, both negatively and positively. The book gives a variety of examples, ranging from historic and current to potential future use of and interactions with the subsurface, where sustainability plays a major role. The book further aims to raise awareness about assets and threats provided by the last frontier for human development on the solid Earth.

The subsurface has provided shelter, water, food and materials to humankind in the past, as described in Chapter 2. Also in modern times, the Earth provides all materials for daily life and forms an essential component in the human life support system. Current use of the subsurface for sustainable solutions to problems of today's society is described in Chapter 3. The need to go underground to explore and exploit the 'empire of darkness' to the benefit of future generations urges us to find new solutions to break down psychological, technological and economic barriers, which hamper such development now and to make the underground a more attractive realm. Examples of potential future use of the subsurface and some tools in subsurface management are discussed in Chapter 7.

To understand the impact of subsurface development on the Earth, at least something about the composition of the Earth should be known. Our knowledge of the Earth properties, in particular the geotechnical properties, has expanded significantly over the past decades, as outlined in Chapter 4. It also gives an overview of the technical aspects of interactions between humans and the subsurface. The subsurface is a relatively poorly known domain that cannot simply be observed or analysed. However, subsurface data and information systems are rapidly growing. Communicating such information to engineers, municipal authorities and hence to citizens is vital, as discussed in Chapter 5. Even if the subsurface would be fully transparent, lacking or conflicting legislation, regulations and procedures may hamper any further underground development and may lead to conflicts or unsustainable approaches. Chapter 6 explores such aspects and compares legislation about the subsurface and management tools in various countries.

Except for mining and oil and gas extraction, almost all past and present underground activities as described in this book, were restricted to the 'shallow subsurface'. With a few exceptions, no particular attention is given in this book to the topmost metres.

As this book discusses the interaction of humans and their subsurface it is relevant here to determine (potential) roles and functions of the subsurface for humans. Several such classifications have been made in the past (Van der Maarel & Dauvellier 1978; De Mulder 1999; Sandberg 2003). Here seven roles of the subsurface are indentified:

(1) source of natural resources
(2) storage of materials
(3) space for public and commercial use
(4) space for infrastructure
(5) medium for foundation for constructions
(6) component in life-support systems
(7) archive of historical and geological heritage

Historic, current, and future use of the underground is discussed here roughly along these lines (Chapters 2, 3 and 7). 'Source of natural resources' (including groundwater) may be referred to as 'extraction'. 'Space for public and commercial use' is combined here with 'space for infrastructure' and 'foundation for constructions'. The 'life-support' and 'archive' functions of the subsurface are discussed jointly as well.

1.5. Earth scientists and the International Year of Planet Earth

Earth scientists are ready to share their daily growing knowledge about the Earth and how this planet works. Raising awareness of a sustainable use of the subsurface and to use Earth scientific knowledge to make this planet a better, healthier and wealthier place for human societies, urged the United Nations to proclaim the International Year of Planet Earth (subtitled 'Earth Science for Society') for 2008 (UN General Assembly 2005). This book aims to contribute to this initiative.

Chapter 2 Historic use

The subsurface has always been essential to human survival and development since human lineage diverted from that of the chimpanzee some 6 million years ago (Brunet et al. 2002). From a geological point of view humans are a relatively young species on the face of the Earth. The subsurface almost literally played a mother role for humankind, a role that is still acknowledged in certain cultures (Fig. 2.1). Newton compared the Earth with an animal or a vegetable. In his Gaia-theory, Lovelock (2003) states that the Earth behaves as a single, self-regulating system comprising physical, chemical, biological and human components. This concept was initially severely attacked, mostly by biologists, but since the mid 1990s, resistance has diminished (Lovelock 2006).

The subsurface provided early humans with most basic requirements: fertile soils for collecting and growing food, water, shelter in caves, and natural resources. Prior to farming, some 10 000 years ago (Mithen 2003), at the onset of the Neolithic, humankind has been strongly reliant on the Earth's subsurface.

This reliance declined with the cultural shift from hunter and food gatherer societies to settlements and the rise of agriculture. Nevertheless, drinking water was mostly collected from natural springs or rivers, but food and building materials for homes were directly or indirectly derived from the subsurface. In addition, holes were dug in the ground for storage of supplies and for burial of the deceased.

Even though human history has been very short, humans are one of the most successful species in populating the environment in the history of life on Earth. From a few thousand individuals some 200 000 years ago, the human population passed 1 billion around 1800 AD and 7 billion in 2011. The levels of consumption and the scope of technology have grown in parallel with, and in some ways outpaced, the numbers (Harrison & Pearce 2000; EEA 2005a). Population growth was supported by the availability of natural resources thus supporting increase of prosperity. In turn, growing use or consumption also lead to depletion of resources and negative environmental impacts locally causing societal decline. The most obvious examples are related to non-renewable mineral resources (e.g. Gold Rushes).

Over the millennia, occurrences of mineral deposits as a source of wealth and/or power created complex political, legal and even ethical problems. Mineral rights, normally provided by national governments, often lead to conflicts with indigenous people claiming environmental damage from exploration and mining.

Humans have always used Earth materials extracted from the ground. The changing styles of tools are archaeological markers for periods of technological development (i.e. Stone, Iron and Bronze Ages). Humans started modifying the soil conditions and ecosystems by cultivating land. Removing parts of the original vegetation triggered desertification in some regions. Draining land for reclamation lowered groundwater tables, resulting in slow, but significant subsidence in soft deltaic areas (Claessen et al. 1987). Mining has punctuated the subsurface even more directly. Over the millennia, humans have also left their mark in the subsurface through intentional burials. The subsurface may also act as an environment preserving remnants of earlier civilizations that have not been left there on purpose (Fig. 2.2).

When humans began settling, urbanization soon followed. Urbanization implies division of tasks between population groups, between residents and non-residents (or rural people). Urban people gradually detached from direct supply of Earth materials, perhaps with the exception of construction materials. They continued storing food and goods underground, but also used that as shelter for harsh climatic conditions and enemies. Towns emerged in places where people conceived the best economic or strategic perspectives, although humans never gave up living in caves or underground.

With time and as quality of life improved, urban dwellers have become increasingly detached from the production of food, drinking water, fuel and construction materials. As a consequence, they lost their understanding and appreciation for the provenance of the Earth materials that make their lives possible and more comfortable. In most 'advanced' societies, the Earth is no longer perceived as a living 'Mother' producing all that is needed for survival, but rather as a rigid substratum ready for exploitation. However, natural disasters can temporarily change this perception, making people and politicians understand that the Earth is a dynamic system rather than rigid. Awareness on the physical aspects and processes of the Earth peaked during the International Geophysical Year (1957–1958) drawing significant public attention to the planet and how it works. Exploration of the atmosphere, outer space, in Polar Regions, and of the Earth's interior was high on the scientific agenda in the late 1950s and early 1960s. Major discoveries in the Earth sciences, including the mechanisms behind plate tectonics, date from that period. Such scientific discoveries were well exposed by the mass media and subsequently

Fig. 2.1. Worshipping Pacha Mama (Mother Earth) in the Andean Region. Ayllus delegates make the offering to the Sun and to the Earth at the opening of the World People's Conference on Climate Change and the Rights of Mother Earth in Tiquipaya, on the outskirts of Cochabamba, Bolivia (photograph: DPA/Reporters; F. Cartagena).

drove many students to embark on studies in university on topics related to geology and geophysics. About two decades later, a major dip in public attention and perception for the subsurface and the Earth in general occurred during the 1980s. In that time, the price of mineral commodities dropped strongly and mining and underground storage were often perceived as environmentally unsustainable, and in turn, negatively impacting students to consider a professional career in the Earth sciences. In the 1990s and 2000s a further drop occurred with the decline in employment for Earth scientists and in public interest for the natural sciences studies as compared to information and social sciences (De Mulder 2005).

In this chapter, the role of the subsurface in the past (until approximately 1975) is described against their main functions for humankind, both directly and indirectly. Following a description of its role in supplying resources, the role of the subsurface for underground space and infrastructure is discussed, and some concluding remarks are made at the end.

Fig. 2.2. Well-preserved remnants of a Roman wooden vessel at about 6 m below the surface in Woerden, The Netherlands (photograph: Reporters, U. Wesselink).

2.1. Extraction

2.1.1. Earth materials

Most tools of pre-historic humans were made of Earth materials. These were also used for, for example human consumption, heating, trade, medicine, construction, and weaponry. Throughout history, the quest for Earth materials drove civilizations into new territories. The Romans conquered Brittany to find tin, the Spaniards and Portuguese went to South America to find gold and silver, the British and French moved to North America and the Belgians, French, Germans and Italians to Africa, all in the search of natural resources. The drive for valuable Earth materials also triggered railway construction in the United Kingdom, the USA, Canada and Siberia in the 19th and 20th century (Harrison & Pearce 2000).

In the context of this publication, two types of Earth materials are distinguished: those derived from renewable and those from non-renewable resources. Materials from renewable resources are normally reproduced by natural, relatively short-term processes and include fresh water, wood and other plant products, animal products and oxygen. If consumption rates exceed the rate of renewal or reproduction on human timescales, a resource is considered non-renewable. Most Earth materials are thus non-renewable.

2.1.2. Renewable resources

Looking only at Earth materials, fresh water is the only resource that at the scale of a human life-time may be renewable. Fresh water resources occur on land at the surface and underground, where the topmost first tens of metres are normally fresh too.

Fresh water resources occur on all continents, both at the surface and in the subsurface, but they are spread unevenly over the globe. In the past, most of the fresh surface waters could be used directly for drinking purposes, as our predecessors did. Today, most drinking water comes from groundwater. Only a small proportion of groundwater is actively involved in the water cycle and most of it is stored in porous or fractured bedrock sometimes several thousands of metres below the surface. Planet Earth's mean annual renewable volume of water is 43 000 km^3 (BGR/UNESCO

2008; Döll & Fiedler 2008), but total volumes of the Earth's fresh groundwater resources are about 200 times higher. That is because most groundwater resources have accumulated in the subsurface over centuries or millennia. Combining population density and groundwater recharge expressed per capita and relating this to water consumption show the regional/national pressures on groundwater resources (Fig. 2.3). In a number of regions current consumption exceeds the (long-term) replenishment of fresh water resources, for example due to over-extraction and 'mining' of fresh water from ancient aquifers. This may threaten the ability of future generations to satisfy their needs and may therefore be labelled as *unsustainable use of groundwater*.

Under current climatic conditions some groundwater systems are non-renewable. These may have been formed in much wetter climates that prevailed 1000 or 10 000 years ago. In the North-Eastern Sahara, for example, the Nubian Sandstone aquifer system underlies an area of more than two million km^2 in Chad, Egypt, Libya and Sudan, and contains 8800 km^3 of fresh groundwater. That is about one hundred times the present annual water consumption of the Nile Basin countries (Hassan *et al.* 2004).

Until Neolithic Times, humans used fresh water for drinking, food preparation and washing purposes only. When farming began to develop, fresh water was also used for irrigation (Box 2.1). Availability of, and access to fresh water has always been vital to settlers. The earliest settlements and indications for urbanization were found in Western Jordan (in Beidha), 10 000 years ago (Mithen 2003). People from early civilizations dug water wells at or close to their settlements to collect groundwater (Figs 2.4 & 2.5). The oldest known water well dates back to 8900–8300 years ago and is found in the currently offshore Neolithic site of Atlit Yam in Israel (Antiquities 2011).

Other types of centrally organized water supply were built in Roman Times (200 BC–500 AD). The excavation of Herculaneum and Pompeii, destroyed by the Vesuvius eruption in 79 AD, shows that water was transported and distributed through a system of aqueducts and lead pipes. Romans also built extensive underground sewer systems for the removal of waste water (Fig. 2.6).

From Medieval Times to the Renaissance, water was mainly distributed by human action through dedicated construction works (Lindley & Sharp 2003). Over the last few centuries, cities have been increasingly supplied by underground water mains (Box 2.2). Most of these systems leak, and often 20% to more than 50% of the supplied water is lost in the ground. In turn, this fresh leak water contributes to the natural water cycle, keeping urban water tables high. Suburban aquifer recharge may thus be compensated entirely for reduced absorption of rainwater in the ground due to extensive pavement and other hard structures (De Mulder *et al.* 2001). However, replacing leaking water mains may cause other problems, for example lowering urban groundwater levels may lead to land subsidence (Lawrence & Cheney 1996).

2.1.3. Non-renewable resources

Non-renewable Earth materials may occur at the surface or underground and should be dug out, extracted or mined. Mining at the surface normally takes place in open pits, although sometimes Earth materials can be simply scraped from the surface, for example by mining beach sands.

Early humans soon learned to develop tools for the extraction of Earth materials from the ground. Underground mining may be via vertical mine shafts or by horizontal or sloping tunnels into the rock face. Earth materials may be extracted in various ways, depending on the type and geometry of the ore bodies (in veins, seams or layers). One of the first mining techniques was the bell pit, consisting of a shaft opening into a small chamber (Fig. 4.34). Other underground mining techniques include the 'room and pillar' type, where for stability reasons, pillars of unmined rock are left in place. Another type is 'stope and retreat', where the hole from which the Earth materials are removed ('the stope') is left to cave-in. In 'long wall mining', up to one kilometre long tunnels are dug in the deposit and the Earth materials (e.g. coal) are (mechanically) removed from one side of the tunnel prior to collapse. Roof tops can be supported by rock pillars or timber. In the late 19th and early 20th century, steam powered machines allowed for much deeper and more complex underground mining as mechanical water and ventilation pumps had a much higher capacity than manually operated pumps.

In mining, Earth materials can be classified as metallic minerals, industrial (non-metallic) minerals and fuel minerals. Metallic minerals are divided into iron (iron ore); iron alloys (manganese ore, chromium, nickel, molybdenum, cobalt, vanadium); base metals (copper, lead, zinc, tin); light metals (aluminium, manganese, titanium); precious metals (gold, silver, platinum); and rare metals (uranium, radium, beryllium).

Industrial minerals are non-metallic rocks and minerals that are used in industrial processes and products, construction, civil engineering works, agriculture, etc. These are divided into construction materials (sand and gravel, limestone, marble); chemical materials (sulphur, salt); fertilizer materials (phosphate rock, potash, nitrates); ceramics (clay, feldspar); refractory and flux materials (clay, magnesia); abrasive materials (sandstone, industrial diamonds); isolating materials (asbestos, mica); pigment and filler materials (clay, diatomite, barite); and gemstones (diamonds, amethyst).

Fuel minerals, or fossil fuels, include lignite (a low rank, brownish-black coal), black coal, oil, and natural gas (Serrano 2005). Below, some examples of historic mining of various types of Earth materials are described.

2.1.3.1. Base metals (copper, lead, zinc and tin)
Metal mining started quite early in history as demonstrated by the remnants of very old copper mines in Central

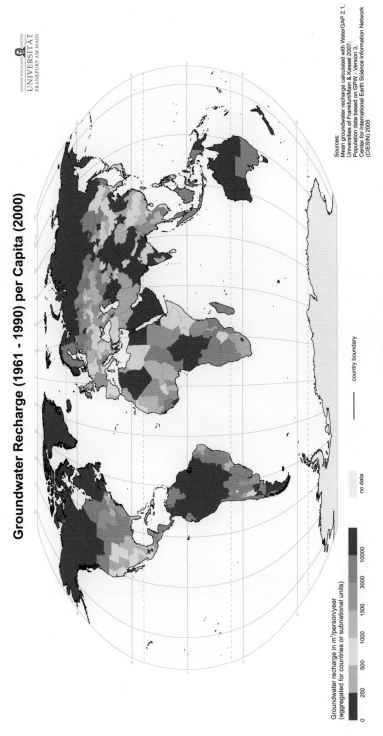

Fig. 2.3. Per-capita renewable groundwater resources around the year 2000, based on long term average diffuse groundwater recharge 1961–1990 (BGR/UNESCO 2008; Döll & Fiedler 2008).

Box 2.1. Water collection and distribution in deserts

The need to find water and to bring it to the people has always been a challenge and a source of innovative solutions. One of these is the 'qanat': underground galleries for groundwater supply, consisting of a horizontal well dug into a dipping aquifer in arid regions and several up to 100 m deep vertical shafts for removal of dug-out ground and maintenance (Fig. 2.4). With their lengths of several kilometres, qanats belong to the world's first major hydrological engineering works that occur in arid regions in China, Morocco, Afghanistan, Peru, Mexico, Spain and Jordan (Fig. 2.5). Extensive qanat systems occur also in Iran where 20 000 of the originally 30 000 qanats are still in use, serving 80% of the water supplies in the Iranian plateau regions (Moran 2000).

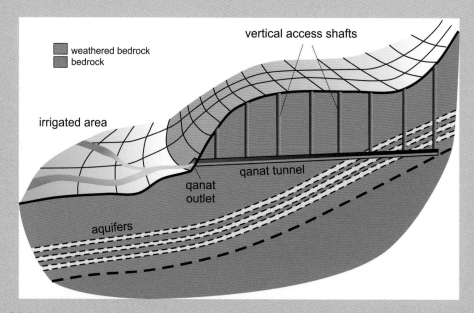

Fig. 2.4. Schematic representation of a qanat: a water supply system found in North Africa, Asia, South America, and Iran. It is made by digging a horizontal tunnel into a mountain or large hill; the vertical shafts are used to help remove excavated materials (modified from Moran 2000; Kaltenborn *et al.* 2010, designer Riccardo Pravettoni).

Water yield in qanats may be as high as 2 m^3/s depending on the size of the intake area of the 'mother-wells'. Their location is determined by aquifer availability and by water table depth, slope conditions, and distance to the settlement. As water demands may vary while qanats produce continuous flows, water management systems were installed, consisting of ponds and gates for storing water at night and in periods of low seasonal use (Sankaran Nair 2004). This technique has been quite significant for farming in desert areas and for societal relationships resulting from customs and laws, for example the codification in the Book of Qanats from the ninth century.

Europe. In Rudna Glava (Serbia), a 6000 years old mine gallery was found at depths of 25–30 m. In Bulgaria, about 300 old mining galleries have been identified, the oldest dating back to about 7000 years (Kaptan 1980).

Copper was one of the metals mined earliest in history. 'Copper' is derived from the Greek word 'kyprios' relating to Cyprus, one of the most important sources of copper in ancient times. Traces of old copper mines and smelting furnaces are also found elsewhere around the Mediterranean; including the Tokat region (Black Sea region, Turkey) dating back to at least the Early Bronze Age. Mine tailings and slag-heaps found in Egypt and in the Sinai Desert witness of production of thousands of tonnes of copper, lasting for a few centuries and suggesting that Bronze Age merchants used standard sized ingots for trading. The extensive use of copper relates to its malleability and resistance against corrosion.

Fig. 2.5. A horizontal waterway connecting a qanat with the historic town of Petra, Jordan (photograph: E. De Mulder).

Bronze is an alloy of copper and tin and production began some 5000 years ago. By adding tin, the alloy's hardness increased significantly while lowering its melting point. These properties made bronze a superb material for weaponry production. Merchants searched Europe for tin occurrences that were much less abundant than copper. Tin was eventually found in shallow occurrences along with zinc, lead, iron and silver in Cornwall, South West England. By the Iron Age, some 1000 years after the Bronze Age, Cornwall had become the world's main tin supplier. Cornish tin mines were an important target for the Romans when they conquered Britain.

Lead mining also started very early in history, as shown by an 8500 year old mine in Turkey. Romans produced

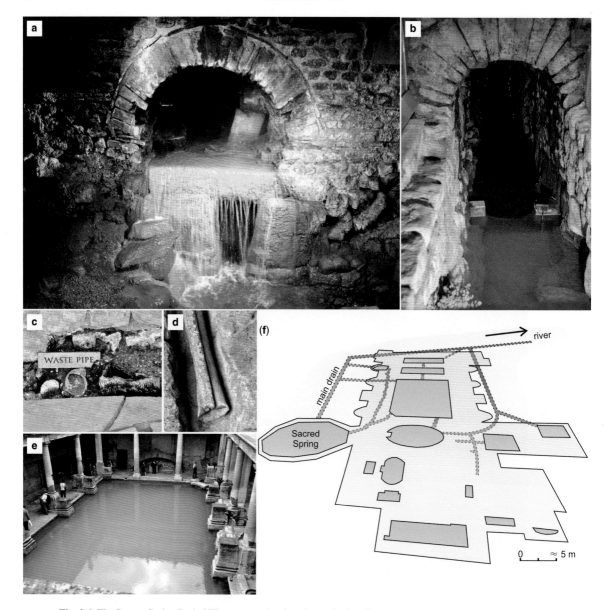

Fig. 2.6. The Roman Baths (Bath, UK), an example of engineered subsurface water and sewer systems built in Roman times. (**a**) Thermal spring overflow with excess iron deposition, spring temperature 46 °C; (**b**) Roman drain to river Avon; (**c**) Original lead sewage pipe embedded in Roman wall in Chedworth Roman Villa (Yanworth, UK); (**d**) Lead piping made from a flat sheet which has been rolled and has a single fold as a seal; (**e**) The Great Bath viewed from the East, depth 1.5 m; (**f**) Water system lay-out (photographs: E.M. Charbon).

lead on a massive scale, with an estimated 18 million tons extracted from 200 BC to 500 AD. Lead was used for cooking utensils, pots, urns and for water distribution (Juuti & Tapio 2005). Some believe that lead poisoning contributed to the decline of the Roman Empire (Nriagu 1983).

2.1.3.2. Iron
In the Iron Age (about 2700–2000 years ago), many small scale iron mines opened all over Europe. Most of the iron was used for weapon production. Iron mining boomed again at various periods during history, such as during the Industrial

Box 2.2. Natural water purification

Apart from a source, the subsurface also acts as a sink for fresh water through infiltration of rain and surface waters. Infiltration promotes natural biochemical reactions between water and sediment particles, resulting in natural purification of ground water. The natural purification capacity of the underground has been applied in The Netherlands since the 19th century. The city of The Hague started collecting such purified waters from the coastal dunes on which parts of that city are built in 1874. Shortly after, water demands exceeded natural supply, resulting in drainage of dune valleys and subsequent loss of biodiversity. Since 1955, pre-treated but non-potable surface water was brought into the dune areas and allowed to infiltrate in the subsurface via ponds and ditches to restore water tables in the valleys (Nordmark 2002). The (now mostly artificial) fresh water zone in the subsurface also acts as a barrier against salt water intrusion by the sea. To reduce environmental impact in the ecologically valuable dune areas by nutrient-rich waters, deep infiltration wells were installed more recently.

---

revolution that started in England in the last decades of the 18th century, during WWII in the United States and during the Cultural Revolution in China (1966–1976).

2.1.3.3. Precious metals

Gold has been one of the most appealing types of all metals throughout history and mining dates back at least 5000 years, when gold was used for decoration in the tombs of the Kings valley in Egypt. However, major gold mining (the 'gold rush') began in the sixteenth century AD by Spanish colonizers in South America. In the USA, the Californian Gold Rush started in 1848 (Rohrbough 1997). Numerous abandoned excavations and small mine shafts can be found in the mountains in the USA, Canada, Australia, Latin America, Philippines, and elsewhere.

Precious metals (gold, silver and platinum) and gemstones played important roles in trading in early civilizations, a development that continued at least until WWII (Box 2.3). Gold is both rare and durable, and became a medium of exchange in ancient times. In ancient cultures, precious metals and gemstones often served as a kind of bank for a tribe or State. Power was largely determined by credibility in terms of

---

Box 2.3. Silver mining for ancient Greek power

Politically significant were the ancient mines of Lavrio, Greece, lying 40 km SE of Athens. Silver and lead mining started here around 1200 BC by using the 'room and pillar' method. In the 5th century BC, the annual silver production was here about 20 000 kg. These mines owe their fame to their contribution in the development of ancient Athens. Their production provided a solid economic base for wealth and power for the City State. Over 2000 shafts with ventilation and hauling, some deeper than 120 m, were found; the usual cross-section was rectangular or square, each covering areas of 2 to 4 m^2 (Fig. 2.7). Excavation of a 100 m deep shaft with simple tools as iron hammers and chisels, typically took 1 to 1.5 years (Kaliampakos & Mavrikos 2004).

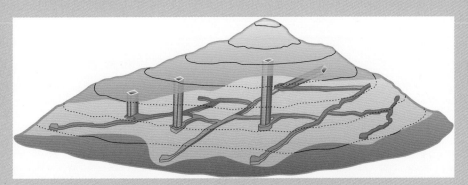

Fig. 2.7. Shafts and adits of the Lavrio mines (Kaliampakos & Mavrikos 2004).

precious metal and gemstone possessions to be used for securing loans or raising armies. The need for capital was a driving force for mining, often at the expense of many slaves' lives. National banks later took over this capital role while keeping large gold reserves: 35% of all the gold ever mined is held by central banks (Lomborg 2001) (Box 2.3).

2.1.3.4. Rare metals
Uranium mining began at the start of the 20th century after discovery of the element/metal by Henri Becquerel in 1896. In 1936, uranium ore was discovered in a mine at Goldfields, Lake Athabasca, Canada. In 1943, the governments of Canada, the United Kingdom and the USA banned private exploration of radioactive minerals in their countries in an attempt to control all uranium sources, but the ban was lifted in 1948. Between 1947 and 1967, most uranium was produced in the USA, followed by Canada, South Africa and France (UIC 2011). Major demands for uranium developed since the late 1960s with the expansion of nuclear power generation (Box 2.4).

2.1.3.5. Construction materials
One of the most popular Earth materials in pre-historic times was flint (silex, chert), primarily used for tool and weapon production. Flint normally occurs in limestone, for example in the white, Cretaceous Chalk cliffs along the North Sea coasts in Northern France and England. For collecting top quality flint, prehistoric humans were prepared to go underground (Fig. 2.8).

Limestone quarrying for cement production has been and still is a major industrial activity in many countries. Limestone may be classified as a construction material, like clay, sand and gravel. Blocks of rock, including limestone, belong to the oldest construction materials for housing and other structures. Together with clay and/or gypsum, lime has been used for millennia as cement or a mortar to immobilize pieces of rock.

The main application of sand has always been for construction and fill purposes (levelling). In construction, it may be used for road pavement, concrete, mortars and glass production. Since humans started building shelter, clay, often with plant materials, was used for roof construction to protect against rainfall. In Roman Times, bricks were produced from clay, as they are today. They also used clay to produce domestic utilities (terra sigilata, or typically red-coloured, stamped earthenware with a glazed surface), as was done in the Neolithic, several millennia earlier. Gravel came into use only in the late 19th century as a component in concrete for more advanced construction purposes. With very few exceptions, construction materials were always mined in open pits or in quarries, and in close proximity to places where the materials were used.

2.1.3.6. Chemical materials
Underground and surface salt mining and seawater evaporation for salt were main sources for salt since ancient times. In The Netherlands salt mining became quite popular in Early Medieval times when marine influenced peat beds were dug out and burnt to retrieve encapsulated salt crystals. Underground salt mining developed later. Over the past few centuries, underground mining of rock salt mainly took place in Europe. The salt mine at Wieliczka, near Kraków in southern Poland, has been a World Heritage Site since 1974. That mine operated continuously since Medieval Times. Miners carved elaborate underground rooms, sculptures and even an underground cathedral in the Miocene rock salt. Solution mining of salt also has a long history. In Ontario, Canada, solution mining in Silurian strata has been in operation since 1866 at depths ranging between 300 and 720 m.

2.1.3.7. Fuel minerals
Underground black coal mining began in China more than 1800 years ago. Until 1949, production was a few million tonnes per year. Production rose since at an unprecedented rate (Wang 2004). During the Industrial Revolution, mining focused on iron ore and black coal in Europe. Extraction of these resources prompted development of many of the current cities in England and Germany. Black coal and iron ore mining reached peak levels by mid 20th century,

---

Box 2.4. Uranium and Aboriginals

Although about 40% of the world uranium reserves are hosted by Australia, there are no nuclear power stations. Most of the Australian uranium deposits were discovered in the Northern Territories on Aboriginal Lands, for example in Kakadu in 1969. In 1985, the rights of Aboriginals for their own lands were formally recognized by the Australian government and these lands were placed under their control. Although mineral rights were excluded from this bill, mining companies had to negotiate and pay compensation fees for mining in Aboriginal land. In 1980, the Ranger uranium mine, surrounded by the Kakadu National Park, became operational. The Commonwealth government granted exploitation rights to mining companies in an area adjacent to what became a National Park and a World Heritage area. The potential environmental and health impacts gave rise to fierce debates. Aboriginal land claims (protected natural environment and areas of traditional livelihoods) and the occurrence of a substantial ore body are an example of competing, justifiable and possibly incompatible use of natural resources (Switzer 2001).

Fig. 2.8. Prehistoric underground (6000 years old) flint mine, near Rijkholt, The Netherlands. The irregularly shaped flint stones (black) are embedded in soft limestone (yellow brown) (Felder & Bosch 2000; © TNO Geological Survey of The Netherlands).

then dropped dramatically in the 1960s and 1970s when deep underground mining became uneconomic for most Western European countries. Black coal mining continued in the currently main producing nations, including Russia, China, USA, India, Australia, South Africa, and Poland.

Brown Coal (or lignite) is a fuel mineral intermediate between black coal and peat. Because of its occurrence, closer to the surface brown coal is mainly exploited in large-scale open cast mining, that may reach depths of hundreds of metres. Its combustible properties were noticed well before it actually served that purpose. In Europe, for example Romans noticed that while constructing aqueducts, close to Cologne, Germany. There, deforestation and the lack of firewood for potteries lead to its use as a fuel upon dewatering the material. Small scale mining started in the 18th century (Kleinebeckel 1986). Germany has been the lead producer and other brown coal producing nations include Greece, Poland, Czech Republic, Romania and Bulgaria.

With the industrialization in the late 19th century, the scale and depth of mines increased. In Germany, the Hambach mine is one of the deepest, up to 470 m (Fig. 3.9). Steam engines facilitated production of (highly dewatered) briquettes of brown coal for fuel and the drainage of the mine. At the turn of the 20th century, most of the brown coal was used for electricity production. For that purpose, mega-sized open pit mines were developed.

While peat mining (digging) in Roman and early Medieval Times was primarily to collect salt, a few centuries later peat became increasingly used as a fuel. Large peat bogs in Germany, Ireland, The Netherlands and Finland were stripped and used for heating the homes of the rapidly increasing population in the emerging cities, in particular in the late 19th century. Unfortunately, large scale peat excavation caused land subsidence, for example in The Netherlands, putting the low and flat delta at risk for flooding by the sea and for salinization of fresh water aquifers.

Sites where oil and gas seeped from the subsurface were quite well known, and in use. Where present, they also used oil shales for fuel. The first oil well was drilled in 1859 in Pennsylvania, USA, but it took more than 60 years before oil became the major fossil fuel for transport vehicles. As an alternative to black coal, oil and natural gas became popular only in the last decades of the 20th century.

2.1.4. Impact of mining

Ancient mining impacted the environment in a quite serious way and their effects may still be visible. Such impacts relate to solids, for example dust, and water, for example surface water, groundwater, occurrence of contaminants and leachates. There have also been physical aspects connected to mining, for example ground stability and the risks of cave ins.

Both open pit and underground mining impact the environment and may disrupt balances between Earth processes. This is particularly true for modifications in groundwater flows to keep active mines dry, but also by abandoning a mining site resulting in rising groundwater levels. This latter modification may have significant geochemical impacts if mine tailings become saturated again and become part of groundwater flow regimes, mobilizing hazardous chemical compounds and bringing these in the biosphere. Chemical traces of mined materials may thus be found at large distances from historic, abandoned mine sites.

Acid mine drainage, painting river sediments orange by its dissolved iron, may also enhance transport of heavy metals like cadmium, lead, arsenic and mercury. This acid mine drainage, where sulphide containing minerals are oxidized to sulphuric acids, may persist for centuries or even millennia, for example 2000 year old Roman mines in Britain still produce acidic water (Fields 2003).

In addition to mine waters, tailings may affect the environment, for example wind may transport fine dusts over large distances. In South Africa, 134 asbestos mines and 400 mining waste dumps produce asbestos dust that is introduced into the surroundings. In Japan, 5500 abandoned mines have been identified as impacting the environment in one way or another. Fields (2003) estimates that some 5% of all abandoned mine sites cause some kind of environmental damage. Gold mining also may have a particular impact. Mercury, used to isolate gold particles can still be found in America and Canada from the gold rushes of the second half of the 19th century. In the past, mining waste was locally stored underground, but this is now causing significant problems. An example is the Giant mine in Yellowknife (NW Territories, Canada), where several hundreds of thousands tonnes of arsenic trioxide from gold mining operations were stored in old tunnels. Leaching by groundwater resulted in the contamination of the surface water, in this case the Great Slave Lake. Containment measures were taken to halt further impact on the environment (Fields 2003).

Fluctuating groundwater levels may also affect the stability of mine sites, in particular of underground mines. Collapsing roof tops of poorly back-filled underground mines often generate land subsidence at the surface. As locations of old abandoned mines are often not well-registered, future generations may be taken by surprise by the sudden appearance of sinkholes (Fig. 2.9; Mather *et al.* 1996).

Mineral resource development has socio-economical consequences as well. Often, local populations do not significantly benefit from the wealth under their feet. One of

Fig. 2.9. Impact of a void migrating to the surface in the outskirts of the city of Glasgow. Collapse occurred over a mine shaft seizing a drilling truck (reproduced by permission of the British Geological Survey; © NERC. All rights reserved. IPR/138-14C).

the reasons for that is the lack of transparency on the distribution of income generated from mineral extraction. This has been and often still is a major problem which directly affects one of the pillars (societal benefit) of sustainability (see Chapters 3 and 6).

2.2. Subsurface space and infrastructure

The subsurface has always provided shelter for people. Today, as well as in the past, many people around the world benefit from the subsurface and live, work or recreate in natural or artificially excavated caves or holes in the ground. In this respect, local topography and geological conditions are important factors.

Caves are formed by geological processes such as river erosion, underground solution (karstification and underground dissolution of salts, coastal abrasion and tectonic movements. Caves, caverns and other underground places host specific microclimatic conditions due to the interactions between the outside atmosphere and the surrounding rock faces. Local geological conditions, with their particular physical and chemical rock properties, determine suitability of underground locations for specific functions to a large extent. Intrinsic properties of the subsurface, such as thermal and chemical isolation capacity, containing pressure, bearing capacity, may favour underground storage. Thermal insulation capacity and the prevailing slow underground processes make the subsurface often an ideal place to store goods against limited energy costs. Such storage may include food, fuel, other strategic materials, or waste.

2.2.1. Human settlement

Closed in by rock or sediment, caves provide shelter and if suitably high above the surroundings, provide comparative strategic advantages in terms of hunting or conflict between rival groups. As a result, and because people have often been staying there for generations, many natural caves contain remnants or traces of the past (Barber & Hubbard 1997). This helps to reconstruct evolution of ancient cultures, and past local to regional environmental conditions (Brown *et al.* 2004). World-famous examples of such caves include Lascaux (Dordogne, France; Fig. 2.10) and Altamira (Cantabria, Spain), with respectively 17 000 and 15 000 year old wall paintings. Human occupation of the Lions Cave (South Africa) with ochre markings on the walls extends back as far as 43 200 years before present.

Some two millennia ago, a large underground city was built under Rome (Italy). These 'catacombs' host a vast, three levels deep cemetery with more than 12 km of underground galleries decorated with paintings, graffiti, stucco work and

Fig. 2.10. Cave paintings of a horse discovered in Lascaux, Dordogne, France, in 1940. These wall paintings date back 17 000 years, just after the last glacial maximum, when the ice caps of the last Ice Age began to retreat (photograph: Fresh Images, Robert Harding Production).

mosaics. The galleries are lined with carved burial niches, frequently two or three levels high (Rome 2011).

In Central Anatolia (Turkey), two underground towns (Kaymakli and Derinkuyu) were built in volcanic tuffs, 8 to 10 floors deep and interconnected by several kilometres of tunnels. The insulation capacity of the surrounding rocks protects people from extreme cold and hot weather (Preethi 2001; Galipeau & Besner 2003). In the Byzantine Period (approximately 400–1450 AD) in Cappodocia, Turkey, dwellings, churches and monasteries were carved in the soft rocks serving as hiding places for early Christians (Erdem & Erdem 2005). Since the 7th century AD, many generations lived underground in the Iraqi desert (Box 2.5).

Box 2.6 relates to an example in Germany on subsurface use (Fig. 2.11). Some 80 000 people live in caves in Southern Andalusia (Spain); and there are also many cave homes in the Loire and Seine valleys (France). These dwellers are called Troglodites, named after cave people living at the borders of the Red Sea in Egypt in pre-historic times. Along the Loire River, cave homes were built in places where limestone was removed and used as construction material for the numerous castles and mansions in that region (Fig. 2.12).

In Central Australia, 3500 people live underground in Coober Pedy, a former opal mining site. There, summer temperatures may exceed 55 °C at the surface, but the underground dwellings remain relatively cool in summer and warm in winter. Cave homes are built fast and cheap: boring machines may excavate a four-bedroom home in one-day time, at prices 20–30% lower than conventional housing. With stable temperatures of 25 °C all year round, the real savings are in energy consumption (Gluckman 1995; Fig. 2.13).

Protection against war activities has been one of the main reasons for underground construction in historic times. The Great Wall in China, with its construction started during the Qin Dynasty (221–206 BC), also contains underground spaces. In Europe, many bastions with vast underground networks were built between the 15th and 17th centuries, one of the most famous being the 16th century bastion of Torino (Italy). Gibraltar holds some 50 km of tunnels and chambers constructed between 1782 and 1968 (Rose 2001). WWII examples include the Maginot Line along the French borders with Germany and Italy (Fig. 2.14). Other examples are Adolf Hitler's 'Adlernest' in Southern Germany, and the 'Atlantik' Wall, built to protect the Germans from anticipated invasions by Allied Forces, along the West coast of France. After WWII, strategic underground complexes were built in the USA, for example the Pentagon and NORAD Cheyenne Mountain Complex in the USA. Such underground complexes may have the infrastructure and extension of big cities and may host and supply many thousands of people with food for many months.

2.2.2. Storage

Containment and isolation capacities of the subsurface have been extensively used to reduce environmental impact of human activities. Former quarries and mines did and still do play a role in this respect. Early evidence of grain storage comes from China, some 5000 years ago (Bergman 1986). More extensive underground structures, including

---

Box 2.5. Underground Al Najaf City

Al-Najaf City lies on a plateau in the very dry Western Desert of Central Iraq, with no surface water resources. This area was first inhabited in the 7th century AD around a holy shrine. Subsequently, a city developed at the site. Early Arab settlers faced major difficulties in the region, including water scarcity, harsh climatic conditions (desert storms, extreme temperature variations), and attacks from neighbouring Bedouins. Nevertheless, people did not give up and coped with these hostile conditions by adopting innovative solutions, in particular the use of underground space. Water scarcity and transportation problems were overcome by digging canals from the Euphrates River, only 12 km from the city. One of these canals was dug under Al-Najaf City, branching into a network of smaller tunnels. Near the western edge of the plateau, these tunnels converge into a single conduit supplying water to the adjacent Bahar Al-Najaf valley, via an opening in the plateau's rock face. Houses, mosques and schools accessed the underground tunnels via shafts providing water by rope and buckets.

To cope with the climatic conditions, buildings and public places were built halfway underground. Underground caverns were dug at 8, 16 and 27 m below the surface, each subdivided into chambers, cellars or storage depots. These were ventilated by vertical shafts, open at roof top and penetrating shallower levels to reach deeper ones. During the day, caverns received sunlight through a central shaft, opening into the internal yard of the house. Residents could rest at three successively cooler levels underground, or hide there when threatened by hostile tribes.

The subsurface of Al-Najaf City hosts a unique underground cemetery. Again, caverns were dug at three levels reaching outside the city. From the cavern walls, tunnels were driven to serve smaller niches where corpses were laid and then sealed by brick and mortar. A single wall may include several corpses stacked up to the tunnel roof. Today, Al-Najaf is a dying city: rising ground water tables are forcing citizens to leave their underground residences for surface dwellings.

> **Box 2.6. The underground City of Oppenheim, past and present**
>
> The German city of Oppenheim is underlain by an extensive network of cellars and corridors, together representing an underground city (Ehlke 2003). Almost all of the 500 houses within the former city walls were connected by tunnels, sometimes three levels underneath each other. The underground system dates back to a period between the 13th and 16th century AD and covers an area of 32 ha. Excavation was controlled by geological conditions. Digging started from the surface, in 7 to 8 m thick loess and loam deposits; good for rapid excavation but with stable cavity walls and was followed by excavating deeper limestone and marl beds. The walls of corridors, typically 1–1.5 m wide and 2 m high, were made of fractured rock for stabilization. Initially, the arching roofs were concave in shape but later more triangular roof types were used showing increased knowledge on construction behaviour. No historic records exist on accidents during construction, but locally in the weakest spots in the ceilings, monitoring devices were installed, consisting of a square slate fixed to the ceiling that flakes off when deformed under stress. This simple visual device is used in the mining industry elsewhere in Europe. The extent, the characteristic shapes and construction differences in Oppenheim's underground system all point to the presence of special skills and expertise in underground construction works.
>
> The reasons why this complex cellar- and tunnel-system was built were never fully understood, but one of these must have been the creation of (cool) storage rooms in a crowded city. The system grew over time and underground wells, cisterns for water storage and drainage became incorporated. The corridors also provided escape routes and shelter, not only in the 17th century when Oppenheim was destroyed, but also during WWII. All written information about construction and ownership of individual cellars was lost when the city was repeatedly destroyed, most prominently in 1689. Also documents kept by the church (one of the owners of the labyrinth system) were lost when the cathedral in Worms burned down in the same year.
>
> Some cellars and corridors survived until modern times but the system as a whole declined also due to groundwater penetration and related loss of stability. Corridors were partially filled with waste materials and over time roof tops collapsed resulting in sinkholes at the surface. Between 1970 and 1985, 27 such sinkholes were recorded, and in 1986, a police car sank into a sinkhole in the street. Because of the deterioration of the underground system and subsequent risks to existing buildings, an action plan was eventually prepared. Cellars and corridors were mapped and remedial measures for stabilization were taken. Also, actions were taken to preserve this cultural heritage and safeguard it from further decay by revival of its use and giving it new functions, as restaurants, theatres and tourist attractions (Stadt Oppenheim 1993). Figure 2.11 shows an example of the underground corridors and a preserved cellar. A geo-information system with detailed subsurface information proved to be of great value for further development of the city and its underground legacy. Oppenheim has also been a challenge for legal and financial reasons as the underground system no longer follows the borders between private property and public space.

tombs, were constructed there during the Qin Dynasty. Examples of multiple uses of the subsurface include salt caverns for storage of oil and natural gas, compressed air and waste. Underground storage capacity may even be economically more valuable than the extracted salt (Manocha 2001). Initially, such artificial solution caverns were used for storing extremely salty fluids (brine) and wastes as a by-product from industrial processes. Later, more options for storing materials in these caverns were applied, starting in Canada during WWII (Thoms & Gehle 2000), followed by the USA, where salt caverns were used to store strategic petroleum reserves. Natural underground cavities resulting from long-term dissolution of rock salt have been used to store hydrocarbon products, liquefied petroleum gas and petrochemicals since the 1950s. Storage of crude oil was not invented before the Suez crisis in the 1950s that promoted building such a facility in the United Kingdom. That was used as a model for storing natural gas in the USA and Canada in the 1960s, and a decade later in France and Germany (Thoms & Gehle 2000). Since 1959, alkali wastes and organic residues were dumped in depleted salt caverns in Holford, United Kingdom.

Other underground activities and related infrastructure include hydroelectric and nuclear power generation. Many such structures were built shortly before and after WWII. Since the 1950s, nuclear power stations were built, often with underground extensions. A special case is the very deep power station at Krasnoyarsk (Russia), where chambers 300 m long and 60 m high were excavated to depths of 250 m in hard rock (Leith 2001).

2.2.3. Stable construction

A safe and stable environment to live and work in is not only relevant for and related to subsurface construction. The characteristics of the underground, also impact buildings and infrastructure at the surface. The weight of a building, for example, has to be transferred to and evenly spread in the underground to create a stable construction without unacceptable deformation of the structure due to possible

Fig. 2.11. Underground corridors under historic Oppenheim, Germany (photographs/drawing: Stadt Oppenheim).

Fig. 2.12. 'Troglodite-dwellings'. Louresse-Rochemenier, Loire Valley, France (photographs: E.M. Charbon).

Fig. 2.13. Underground house in Coober Pedy (photograph: Fresh Images, D. Wall).

Fig. 2.14. Entrance to Ouvrage Schoenenbourg along the Maginot Line in Alsace, France (photograph: E.M. Charbon).

settlement. Conditions related to the underground, like earthquake hazards and slope stability, also need to be considered when building at the surface. The sciences of soil mechanics, engineering geology and foundation engineering evolved in the early 20th century although some fundamental mechanisms were already described in literature in the 18th and 19th century.

In the past, builders used local and empirical knowledge to design and construct houses, bridges, etc. The type of construction and materials used depended on the natural environment including the characteristics of subsurface materials. Coastal and river areas with unconsolidated sediments have significantly different characteristics than hard rock and mountainous areas. Deltaic areas belong to some of the most densely populated areas in the world. Their subsurface consists of relatively recent deposits of gravel, sand, clay and peat with high groundwater tables. The bearing capacity of such deposits varies and may be very low and must be taken into account when designing and constructing buildings. Often, the underground was excavated until a layer with adequate bearing capacity on top of which the so-called building footings were constructed.

Over time, other techniques evolved, for example the use of pile foundations to transfer the load to deeper layers without the need to remove the top layers. The first known bridge (Pons Sublicius) across the river Tiber in Rome is believed to have been built on wooden piles in 621 BC (Tyrrell 1911). Wooden piles have also been used widely in The Netherlands, Germany, Poland and Scandinavian countries, for example, in the soft soil of Amsterdam and Venice (Italy). In Amsterdam, the majority of wooden piles date back 70 to 150 years relating to the major city expansion after the industrialization period. However, in some cases underneath historical buildings wooden piles may date back as much as 300 years (Klaassen 2009). To prevent their deterioration from biodegradation, high groundwater levels should be maintained. In the USA in Chicago wooden piles were first used for building purposes from 1883 onwards (Randall & Randall 1999). By 1915, an ASTM (American Society for Testing and Materials) standard for wooden piles was published. After World War II, wooden piles were mostly replaced by concrete piles.

2.2.4. Space for infrastructure

The subsurface is most frequently used for traffic, water transport, sewerage, electricity, and for communication lines. Tunnelling for transport started earlier in Europe and North America than in other continents and was linked to the rapid growth of railway systems in both continents during the second half of the 19th century. This included the metro systems in major cities such as London, Paris and New York.

Tunnels for European railways developed so rapidly that in some cases more tunnels were made than needed. Fierce competition between rival underground railway companies (before London Transport was created) resulted in a number of abandoned stations due to their mutual proximity. Some tunnels were re-used in the WWII as air raid shelters. In the USA, railway tunnelling collapsed in the early 1970s together with the decline of public railways systems as they were out-competed by the automobile. Underground tunnelling for transport in Asia started in the 1930s in Japan, where the first long railway tunnel (>5 km) was built in 1931. The first long railway tunnel in China opened in 1966. In Europe, the first major road tunnel (5240 m) was completed in Spain at Viella, in 1948.

Although clear advantages can be pointed out from underground construction of infrastructure, the examples mentioned also show that careful planning is needed to prevent wasteful use of underground space and premature abandonment. Sustainable development of the subsurface requires a possible sequence of uses to be considered, for example quarrying for materials, transport infrastructure and final use for storage purposes.

2.3. Concluding remarks

Some lessons for the present and the future may be drawn from this summary of past developments in the subsurface. In the past, the Earth was considered as an integral part of their natural surroundings. Until the Neolithic or Holocene, humans almost entirely relied on the Earth and its subsurface. Humans adapted their ways of living accordingly and had no, or very limited, options to influence or control any of the Earth processes. Technological development resulted in increasing demands for Earth materials. In turn, mining generated more knowledge about occurrences and properties of Earth resources, and about how Earth processes work. Such knowledge brought more wealth and, with respect to groundwater, better health conditions. Wealthier societies required even more Earth materials to sustain their standard of living, resulting in further development of the subsurface. More knowledge about the composition of the Earth and its processes gradually provided a basis for a better understanding of human influence on Earth processes.

This reliance on the Earth and its materials was not perceived by the average citizen until the mid 1970s when population and urbanization began to merge with Earth-based infrastructure. Most people were unaware of the environmental impact of their activities and interference with Earth processes, and often considered the subsurface as an adequate sink for waste materials.

As described in the next chapter, this perception has changed rather dramatically during the past three decades as a result of a worldwide paradigm shift in environmental awareness. In a way, humans regained an awareness of the relevance of the subsurface for society and of the necessity to live in harmony with the natural environment, as was common in the past.

Chapter 3   Current use

The previous chapter discussed humankind's use of the underground since his earliest encounters on planet Earth. They relied heavily on what the Earth produced and the subsurface played a key role for their subsistence. Over the centuries, that role diminished and in more recent times most people became almost totally unaware of their subsurface.

That paradigm changed in the mid 1970s when concerns grew about the planet's environment and its impact on health and well-being. Two publications played a decisive role in this rather sudden increase in environmental concern: 'Silent Spring' (Carson 1962) and 'The limits to growth: a global challenge' (Meadows et al. 1972). These placed 'the environment' on the political agenda which prompted some governments to take measures to protect the environmental compartments of the atmosphere, (surface) water and the subsurface. In these political waves the subsurface was not ignored. Pressured by a deeply concerned public opinion, governments became reluctant to grant new mining licenses and other functions for the subsurface were reconsidered as well. On the positive side, this concern also prompted unprecedented geological research programmes to find solutions for example for permanent storage of hazardous chemical and nuclear waste materials and (later) for underground storage of greenhouse gasses.

This shift in public perception marks the start of the period discussed in this chapter, from the mid 1970s until today (2011). Extraction of Earth materials from and storage in the subsurface over the past few decades will be reviewed briefly first. Next, rapid expansion of subsurface space for public and infrastructure activities is discussed, followed by other societal functions relevant in the subsurface. Chapter 3 closes with a discussion about sustainability of current use of the subsurface.

3.1. Extraction and storage

3.1.1. Renewable resources

As discussed in Chapter 2, groundwater is the only renewable Earth material resource. With about 10 000 billion tonnes, the worldwide annual recharge of groundwater greatly exceeds global consumption (600 and 700 billion tonnes per year, Struckmeier et al. 2005). But groundwater cannot be considered a renewable resource everywhere on Earth. In arid and semi-arid regions precipitation is often too little to adequately recharge the aquifers. If deep wells tap deeply seated aquifers, not actually connected to the water cycle, these will be depleted eventually.

Groundwater is not just a resource but also an essential component of the subsurface 'life-support system' (see Section 3.3.2). Many of the two billion people who fully rely on groundwater use this for drinking water. That is only 10% of all fresh groundwater consumed, most of it (70%) is used for agriculture (irrigation), followed by industry (20%) (Struckmeier et al. 2005).

Groundwater occurs in porous sediments or rocks in the subsurface (aquifers, or reservoirs) (Fig. 4.10). It took several millennia to saturate these aquifers with fresh water by precipitation, by rivers or lakes. If small quantities are withdrawn the reservoirs will normally be recharged by rainfall or otherwise.

Under normal conditions, groundwater is safe and clean and well suited for human consumption. In some areas, however, groundwater may flow through aquifers with natural mineral contents that may affect human and animal health (Selinus et al. 2005) (Section 4.3.9). Aquifers may also be contaminated by, for example, leaking tanks with chemical liquids.

Another groundwater-related threat concerns over-exploitation. This frequently occurs in urban and industrial areas, for example in 60% of the European cities (EEA 1998). That may not only result in depletion or in unsustainable use of groundwater resources, but also in land subsidence. This man-induced hazard develops gradually but may have significant impact. Cities such as Bangkok, Venice, Mexico City and Jakarta are examples of such 'sinking cities' due to, among other reasons, excessive groundwater extraction (Lawrence & Cheney 1996). This process may be retarded and halted through strict regulation and water management as was achieved, for example in the city of Shanghai (UN ESCAP 1985).

Land subsidence may be registered by Earth observation techniques that may identify very small (1 mm per year) modifications at the Earth's surface, for example caused by compaction of over-exploited aquifers or uplift related to aquifer rebound from groundwater recharge (Fig. 3.1). This microwave imaging system (Interferometric Synthetic Aperture Radar: InSAR) is a highly effective spatial technique using data acquired by the European C-band ERS1/2 and ENVISAT-ASAR satellites, 800 km above the Earth's surface. Today, this technique is used more and more often as an early warning system for other geohazards such as landslides, volcanic activities and earthquakes (Terrafirma 2011) (Section 5.1.4).

3.1.2. Non-renewable resources

Demands for Earth materials rose dramatically over the last three decades due to unprecedented population growth. For example, aluminum production doubled in the last quarter of the 20th century while copper production increased more than 6 fold between 1950 and 2000 (USGS 2001). Similarly, production of other Earth materials, as industrial minerals and fossil fuels, increased exponentially. Such high demands, coupled to continued world population growth and the inherent non-renewable character of such commodities, inspired the Club of Rome to predict rapid depletion of most of the world's Earth material resources anticipating exploding commodity prices as a result (Meadows *et al.* 1972).

However, the opposite happened and prices dropped in the 1980s following a long-term historic and general downward trend for raw materials (Fig. 3.2). A similar trend was observed by the International Monetary Fund for metal prices in the period 1957–2000 (IMF 2012). Despite their sharply increased production the world's registered reserves of most commodities increased (Crowson 1998).

This paradoxical situation is mainly attributed to new discoveries. A well-trained generation of exploration geologists, equipped with new tools including satellite images and 3D seismic surveying techniques, and assisted by fast data handling and modeling methods, identified many new and large resources of almost all metals and minerals. These discoveries were accomplished in times when exploration budgets in the mining sector generally dropped.

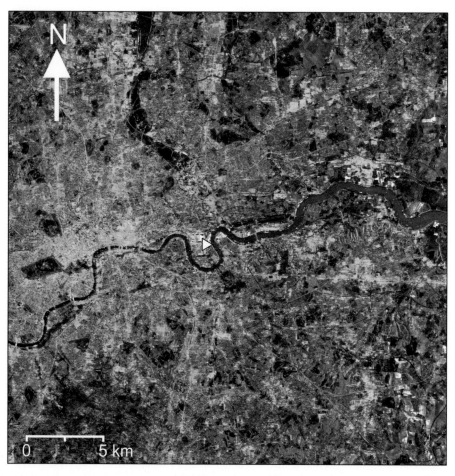

Fig. 3.1. Earth surface modifications in London, UK. Yellow colours refer to annual subsidence of 1 to 2 mm a^{-1}, red stands for places sinking 3–4 mm a^{-1}. Note the subsiding area in the lower left corner due to groundwater extraction and the ribbon like structures further north coinciding with recently constructed electricity and metro tunnels. Blue colours refer to 2 mm a^{-1} ground heave where paper, printing and brewing industries withdrew decades ago and stopped groundwater extraction since. In green coloured areas minor or no modifications occurred. Results are based on an average of over 200 measuring points/km^2 between 1995 and 2000 (courtesy: Fugro NPA Ltd (FNPA) and Tele-Rilevamento Europa, TRE).

Normally, exploration budgets are positively correlated to commodity prices (Goulden 2006).

After two decades of decline, commodity prices rose sharply since 2002. This was particularly true for energy and metal resources and was attributed to the strongly increased demands in rapidly industrializing countries as China and India. Due to the global economic crisis following the mid 2008 credit crises, commodity demands and prices decreased again.

Higher demands make mining of lower grade ore bodies more viable resulting in larger extraction sites. New processing techniques resulted in higher net yields as well. While 3% copper ore grades were normal mining targets at the beginning of the 20th century, today often 0.5% grades are produced, leaving 99.5% of waste materials behind (Harrison & Pearce 2000).

A particular big extraction site is the world's largest open cast copper mine Chuquicamata, near Antofagasta in Northern Chile (Fig. 3.3). The mine is elliptical in shape, with a surface area of almost 8 km^2, and is 900 m deep. One of the world's deepest mines is the TauTona Gold Mine (or Western Deep), near Carletonville, West of Johannesburg (South Africa) that reached a depth of 3900 m below the surface in 2008 and is scheduled for further deepening to 5 km (Yefim Cavalier 2003). At 3900 m, rock face temperature in the TauTona mine is 60°C requiring major cooling facilities (TauTona 2011). Rock pressure is another problem in deep mines. For example, at 3.5 km rock pressures are some 90 MPa (for reference, normal concrete has an unconfined strength of around 20 to 40 MPa), and when rock is removed through mining this pressure in the surrounding rock becomes a multiple. These high pressures may cause a phenomenon known as rock bursts which is the sudden explosive failure of the rock mass. The effect may be more severe due to the combination with the cooling of the rock. Rock bursts are responsible for many fatal accidents in deep mines every year (Yefim Cavalier 2003). Figure 3.4 shows miners working at great depth in another South African gold mine. The world's largest zinc and lead mine is located in Alaska, in the Arctic Circle.

In 1997, some 1 Mt Nickel was produced worldwide, most of it in the Russian Federation. This resource is mainly extracted from large and deep open cast mines in the NW of the country.

Alluvial diamond exploitation in the southwestern Namibian deserts also covers wide areas but is shallow and part of the extraction is offshore. The Namibian diamonds are derived from weathered rocks (Kimberlite pipes) in central Southern Africa and were transported downstream to the west by the Oranje River and its Tertiary and Quaternary predecessors. Detailed geological studies of these river deposits and their fossil fluvial environments led to accurate predictions of distribution patterns and to high yields of diamond production. Seven percent of the Namibian GDP depends on diamond production in 2008 (IHN 2011; Namibian diamonds 2012).

Except from areas where rock salt and gypsum deposits are exposed at the surface, as in Western Iran, most of the world's salt mines are underground. There, extraction is either by excavation or by dissolution. As in all underground mining, extraction may cause land subsidence due to collapsed underground cavities (Fig. 3.5).

The world's largest volumes of mined materials are construction materials. Their production goes up and down with economic growth (Vagt & Irvine 1998). Volumes dropped during the economic recession period that affected many OECD countries (OECD is the Organization for Economic Cooperation and Development) in the 1980s while economic revival in the mid 1990s is well reflected in increasing production rates of sand, gravel, clay and limestone (Fig. 3.6).

Construction materials are normally dug from excavations at or near the surface. Like other mineral resources, their occurrences are determined by geological processes that did not necessarily produce such Earth materials in places where they are needed most by society today. As the price of such bulk materials is strongly controlled by transport costs, distant resources are often too expensive. Environmental constraints apply but may be overruled by economic factors. If building activities are halted due to lack of (cheap) construction materials, certain environmental (landscape) damages may be tolerated. Alternative solutions may be found by digging deeper, but that comes with a price as well. Where groundwater tables are high, excavations are easily flooded and exploitation is mainly by dredging.

As construction materials outnumber all other Earth materials in volume, impact of such excavations on the

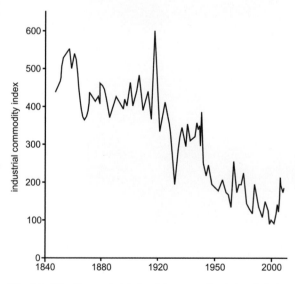

Fig. 3.2. 'The Economist' industrial commodity index, inflation adjusted, with the year 2000 = 100 (data from The Economist 2005, 2009).

Fig. 3.3. Chuquicamata mining complex, the world's largest open-cast copper mine, in Calama, Chile, on 16 September 2005. Surface area: 8 km^2 and 900 m deep (photograph: DPA/Reporters, V. Rojas).

landscape and local groundwater conditions may be quite significant. Almost all cities deal with these problems as they need such materials for urban development and maintenance. Proper understanding of the regional geological conditions in combination with the use of Geographical Information Systems (see Section 5.2.6) may provide

Fig. 3.4. Deep underground mining (2600 m depth, South Deep Gold Mine, Gauteng, South Africa) (photograph: South Deep Rock Engineering Department).

proactive municipalities with tools for designating adequate excavation sites while anticipating successive urban sprawl (De Mulder *et al.* 1997).

Energy materials have their own dynamics. Fossil fuel (coal, oil, natural gas, tar sands, etc.) production rose, uninterrupted since 1975, due to the sharply increased energy

Fig. 3.5. Local land subsidence due to dissolution mining in salt domes in Twente, East Netherlands (January 1991; copyright Tubantia).

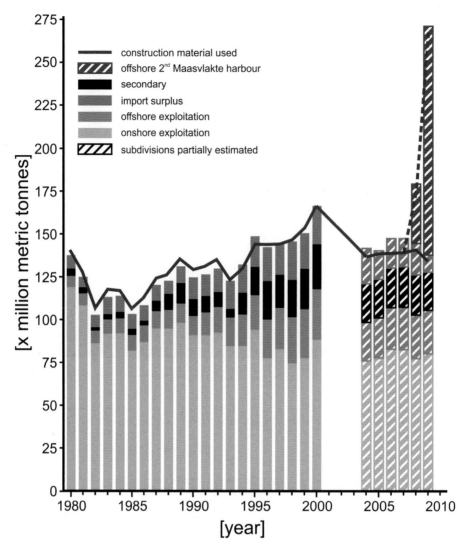

Fig. 3.6. Consumption and supply development of all natural construction materials from the subsurface of The Netherlands between 1980 and 2009 (2nd Maasvlakte is a very large extension of Rotterdam harbour; data: 1980–2000: De Mulder et al. 2003; 2004–2009: Compendium 2011; data 2nd Maasvlakte from Port in Action 2010).

demands. This was in particular true for economically rapidly expanding countries in SE Asia, followed by China and India since approximately 2000. Of the 2009 world's annual total coal production (including lignite) of 6.8 billion tonnes, almost 48% is extracted in these two countries (Worldcoal 2011). Other major coal producing nations are the USA (about 1 billion tonnes a^{-1}), Australia, Russia and South Africa (Taylor et al. 2005).

Between 1975 and 2009, oil production increased significantly and natural gas production roughly doubled. Since 1980, the world's proven oil reserves almost doubled from 667 to 1258 billion barrels at the end of 2008 (BP 2010).

The Reserves to Production (R/P) ratio of oil increased strongly in the 1980s and stabilized at approximately 40 years by 1990 (Fig. 3.7).

Natural gas burns cleaner than oil. As of 2005, natural gas constituted 23% of the world energy production. Natural gas resources are widely spread over the globe but in remote areas it is often flared. Instead of flaring natural gas (in Africa alone for 500 million US$ per year), this energy resource may be used to address local, small-scale energy needs (Sinding-Larsen et al. 2006). Application of natural gas as an energy resource may be economic only if transported by pipeline to larger concentrations of consumers.

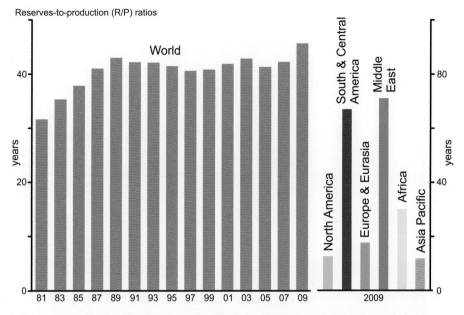

Fig. 3.7. Reserves to production ratios for oil, expressed in years of future consumption; left: world, right: per region for the year 2009 (data: years 1981 and 1983 from BP 2006, years 1985 through 2009 from BP 2010).

Only when prices are high, compressing natural gas to Liquefied Natural Gas (LNG) may be profitable.

As oil and natural gas occurs in the pores of sandstone layers under high pressures, its release from these rocks normally causes the sandstone to shrink. In turn, that may cause local subsidence at the surface. In low lying areas close to the sea such subsidence may cause water management problems (Fig. 3.8).

Very large volumes of potential hydrocarbon resources occur in tar sands, mainly in Canada. Again, exploitation may only be economic when prices are relatively high (>60 US$ barrel^{-1}). Natural gas trapped in shales (shale gas) and gas hydrates (pressurized, low temperature solid cubes of natural gas in deep-seated marine sediments) may also represent significant future hydrocarbon energy resources (see Chapter 7).

Uranium is another fossil fuel with large resources worldwide. Due to an almost worldwide moratorium on expansion of nuclear power generation since the Chernobyl (Ukraine) nuclear power plant accident in 1986, uranium production temporarily stagnated but rose again by the end of the 1990s. Canada and Australia are today's main uranium producers. The world's largest consumers are countries that rely heavily on nuclear power for electricity generation. Figures for 2007 are: Lithuania 80%, France 78%, Belgium 55%, South Korea and Switzerland 40%, Japan 26%, and the USA 20%. By 2009, fifteen percent of the world electricity production was nuclear (Nuclear power 2011). Highly radioactive nuclear waste is a detrimental side product of nuclear energy production as it is highly toxic and remains radioactive for extremely long periods (10 000–100 000 years). The tsunami in Japan and resulting nuclear disaster of 2011 triggered the closure of a number of nuclear power stations throughout the world and consequently it is likely that production of uranium will reduce too for some time.

3.1.3. Environmental concern and reclamation

Large-scale mineral exploitation impacts local landscape and environmental balances. Drifting on the waves of increased public environmental awareness, more and more governments considered mining as unwelcome, regardless of the strong demand for its products by the same public. This negative perception was reflected in policies to constrain mining and related activities. Except for Spain, Finland, Sweden, Poland, and Ireland, active metal mining became almost extinct in the European Union since the late 1980s.

Impact on the landscape is greatest with open pit mines. In the industrialized world there are many examples of derelict and abandoned mine sites frequently accompanied by dump sites of tailings as a by-product of mining. As many of such areas were subsequently transformed into industrial centres, they were often affected again by soil pollution and turned into waste dumps. This problem became well-acknowledged in OECD countries and in the early 1980s, the first measures were taken to minimize and mitigate such impact. For example, 170 km^2, up to 425 m excavated lignite mines were reclaimed and turned into farmlands, leisure areas and recreational parks in the Lower Rhine Valley, near

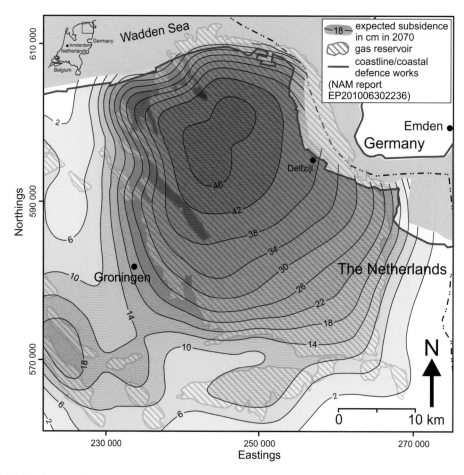

Fig. 3.8. Total expected land subsidence in 2070 upon termination of exploitation of the Groningen natural gas field in the NE of The Netherlands, adjacent to the North Sea. As the surface is close to sea level specific measures have to be taken to prevent the area from flooding (data: Bodemdaling door Aardgaswinning, EP201006302236, NAM B.V. 2010).

Cologne, Germany. Most of the recovered top soils were laid aside during excavation and were re-used in the new landscapes as a post-mining operation (Fig. 3.9). However, such abandoned lignite mines may also provide examples of poor reclamation practises when rising groundwater tables mobilized and deposited chemical waste products, threatening ecosystems in the wide surroundings of, for example the Bitterfeld area, in Germany. In a positive sense, the area turned into a test-site for new clean-up technologies and an *in situ* pilot plant for treatment of polluted groundwater afterwards (Bitterfeld 2011).

To cover reclamation costs after mining, special taxes are levied for construction materials in France. Many engineering geological, hydrological and geo-environmental studies were conducted to convert abandoned open-pit mines into stable lakes and landscapes, for example in Chogart, Iran (Osanloo & Gholamnejad 2005).

In the USA, reclamation of abandoned mines may be a joint State and Federal responsibility. The Tennessee Land Reclamation Section, for example, is responsible for reclaiming abandoned coal mine sites, using State money and federal grants from the U.S. Department of Interior's Office of Surface Mining. With an annual operating budget of approximately 1.4 M US$, the Section administers around 10 reclamation contracts annually (Tennessee 2011).

Excellent examples of reclaimed mine sites can be found in Peninsula Malaysia where in the 1980s and 1990s many abandoned tin mines in the Kinta tin belt were reclaimed and transformed into residential areas, shopping malls, parks, golf courses and other leisure areas (Malaysia 2011).

Apart from impacting the landscape, processing and storage of waste mining products (tailings) may cause considerable environmental damage. In 1998, part of a waste reservoir at the Aznacollar Zinc mine collapsed, spilling

Fig. 3.9. Reclaimed lignite mines; Hambach, near Jülich, West of Cologne, Germany. The right hand side of the photo shows the active open-pit mine, the conveyor belts for transport of lignite to the power station are in the centre. The area on the left hand side has been reclaimed (photograph: W. Tegtmeier).

nearly 7 million cubic metres of toxic waste just outside Europe's largest nature reserve, the Doñana National Park, causing Central Spain's worst ever ecological disaster (Fig. 3.10). Tailings may cause yet another detrimental effect, due to their inherent geomechanical instability which may impact any development on top of such tailings.

Other detrimental impact on the environment may be caused by black coal mining which may evoke coal fires. These occur in many of the shallow coal belts worldwide (Stracher 2007). In China, for example, coal fires occur along the entire 5000 km long coal belt in the North. Once lit, often by spontaneous combustion, they are hard to fight and may burn for as long as oxygen is available. Such coal fires contribute significantly (2–3% by China only) to the world's total annual output of CO_2 (China coal fires 2011) (see also Section 4.7.4.2). Mining may cause casualties as well. Through cave-ins, floods and gas explosions about 6000 coal miners are killed every year, 80% of them in China. Due to improved safety conditions these numbers show sharply downward trends.

Public environmental concern was politically translated in national environmental policies, initially mainly focused on environmental protection. At a global level, the UN created the United Nations Environmental Programme (UNEP) in 1972, which issued guidelines for sustainable mining. UNEP was strongly involved in the Earth Summit (UNCED), in Rio de Janeiro in 1992 and in the World Summit in Johannesburg in 2002. There, mining was an issue of intensive debates as the mining industry was making efforts towards more sustainable mining. In 2003, the International Council for Minerals and Metals (ICMM) launched its 10 Sustainable Development Principles, amongst others addressing environmental performance, land-use planning, good governance, human rights, transparency, and local communities (ICMM 2011). Following the World Summit in Johannesburg, many governments appeased their resistance

Fig. 3.10. Flood level (half way door) marked on pumping house as a result from the Aznacollar mine incident (photograph: D. Van Ree).

against development, including mining, on the condition that this should be environmentally sustainable.

3.1.4. Storage

For strategic purposes, spatial and energy management reasons, more and more often fossil fuels, drinking water and compressed air are being temporarily stored in the subsurface. In Norway also frozen food is stored underground. From comparative studies it appears that underground storage is cheaper than at the surface, in particular for underground storage tanks for liquids exceeding 5000 m^3 (Broch 2007). Saving energy costs drive underground food storage in Korea as well (Shin & Park 2007). For comparable reasons, temporary storage of warm or cold water has become an increasingly widespread option, mainly in European countries.

Driven by current international commitments reducing greenhouse gasses in the atmosphere, initiatives were taken to store excess carbon dioxide in the subsurface. Depleted underground gas reservoirs may quite well serve this purpose (Torp 2003; Box 3.1).

Underground ('geological') disposal of nuclear waste has been a topic of fierce public debate since the early 1980s. For addressing this problem by permanent underground disposal proper geological expertize is required, boosting geological research in many of the OECD countries. The main advantage of underground disposal is the isolation capacity of targeted host rocks, for example granite, consolidated clays, shales, and rock salt deposits. To test isolation capacity of rock salt, two nuclear devices, equivalent to 5 kt of TNT each, were brought to explosion in salt deposits in New Mexico and Mississippi in 1964. No leakage of radioactivity has been recorded by test devices since (Thoms &

Box 3.1. Carbon dioxide storage in the subsurface

As 'traditional' methods for reducing CO_2 emissions will probably fail to comply with the requirements set by the Kyoto Climate Treaty, underground storage of carbon dioxide from fossil fuel combustion is seen as a more feasible option in some countries today. Once CO_2 is freed through burning fossil fuels, sequestering it from the atmosphere is like returning it to where it was stored for hundreds of millions years. CO_2 sequestration (or: Carbon dioxide Capture and Storage: CCS) is normally realized in two steps. First, it is captured from power plants exhausts or from other energy-intensive industries. After compression and transport, CO_2 is injected in natural underground reservoirs. Such reservoirs may include depleted or disused oil or gas fields, deep saline aquifers, and deep-seated coal seams unsuitable for mining. In terms of storage capacity, deep saline aquifers sealed by impermeable rocks may have the best potential worldwide.

To investigate technological feasibility and environmental sustainability of such underground storage, the European Saline Aquifer CO_2 Storage (SACS) project was conducted at the Sleipner field in the North Sea where the world's first commercial carbon dioxide storage in deep saline aquifers has been in operation since 1996 (Eiken *et al.* 2000). Every year, one million tonnes of CO_2 is injected in the Utsira Formation in this field (Fig. 3.11). Through repeated seismic surveys migration of this injected gas in the aquifer is monitored (Fig. 3.12). Upward migration is hampered by shale beds in the aquifer prompting dissolution of CO_2 in the adjacent reservoir's formation water (brine). Seismic and gravity surveys are being used for calibration of reservoir simulation models. In turn, predictions from such models are verified by the measured volumes of injected carbon dioxide.

Fig. 3.11. Schematic view of CO_2 storage in the Utsira Formation in the Sleipner field (copyright Statoil).

Leakage to places beyond the confined underground storage facility is perceived as one of the main potential problems of sequestration. Leakage may occur abruptly (through well or reservoir failure by earthquakes) or gradually (via undetected faults, fractures or wells). The impact of elevated CO_2 concentrations in the shallow subsurface may include lethal effects on plants and underground life forms and contamination of groundwater (IPCC 2005). Simulation models in the SACS Project seem to show that CO_2 can safely be stored in a deep saline aquifer for at least several millennia. In the reservoir, CO_2 undergoes chemical reactions eventually and gradually resulting in immobilizing and neutralizing carbon dioxide. Scientific knowledge built up in this way serves as a basis for appropriate risk management, including assessment, monitoring and remediation (Arts *et al.* 2004). The effectiveness of the available risk management methods should still be demonstrated. Research at the Sleipner CO_2 injection demonstration project has resulted in a Best Practice Manual which may be applicable to other projects on safe and environmentally sustainable underground CO_2 storage (Holloway *et al.* 2004). Ten years after the start of the Sleipner project, many other demonstration projects of CO_2 storage have been implemented.

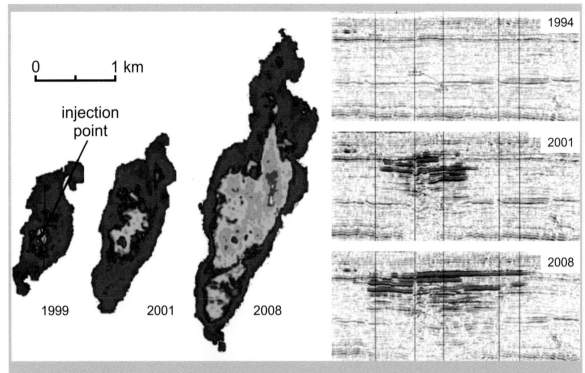

Fig. 3.12. CO_2 plume in map view (left) and time-lapse seismic images (right) from the Sleipner field showing vertical slices through the expanding plume from 1994 to 2008. The total height of the plume is about 250 m. The topmost reflector in the seismic images represents the top of the Utsira Formation; the base of the sand formation corresponds to the lowermost reflector, well visible in the 1994 image. CO_2 injection point lies at the centre at about one third from the bottom (visible as of the 1999 image) (courtesy Torp; copyright Statoil).

Gehle 2000). Eventual migration of the waste material into the biosphere, through groundwater, is considered the main risk.

Safety of geological disposal of radioactive waste in proper host rocks is widely accepted in technical communities and several nations decided to move forward with this option (IAEA 2003). Application of fast-neutron reactors and recycling of spent fuel by pyrometallurgical processing will greatly reduce the creation of long-lived reactor waste materials and would thus make the nuclear waste problem more manageable (Hannum et al. 2005). Underground storage of residual nuclear waste products would contribute to managing this safety problem.

3.2. Subsurface space and infrastructure

The subsurface provides society more options than to extract or to store materials. Such other options include the use of underground space for commercial or societal purposes and as a means for transportation. Using underground space seldom develops from a perceived lack of space at the surface only, but rather from a mix of unique opportunities. Such opportunities may be provided by, for example, the occurrence of such (abandoned) underground space like former mines under cities (e.g. Kansas City, Coober Pedy City, Fig. 2.13), or the presence of dense underground infrastructure networks (e.g. Montréal, Box 3.3). In most cases so far, underground development has seldom been the result of long-term forward urban planning, perhaps with the exception of building line-infrastructure networks. Such planning may develop once some 'autonomous' underground development took place triggering municipalities to support and facilitate other underground development.

Since 1975, underground infrastructure grew significantly both in length and in volume but still lagging far behind such development at the surface. Most such underground developments occurred in and around big cities, particularly in the currently (2010) 11 megacities with more than 10 million urban people. First, developments in tunnelling for infrastructure purposes are described for various regions, followed by some developments in underground space.

3.2.1. Tunnelling

Tunnelling is one of the most prominent and appealing types of underground development. It is directly related to urbanization, mobility demands, traffic congestion at the surface, and more indirectly, to environmental concern. As safer, faster and cheaper techniques (e.g. tunnel boring) became available and traffic congestion became a more pressing issue, tunnelling boosted in several parts of the world since the mid 1970s.

In Europe, tunnelling started relatively early, particularly for rail infrastructure. For a century, construction of longer tunnels developed slowly until about 1975. By then, 50 longer (>5 km) railway and road tunnels existed. From 1975 onwards, that number rapidly increased. In the next ten years only (between 1975 and 1985) 33% more longer (>5 km) road and railway tunnels were completed than in the 25-year period before (Fig. 3.13). This trend continued over the next few decades and is still rising (see Chapter 7).

Before 1975, European tunnel construction mainly took place in the Alpine region, notably in Switzerland, Austria, Italy and France. There was one notable exception: Norway. There, out of their current 10 longer railway tunnels, four were built and completed before 1950, and three in 1944, during WWII. In Norway, emphasis shifted from railway to road tunnels in the 1970s. The oldest longer Norwegian road tunnel dates back to 1966, but 19 out of their present twenty-three longer road tunnels were completed since 1985 and ten of these only since 1995, including the world's longest (toll free) road tunnel in Lærdal (24.5 km). It took 5.5 years to build that tunnel while producing 2.5 million m^3 of solid rock waste material (Fig. 3.14).

This demonstrates Norway's policy to invest much of its oil and natural gas revenues in infrastructure, a long-term sustainable investment. Italy, the other leading European (longer) tunneling nation, invested more in public transport. They completed 50 longer traffic tunnels so far, 43 of which are railway tunnels as are all, longer tunnels currently under construction.

The Euro Tunnel, crossing The Channel and connecting Britain with France, was a landmark in European tunnelling history. Work on the second longest railway tunnel in the world commenced in 1988 and completed in 1994. In only three years giant tunnel boring machines cut their way through the Chalk marls. It has three 50 km long tunnels (of which 39 km are sub-sea); two of the tubes accommodate Eurostar rail traffic while a smaller tube in between acts as a service and emergency escape route. Almost 4 million m^3 of chalk were excavated at the English side, much of which was dumped below Shakespeare Cliff near Folkestone to reclaim 90 acres (360 000 m^2) of land from the sea. This area was converted into a park.

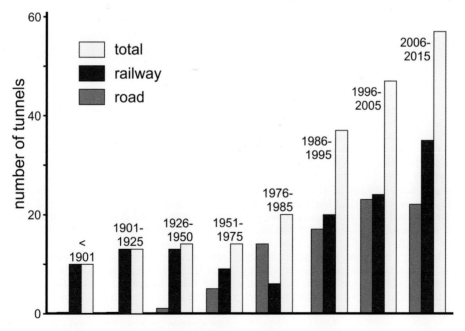

Fig. 3.13. Completion of longer (>5 km) railway and road tunnels in Europe. Note the change in horizontal scale between 1951–1975 and 1976–1985 (data from: Road Tunnels 2011 and Railway Tunnels 2011).

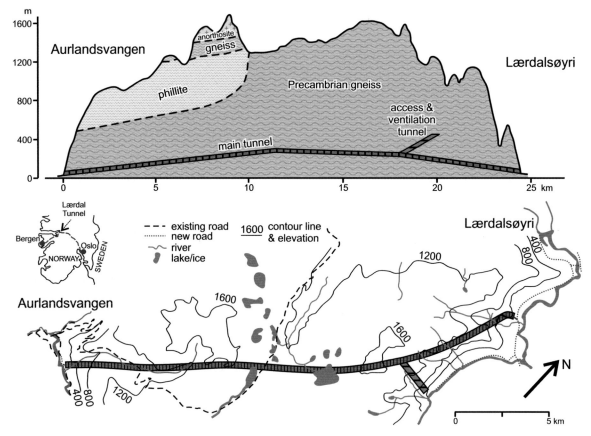

Fig. 3.14. Position and cross-section of the world's longest road tunnel; Lærdal Tunnel, Norway (modified from Broch & Grøv 2008; Geology after Engineering 2006).

In Canada and the USA, tunnelling activities were less common than in Europe and date back mainly to a period well before 1975. In North America, railway construction was responsible for rapid and widespread tunnelling activities in the late 19th and in the first quarter of the 20th century. The most recent long (>5 km) railway tunnel in the USA was built in 1970 while the oldest (the 7.6 km long Hoosac tunnel) was completed in 1876. The longest North American railway tunnel (12.5 km) runs between Seattle and Saint Paul and came into operation in 1929. The longest US road tunnel measures 4.2 km and lies in Boston's Harbor area (Box 3.2, Fig. 3.15). Most of the long road tunnels (2–5 km) in the USA are rather old and date from the construction peak between 1940 and 1960, or from the 1920s and 1930s. A more recent (1979) tunnel is the world's highest road tunnel, the Eisenhower Memorial in Colorado, situated at 3.4 km above sea level.

In contrast to the USA, major railway tunnelling continued in Canada into more recent times where the 14.7 km long Mount MacDonald tunnel was opened in 1988 and two other longer tunnels in 1983. The longest road tunnel in Canada measures 1.5 km and is (partly subsea) situated between Detroit and Windsor (Tunnel Canada 2011). After a peak in 1975–1985, when 108 tunnels were completed out of the 173 ever built tunnels in Canada, tunnelling declined there and only 8 longer tunnels were completed in Canada between 1990 and 2000. The Montréal example (Box 3.3), however, demonstrates that underground construction developed in quite different directions in Canada.

In Asia, major tunnel construction started in the 1930s but really took off with the construction of the Shinkansen rapid railway tracks in Japan in the late 1960s and peaked in 1975 with the completion of 16 longer railway tunnels, 3 of which longer than 10 km (Railway Tunnels 2011). With 92 longer (>5 km) rail (80) and road (12) tunnels completed by 2011, Japan is the world champion in longer tunnel construction. It hosts the world's longest railway tunnel: Sei-kan, 53.8 km, a subsea tunnel completed in 1988. An economic recession caused a sharp decline in railway tunnelling in 1985–1995. Afterwards, the numbers went up again (Fig. 3.16).

In China major tunnel construction began later and has expanded rapidly since the late 1990s and accelerated from 2000. Initially only longer train tunnels were built but that has changed since 2005 when the first longer road tunnels were opened in China. With some 30 longer (road) tunnels in construction in 2011, China has taken over the position of the world's leading tunnelling nation. In metropolitan areas, some 180 km of underground metro lines are added to the system annually (Qian & Chen 2007). The 18 km long tunnel through the Zhongnanshan Mountains in Shaanxi Province, the second largest road tunnel in the world, was completed in 2007. Out of the world's ten longest road tunnels, three are in China (Road Tunnels 2011). Where Japan invested mainly in rail infrastructure, China puts more emphasis on the construction of road tunnels today.

Korea's underground transport infrastructure grew rapidly within the past decade. In 2007, the total length of constructed railway tunnels was 241 km, almost 8% of all railway tracks. Construction of high speed railway lines boosted tunnelling as well: 46% of Korea's 191 km high speed railway lines in 2007 are underground (Shin & Park 2007). Korea has a long highway tunnel (Chang Su: 6.8 km) and a long high-speed railway tunnel (Sangchon: almost 10 km) under construction.

Other Asian nations with longer traffic tunnels include India with the 6.5 km Karbude railway tunnel in the Konkan railway line and the 8.8 km long Rohtang road tunnel in Kashmir, completion anticipated in 2015. Apart from some highways, underground works completed or in construction include urban metro systems (Delhi, Calcutta), and tunnels for irrigation works. Singapore has one 8.6 km, partly subsea road tunnel (KPE), opened in 2008, and Vietnam recently completed a 6.2 km road tunnel.

Very few longer tunnels or other major underground constructions other than mines have been built in Australia. A few more occur in New Zealand where three longer (8.5–8.8 km) railway tunnels were built between 1923 and 1980. In Africa, some 2–5 km long tunnels occur in the Republic of South Africa where major tunnelling activities started in the 1960s. The Orange River Project and the Hex River railway tunnel attracted many overseas consultants and stimulated local engineering expertise. Not many longer tunnels have been constructed in South America, the longest in Brazil: an 8.6 km long railway tunnel (Ferrovia do Aço), completed in 1989. Another long (7 km) railway tunnel is in Venezuela (Tarzon). A long (6.3 km) road tunnel is in construction in Colombia, between Tobiagrande and Puerto Salgar.

Tunnelling is also used to construct pass ways for hydropower generation in mountainous areas. In the Himalayas tunnels are constructed for water transport by pressured pipes and underground power stations are built. In Bhutan, a country with very low energy demands, major hydropower projects are under construction to supply major North Indian cities with energy.

Box 3.2. The Big Dig …

The Central Artery in Boston (Massachusetts) was one of the most congested urban highways in the USA. The elevated 6-lane highway carried almost 200 000 vehicles per day causing up to 16 h day^{-1} traffic jams and an accident rate of 4 times higher than the national average on other urban interstates. Moreover, the highway also separated Boston's North End and Waterfront neighbourhoods from the downtown, limiting these areas' ability to participate in the city's economic life. A solution was found by replacing it by an eight-to-ten-lane underground expressway directly beneath the existing road which is culminating at its northern limit in a 14-lane, two-bridge crossing of the River (Fig. 3.15). After the underground highway opened to traffic, the old, elevated highway was demolished and in its place 300 acres of parks and open space is being created. In addition, the Massachusetts Turnpike had to be extended from its former terminus south of downtown Boston through a tunnel beneath South Boston and Boston Harbour to Logan Airport. The 3.5 miles long Central Artery/Tunnel Project with its numerous connecting roads is nicknamed 'The Big Dig' as it was one of the most complex, technologically challenging and with its price tag of 15 billion dollars, the most expensive highway projects in American history. For the tunnel, more than 16 million cubic yards of soft river deposits had to be excavated and trenched up to 120 ft deep before bedrock was reached (Salvucci 2003).

Realization of the Big Dig was based on adequate consensus on broad sustainable goals for the region. Solid planning and meeting environmental targets were a substantial component in the success of this enterprise. A quarter of the total budget was spent to mitigate environmental problems, traffic, and community outreach. Millions of cubic metres of excavated soil were used to cap an old landfill in Boston Harbour while halting leaching of toxic wastes into the harbour and converting the island into a public park. That saved cost of expensive disposal at distant sites. Where highway projects once bulldozed bisected cities in the name of mobility, the Central Artery/Tunnel project reunited neighbourhoods and preserved the fabric of a city, even as it made dramatic improvements in its transportation system.

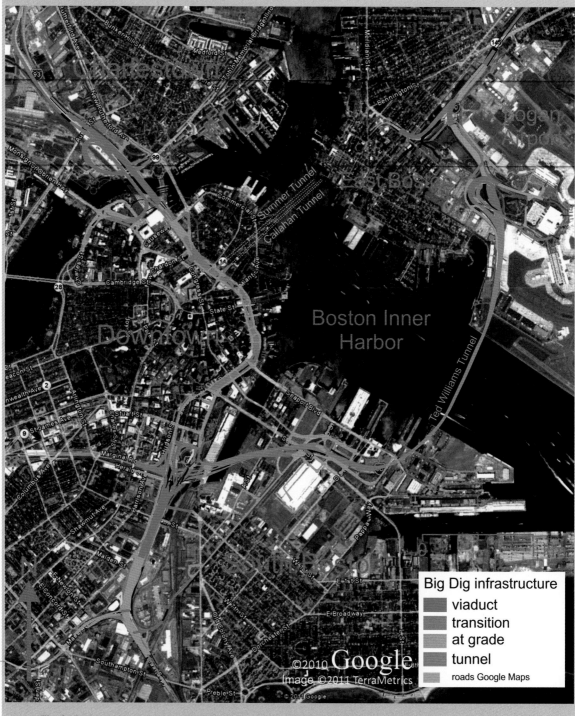

Fig. 3.15. Urban highway tunnelling in the USA: the 'Big Dig' in the Boston Harbour area (modified from massDOT 2011; aerial photo: Google Earth/TerraMetrics).

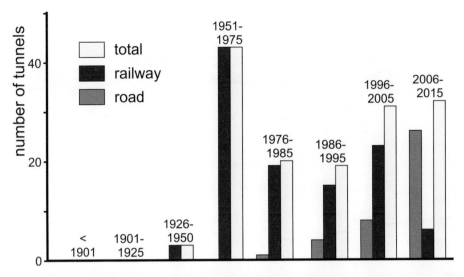

Fig. 3.16. Completion of longer (>5 km) railway and road tunnels in Asia (mainly Japan and China). Note the change in horizontal scale between 1951–1975 and 1976–1985 (data from: Road Tunnels 2011 and Railway Tunnels 2011).

3.2.2. Underground space

Underground (or: indoor) cities are a relatively new phenomenon. They grow rapidly, in particular in Asia. In 2007, Beijing had 30 km² of underground space in use, with an annual growth of 10%. According to Qian & Chen (2007), China counted more than 200 cities with an underground floor area of over 10 000 m² in 2007. Shanghai Expo (in 2010) promotes subsurface space development by the construction of an underground area of 400 000 m² (Yu et al. 2009). Development of underground public and/or commercial facilities often spreads from nodal points in existing infrastructure. This was evident in Toronto and Montréal (Box 3.3, Fig. 3.17) and in many cities in China.

Indoor cities may also develop from depleted or abandoned mines, for example in Coober Pedy City (Southern Australia), where a full underground town developed from opal mining in the early 20th century (see Chapter 2 and Coober Pedy 2011).

In the USA, underground construction activities normally develop at local levels in municipalities. This is usually economically driven and initiated by private companies.

Box 3.3. Underground Montréal

Urban dwellers of Montréal, Canada face a harsh climate. Every year extreme arctic winter conditions alternate with hot and humid summers. Climate has been one of the prime drivers of the successful development of an underground city that began in 1958 when three visionary architects and urban planners started to design a plan for redevelopment of Place Ville-Marie and its surrounding lands owned by a train company. Boundary conditions of this large-scale plan were multi-functionality (shopping centres and offices) and completely separating pedestrians from road traffic. That was realized by underground passages, car parking lots and shopping galleries. When the new Place Ville-Marie opened its doors in 1962, an underground pedestrian network of 500 000 m² was realized. Two parallel subway lines with ten stations, and a few hundred metres from the Place, were developed next. Access to these subway stations was through lobbies of buildings connected with spacious mezzanines and shopping galleries, decorated with plenty of art works providing a pleasant resort for the Montréalers, not only when the winter blizzards blew. When the subway started running, in 1966, ten major buildings were connected with train stations. By then, the indoor, underground city had 240 boutiques and department stores, 4000 indoor parking places, 2200 hotel rooms, 36 restaurants and 4 cinemas. By November 2003, the indoor pedestrian network of corridors, squares, cinemas and shopping galleries had extended to 32 km (Fig. 3.17). It is connected with ten subway, two train and two bus stations, and gives access to 62 real estate complexes with in total 4 million m² of floor space including 80% of all office space in the city centre, to 1600 apartments and 10 000 car parking lots. The network comprises 155 entry points giving access to 500 000 pedestrians daily.

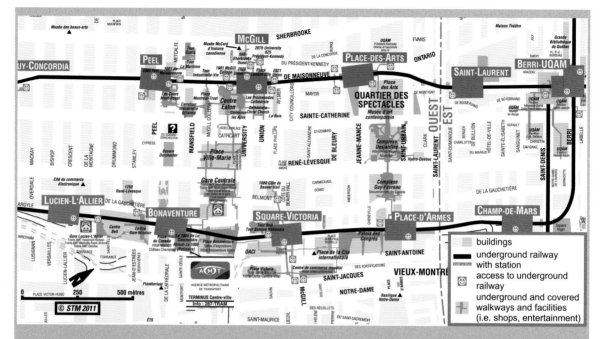

Fig. 3.17. Map of Montréal city showing the network of underground passages (copyright STM©, 2011) (courtesy: Jacques Besner).

Not only climatic conditions attributed to the development of one of the largest underground cities in the world, positive interactions between the municipality and urban developers generated considerable synergy as well. Connecting passages between subway stations and nearby high rise buildings by the private sector, public access to and free passage of private property was negotiated for municipal building permissions and numbers of indoor car park units. In many cities, such public private partnership constructions prove indispensable to successful underground development (after Galipeau & Besner 2003).

Payback and permitting times are relatively short. In contrast to most European countries with lengthy decision-making processes, once the idea for underground construction is born, construction may start fast in the USA.

3.3. Other functions

Excavation, storing, tunnelling and underground cities are probably the best known applications for underground development. In this section some other options for use of the underground are discussed, including bearing capacity, food supply, life support and storage of cultural, historical and geological records are briefly reviewed.

3.3.1. Bearing capacity

The period 1975–2007 witnessed a very significant increase in urbanization. Worldwide, the urbanization rate rose from 37.9–50% (UN Population Division 2008a, b). Seventy-five percent of the Europeans live in cities (EEA 1998). In 1997, cities covered 25% of the European Union land area. Although the general population growth in the European Union was only 2.5% over the past 2 decades, its urbanized areas expanded 11%.

To accommodate 2 billion more people in the past 35 years, cities have expanded at an unprecedented rate. They even (or in particular) did so at places previously avoided for their poor foundation conditions, for example in swamps, river plains or at waste dumps. Where expanding cities encroached upon mountainous areas, steep slopes were domesticated as well, for example the Blue Mountains in Sydney. Such slopes may consist of rock, but foundations on such slopes are often unstable and building there may generate landslides. Reduction of landslide risks normally requires expensive geotechnical measures that obviously are lacking in poor urban regions. Urban sprawl in flat deltaic or river plain areas may face other geotechnical problems, such as very soft subsurface conditions where thick peat or clay beds occur in the subsurface, as for example in The Netherlands. Here, long foundation piles are used to support housing and other structures. The quality of piled foundations may vary from place to place and is often related to the social economic status of the suburb and its inhabitants. As houses can be built even on water (Houses on water 2011), such foundation problems in soft grounds may be technically overcome.

Intensified urbanization often causes densification of city centres and may encourage construction of high-rise buildings, adding to their visible skyline and perceived positive image. Such sky scrapers also need foundations and the higher the building, the more basement floors below it. As shown in Box 3.3, this links the two vertical dimensions in urban building practices and may stimulate the interest of engineers for underground constructions.

Bearing capacity may also be considered one of the factors controlling sustainable urban planning and management. Zones with extremely poor foundation conditions may therefore best be assigned to public parks or to other recreation purposes, whereas those with strong subsurface conditions may be better used for housing or for industrial plants. This role of the subsurface will be discussed in more detail in Chapter 7.

3.3.2. Life support

Also with respect to the life support function of the subsurface public perception has remarkably shifted. Before the mid 1970s, the public and politicians considered the subsurface as a rather inert medium ready to absorb all kinds of waste materials with no or no known negative impact on health and well-being of humankind, plants or the animal kingdom. Around that point in time, laboratory tests on soil and groundwater started to alert people and politicians on deteriorating ecological conditions of the subsurface, jeopardising its life support function and threatening public health and safety. This was confirmed by a flood of environmental studies conducted since. As the distribution of contaminants through the subsurface is controlled by groundwater flow regimes, understanding their behaviour in relation to the often-heterogeneous geological environment is critically important to predict flow paths of contaminants in the subsurface and the integrity status of the human life support system. The past 30 years witnessed a paradigm shift in perception of the life support function of the subsurface. Before, it was ignored if understood at all; now, the subsurface is often seen as an essential element in survival of humankind. This understanding has brought the subsurface into the public eye as an integral component of the environment.

3.3.3. The subsurface as an archive

The subsurface holds an important storage function for cultural, historical, and geological records. The subsurface may therefore be considered as an immense archive, accessed only very marginally yet. Although this perception is not new (Winchester 2002), the past 30 years revealed some innovative views on this archive function. Many cultural and historical remains were discovered since. The Malta Convention of Valletta Treaty (1992) forces contractors in the European Union to give priority to physical protection of archaeological heritage materials over construction (see Chapter 6). Today, more and more archaeologists tend to keep archaeological material provisionally stored in the safe and controlled environment of the subsurface rather than excavating it all at once.

The subsurface also played a role in storing seeds of all kinds of plants from around the globe. This is done in a man-made tunnel in the mountains in the permafrost area of Spitsbergen. Here, gene conservation is brought into practice through seed banks saving formerly genuine types of food producing plants from extinction and as a contribution to the protection of biodiversity (Gene conservation 2011).

Geoscientific research proved that rock cores may store detailed (century scale) time records. These time records may be extended from the last 3 million years (as known in 1975) to over 100 million years in 2005. The accelerating and detailed insights in climate conditions during the past half million years are to a large part attributed to specific information about the Earth atmosphere derived from gas bubbles captured in polar ice sheets, validating climate models in producing more realistic forecasts (Siegenthaler *et al.* 2005).

3.4. Concluding remarks

This chapter described the greatly increased use of the subsurface since the mid 1970s. Both mining and constructions for underground transport facilities took off at unprecedented scales. Surprisingly, this massive exploitation of the Earth resources did not lead to worldwide depletion. Supported by new technologies, geoscientists identified even larger resources postponing inevitable depletion of most commodities into a more distant future. Simultaneously, tunnelling costs went down while production and safety increased. Almost all of the 2.5 billion more people on this planet since 1975 found their homes in cities which expanded enormously.

All this had a major impact on the environmental conditions resulting in a more than 60% increase in the global ecological footprint (0.8 in 1975 and 1.4 Earths in 2009 (Wackernagel *et al.* 1997; Footprint 2011). Simultaneously, public environmental awareness rose quite significantly and became more and more reflected in national and international policies. Extensive numbers of Environmental Impact Assessment studies have increased the knowledge on the interaction between subsurface development and the environment. 'Sustainable Development' has become a leading issue for policy makers since the Johannesburg

Summit in 2002, by making environmental concern as relevant as economic and social development.

However, the geosciences have also taught us that the planet is and has always been dynamic and that the many interacting, often slow Earth processes are and will never be fully in balance. Change is the rule and humans impact such processes in a very complicated but not always 'negative' way. As understanding of how humans interact with Earth processes is expanding, this knowledge can be used to properly manage the interactions between humans and the subsurface. The technical aspects hereof are discussed in the following Chapter 4.

Chapter 4   Technical challenges and assets

Underground construction is much more complex than building constructions at the surface. At the surface, most site-characteristics, such as topography, nature, and drainage, may be determined relatively easy by visual inspection but underground information may only be obtained indirectly, by excavation, boreholes, *in-situ* testing, or by using geophysical methods. A second major difference arises from the loading conditions. For surface structures, loading is normally perpendicular to the surface. In the underground, loading conditions of the subsurface materials are more complex (Fig. 4.1) and in three dimensions, and only recently are the mechanics of subsurface loading becoming understood.

Thirdly, underground constructions behave quite differently in cases of failure. At the surface, the potential failure of a building under construction usually displays visual warning signals such as cracks in stone or concrete or by excessive settlement well before collapse occurs (Fig. 4.2). Normally, such warnings are easily spotted whereas signs of imminent failure of underground constructions are far more difficult to notice and often only recognizable to experts. A fourth difference concerns how to collect the necessary expertise. From experience it is known that specific near-surface layers (e.g. peat or wet clay) are soft and are not very suitable as a foundation for construction. Except for miners and underground workers, most people do not have such knowledge and expertise on the underground.

Ground properties determine options for underground excavation. Generally, stronger and cohesive materials support excavations better than weak, compressible materials. However, stronger intact materials are normally more difficult to excavate than weaker materials. Extraction of materials from the subsurface disturbs the existing stress–strain equilibrium in the underground materials and a new equilibrium will be established. Extraction may not only modify underground conditions but may affect stability at the surface as well. Vice versa, surface structures may influence the subsurface.

4.1.   Surface v. underground engineering

Although surface and subsurface engineering may differ, they have many features in common and are often applied in the same project in combination or successively.

A certain defined depth boundary between surface and subsurface does not exist; many structures on the surface have foundations or parking garages down to depths at which tunnels are made by underground excavation. Projects may start with an open excavation from the surface and laterally continue as an underground excavation. Construction techniques are not much different either and many techniques originally developed underground are used at the surface and vice versa. Moreover, technically many projects can just as well be constructed in a surface excavation as by an underground excavation. The decision how to build depends then on other factors such as economics, the type of skills of available labour, interruption of surface traffic, public nuisance and possible social unrest, damage to (historical) buildings, and environmental aspects.

4.1.1. Mining v. civil engineering

This book emphasises civil engineering, while mining is considered to a much lesser extent. There are two fundamental differences between civil engineering and mining. The first is that most mining excavations, whether from the surface or underground, are constructed for temporary purposes only. They have to last as long as extraction takes place and are allowed to collapse or are backfilled immediately after. Apart from semi-permanent structures, such as slopes along an access ramp in an open pit or shafts in an underground mine, a lifetime of days or weeks is quite normal. In contrast, fifty or more years are normal lifetimes for structures in civil engineering. A second difference results from the users of the underground space. Miners are used to deal with risks of mine excavations. They are aware that such excavations may not be very stable and prone to collapse in due time. Routinely and mostly instinctively, they check whether the space they are in is still safe. Users of underground civil constructions just expect a stable and safe environment, and do not consider checking its safety. The safety standards for civil and mining engineering reflect these differences (e.g. the ESR values in Section 4.6.1.2). The differences provide an opportunity, but also a possible disadvantage. The necessity to check the actual and short term behaviour of the ground and experiencing not very stable excavations adds to the knowledge of the miners on ground engineering and many develop a better feeling for ground behaviour than civil engineers who encounter ground prone to collapse far less often. This

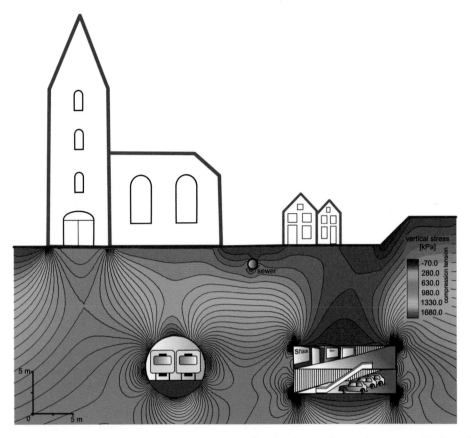

Fig. 4.1. Illustration showing the far more complicated stress loading of underground excavations compared to the surface foundations. The stress contours are the result of the combined effect of the weight of the subsurface materials and the influence of the structures on and in the subsurface. The subsurface is homogeneous with horizontal elasticity properties different from vertical. (From Hack 2012; Examine 2D, Rocscience 2011).

expertise is useful in civil engineering too. The disadvantage may be that former miners working in civil engineering follow practices that are standard in mining but are not sufficient in civil engineering. However, this is controlled by relevant legislation, building codes and laws effective in most countries.

This chapter starts with a very brief description of the fundamentals of geology and origin of ground materials (Section 4.2), followed by geotechnical properties of the subsurface (4.3). Section 4.4 describes and discusses the basics of excavation and support techniques, and 4.5 presents examples of underground excavations and the problems encountered in various environments. Sections 4.6 and 4.7 provide a brief outline of the design methodologies for underground excavations, working practices, and special considerations with respect to hazards and risks. Expected future developments and general remarks conclude this chapter (4.8).

4.2. Geology and ground materials

Geoscientific knowledge of rock and soil properties and geology is essential for functional and safe underground construction. This knowledge builds on the understanding of geological processes that shaped the present configuration of the planet and how the Earth works. Some layers and subsurface materials require very little effort to make and support underground excavations whereas others need advanced techniques. 'Ground' and the structure interact, knowing their mutual interaction is critical for economic and safe designs. 'Ground' is a very general term and may be described as any natural or man-made material present at the construction site. Natural materials may be subdivided in 'soil' and 'rock'. Man-made grounds in fills or dumps may behave mechanically and chemically comparable to natural ground where made of rock or soil but, for example,

Fig. 4.2. Occurrence of weak and inhomogeneous foundation layers indicated by differential settlement at old structures, as for example the Oude Kerk (Old Church) in Delft, The Netherlands. The foundation of the 75 m high tower is on a strongly compressible layer. During construction (between 1325 and 1350 AD), imbalance of the tower caused tilting of the tower foundations. The tower would have collapsed if completed as planned. However, the builders avoided collapse by constructing the top of the tower in a (by then) more upright angle causing the lower part to be more leaning than the top (photograph: D. Van Ree).

dumped household waste may behave quite differently from any natural material. Various definitions for rock and soil exist in geology, engineering geology, and geotechnical engineering. Sometimes, an unconfined compressive strength value (Section 4.3.2) for intact material of 1 MPa is used as the boundary between rock (more than

1 MPa) and soil. As this boundary is rather arbitrary, the authors favour a distinction where 'soil' consists of loose, not cemented particles, and 'rock' of particles interlocked, cemented or otherwise bound together, resulting in a tensile strength, that is the stress required to pull the material apart. A distinction in ground types may be clear in theory, in practice, however, the difference is often much less pronounced. Mechanically weakly cemented rocks often behave as a 'soil' and vice versa, depending on the site conditions, stress environment, and the engineering application.

4.2.1. Rock–soil cycle

A very brief outline of the most common rocks and soils and the geological processes for non-geoscientists involved in decision making on the (urban) underground is presented below. Materials in the subsurface undergo a continuous cycle from rock to soil and back into rock again on long, geological time scales. At the surface, rocks weather continuously as they are subject to atmospheric processes such as temperature variations and rain. Weathering is a complex process of physical, chemical, and biological alteration of rock and soil over time, influenced by the atmosphere and the hydrosphere, and causes rock material to fall apart into smaller pieces or to dissolve in water. Erosion and transport by water in rivers, glaciers or by wind further reduce the size of rock fragments and complete the degradation of rock into soil. After transport, soil and organic remains are deposited in topographically lower situated areas, on- or offshore. Soils may become cemented again due to partial dissolution and re-crystallization of the minerals in the soil or to mineral precipitation out of groundwater. Over time, accumulated soil materials generate increasingly higher pressures on underlying strata (i.e. layers) resulting in lithification that eventually transforms most of the soft soils into rocks again (Fig. 4.3).

Accumulated layers cause the original soil bed to sink deeper into the Earth crust where higher pressure and temperature regimes prevail. These alter the material composition and new minerals may be formed ('metamorphism'). At

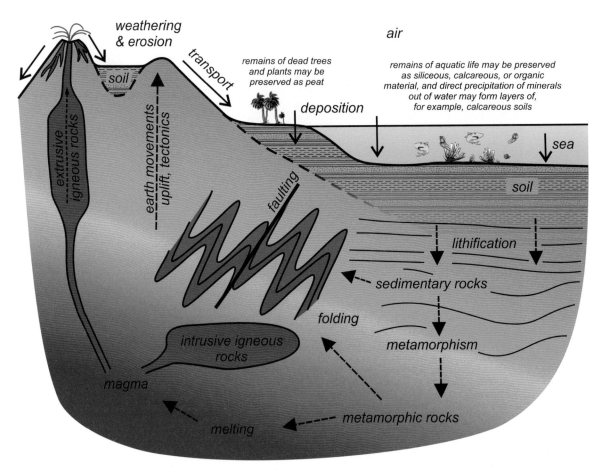

Fig. 4.3. The recurrent cycle of rock and soil formation.

even higher temperatures, the material may melt, intrude in other not-melted material and coagulate as an intrusive material underground, or migrate to the Earth's surface to flow out at the surface, for example, lava flows from volcanoes. After deep burial, all deposits may eventually emerge at the surface again by forces and movements in the Earth (i.e. 'tectonics') where these deposits become subject to weathering processes again, starting a new cycle. Rocks and soils may move to the surface anywhere during the cycle and the variety in composition of soils and rocks may therefore be almost infinitely broad.

Here, only a very brief outline of the most common rocks and soils and the geological processes involved is given. For more extensive descriptions of geology the reader is referred to textbooks such as '*Earth: An Introduction to Physical Geology*' (Tarbuck *et al.* 2010), or '*Historical Geology*' (Wicander & Monroe 2009). A simplified set of descriptions of rock and soil for engineering purposes is given by ISO 14688-1/2:2002/2004 (2002/2004) and ISO 14689-1:2003 (2003). Textbooks such as '*Principles of Igneous and Metamorphic Petrology*' (Philpotts & Ague 2009) and '*Soil and Rock Description in Engineering Practice*' (Norbury 2010) provide details on rock and rock forming minerals.

The subsurface of any site may contain many different rock and/or soil types. They may occur in horizontal, inclined, folded, parallel, or wedging layers, or in irregular volumes. Box 4.1 describes an exposure of rocks and soils and its geological history.

Box 4.1. Geological reconstruction

Figure 4.4 shows an example of a hill composed of various rocks and soils. Granite is an intrusive rock formed after cooling of melted material deep under the Earth's surface. In contrast, limestone (consisting of calcium carbonate), sandstone (cemented sand) and shale (modified clay minerals), are sedimentary rocks normally deposited as horizontal soils and later modified under pressure and temperature, and cemented to rocks. At some point in geological time, layers may have been folded and faulted by tectonic forces.

The million year's geological record in this outcrop may be read as follows: deep in the Earth granite intruded into a country rock. This country rock is not visible in this exposure but may occur below the granite. After cooling, this granite moved upwards to the surface of the Earth where the granite was exposed to weathering processes affecting the top of the granite. Later, the granite moved to a deeper position and sediments (sands, limestone, and clay) were deposited. Again later, tectonics caused faulting, moved the sediments on top of the granite, and tilted the now overlying sediments. This entire complex moved again to the surface and it became exposed to weathering processes. These affected the topmost sediment layers and for a second time the part of the granite that is again exposed to the atmosphere. The groundwater table follows a pattern directly associated with the present-day permeability of the rock strata. Erosion has been responsible for the creation of the valleys on both sides of the hill. Gravity and flowing water transported the weathering products from the higher parts as 'soil' down to the valley bottoms. There may be other, more complex explanations but this is the simplest explanation matching all information available from this exposure and is therefore adopted.

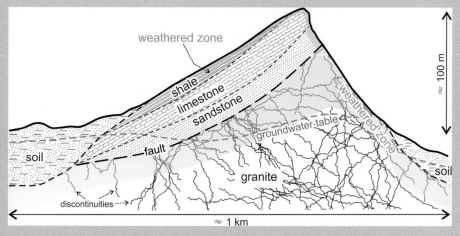

Fig. 4.4. An example of an outcrop with 'soils' and 'rocks'.

Earth materials are composed of minerals, which differ in chemical composition and physical properties. Quartz, a silica oxide, is the most common mineral on Earth and most soils and rocks are composed partly or entirely of this mineral. Quartz is a hard, strong, and abrasive mineral and highly resistant to chemical alteration by weathering. Therefore, some soils (e.g. sands) are composed entirely of quartz grains, as this is the remaining residual product once all other and less-resistant minerals are decomposed. After lithification and cementation of the quartz sands, these may become quartz-sandstones and successively, after metamorphosis, hard, very to extremely strong and nearly always highly abrasive quartzites.

In engineering granular soils are commonly described based on the grain size of the particles; sand has a grain size between 0.063 and 2 mm. Other granular soils are silt (size between 0.002 and 0.063 mm), gravel (2–63 mm), cobbles (63–200 mm) and boulders (rock blocks from 0.2 up to many metres). In a clay the grain size of the particles is less than 0.002 m, however the definition of 'clay' is not unambiguous, as clay may also refer to a soil mainly consisting of clay minerals or refer to plasticity of the soil (see below).

Feldspar, an aluminium-silica oxide, is another mineral commonly found in rocks and soils. When the alteration and decomposition process of weathering has not been completed, the residual material may be composed, partially or entirely, of feldspar grains. Sands with high feldspar content are less abrasive than sands solely composed of quartz grains. Continued weathering transform feldspars into clay minerals.

Clay minerals are the final weathering products of many aluminium-silica minerals and consist of hydrous aluminium-silica with various quantities of water. Many types of clay minerals exist; the most common are illite, montmorillonite, and kaolinite. Clay minerals are small in diameter, flake-shaped and have a relatively large surface area. Wet clay (e.g. soil consisting mainly of clay minerals) behaves plastically, and most wet clays are soft but become firm or stiff with a lower water content. Some clay minerals (e.g. montmorillonite) exhibit a large swelling potential if absorbing water. Soils only composed of clay minerals are not abrasive. Once submerged deeper into the Earth crust, clays may become lithified and cemented to claystones and shales. On geological timescales, shales are transformed into slates, schists, and gneisses when metamorphosed. Claystones and shales are generally not very strong and not abrasive, in contrast to slates, schists, and gneisses. A characteristic of shales is their very thin bedding structure. Deformation and forming of new platy minerals during metamorphism may cause planes with aligned minerals in the rock; so-called 'slaty cleavage' planes in slates and 'schistosity' in schists. These may form planes along which the rock more easily breaks (Section 4.2.4).

Another quite common mineral in the Earth crust is calcium carbonate ($CaCO_3$), the main constituent of shells, foraminifera (very small unicellular organisms with a calcium carbonate skeleton), and coral reefs in the seas and oceans. When the organisms die, their shells and skeletons accumulate on the sea floor and these are subsequently covered by younger deposits. Calcium carbonate also occurs in solution in seawater and may precipitate directly from seawater under specific conditions. Lithification and cementation transform animal remains and direct precipitated carbonate deposits into limestone rock and, after metamorphosis, into marble. Calcareous soils, consisting of calcareous grains, are not abrasive. (Fossil) coral reefs are quite irregular structures and may have variable properties; they may contain voids and holes, often filled by other (e.g. soft soil) material. The strength, of limestone rock may vary from very weak to very strong depending on their structure and on their degree of cementation. Marble is strong to extremely strong. Limestone rock and marble are not very abrasive. In the geological past, conditions have been more favourable for producing calcareous deposits than today, as shown by the widespread occurrence around the globe.

Infrequent at the surface, but rather abundant at larger depths, rocks occur formed by minerals that easily dissolve in water, for example, many sulphates and salts. Most common representatives of this group are gypsum, halite (table salt), and sylvite (potassium chloride). Such rocks behave plastically under surface conditions rather than elastically or in a brittle manner (see Section 4.3) due to easy deformation under pressure by solution and re-crystallization. Generally, they are highly corrosive and soluble in water, not abrasive, not very strong, and notorious for engineering problems.

4.2.2. Cementation and interlocking

Cementation or interlocking between grains or minerals binds grains or minerals in a rock, which is crucial for the strength and behaviour of rocks. The more cementation or interlocking, the stronger and often the harder the rock is. Most common types of cementing minerals include quartz, calcite, clay, and, less frequently, gypsum. Cement and the minerals in soil grains and intact rock blocks are not necessarily identical, for example quartz sandstones are often cemented by calcite or gypsum. The cement bounds in weakly cemented loosely packed granular soils may be broken by a relatively small load exerted on the material causing collapse of the grain skeleton. A similar effect may occur where cement is dissolved by groundwater, for example gypsum or salt cement. Cement consisting of clay particles may also cause a very weak binding between grains that diminishes when becoming wet. Very loosely packed soils with or without very weak, water-soluble cement are commonly known as 'collapsible soils' and may be a serious hazard in underground excavation (Clemence & Finbarr 1981; Houston *et al.* 2002).

4.2.3. Igneous rocks and unstable minerals

Deep under the Earth's surface, pressures and temperature may be high resulting in metamorphic rocks. Under certain conditions, temperatures may become so high that rocks undergo complete melting. After cooling, melted rocks may recrystallize deep below the surface (intrusive igneous rocks) or be transported to the surface and extruded as ash or lava from a volcano (extrusive igneous rocks) (Fig. 4.3). Granite and diorite are examples of rocks cooled and recrystallized deep in the underground, while basalt and andesite are volcanic rocks formed by lava flows. Generally, intrusive igneous rocks and rocks resulting from cooled lava flows are hard, strong, and often abrasive. The cooling process in the Earth takes long, geological times. Minerals formed under such conditions are generally relatively large and stable at the surface. In contrast, minerals in rocks expelled from volcanoes are formed relatively rapidly as lava cools very fast resulting in small minerals some of which are unstable under surface conditions. For the same reason, minerals in volcanic ashes (pyroclastic deposits such as tuff) are notoriously unstable; they disintegrate quite rapidly, dissolve, or weather into clay minerals (Hay 1960; James et al. 2000). These processes generate the rather fertile agricultural soils in volcanic areas. However, these soils may also create major challenges in engineering as their properties can easily change during the lifetime of a construction.

4.2.4. Mass

Rocks and many soils in the subsurface are often arranged in layers or beds. These layers are separated by bedding planes, formed for example when the sedimentation sequence was interrupted, and fractures such as joints, fissures, cleavage, schistosity, and faults caused by tectonic processes, heating, cooling, or by newly formed minerals during metamorphism. Such 'discontinuity planes' divide the ground into blocks or zones of intact material. Blocks or zones of intact material and discontinuity planes are jointly designated as 'rock mass' or 'soil mass' or if undifferentiated: 'groundmass' or 'discontinuous groundmass' (Fig. 4.5).

4.3. Ground properties

The strength and deformation behaviour of ground materials is complex. The range of differently behaving materials and different configurations of the materials is very large and the influence of fluids and gases complicates the behaviour even further. As a consequence, concepts for generic characterization are also complex and new concepts are developed quite regularly. Frequently, actual ground behaviour deviates from that as theoretically forecasted. Descriptions of strength and deformation as given in this book are simplified while sophisticated theories, such as 'critical state', or details of rheological behaviour have not been incorporated here. In the following sections, some of the most relevant soil and rock properties for creating underground excavations are described briefly. Descriptions that are more detailed can be found in Bell (2007), Wyllie & Mah (2004), Hoek & Brown (1990), Hunt (2005), Price et al. (2009), Schofield (2005), and through GeotechLinks (2010).

4.3.1. Strength, deformation, and failure

Material failure is commonly used to describe the breakdown of material if external forces are higher than the material strength and cause it to break. In ground mechanics, however, the concept of failure is less clear. Some ground materials may 'fail' in the sense that these break when their strength is exceeded but many other ground materials do not. They just deform under a load by the relative

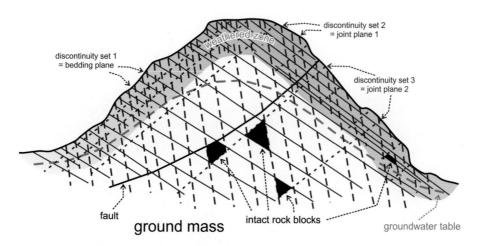

Fig. 4.5. Discontinuous groundmass. In the weathered zone discontinuity spacing and size of blocks reduce in most rock masses (from Hack 2012).

movement of grains in a soil or blocks of rock in a rock mass. Moreover, grounds in a particular situation may do both, that is grains or blocks move but break as well. That may happen simultaneously or subsequently. The rate of deformation, that is the deformation in time, may reduce or increase, occasionally resulting in breaking of the material after a long period. Hence, 'failure' as used in geotechnics is often not related to the actual breaking of ground material but rather to the practical functionality of the application. For example, failure of the ground is considered to have occurred when the construction does not fulfil its design criteria anymore due to the behaviour of the ground.

4.3.2. Intact ground strength

If an underground opening is made, stress patterns are changed and the underground material deforms. Under stress, the material and cement may break, and individual particles, crystals, or grains may break or move relative to each other in shear (Box 4.2).

Strength properties of intact material vary widely depending on factors such as mineral type and size, interlocking, cementation, and geological history. Therefore, it is hardly possible to generalize strength values (Table 4.1). The strength of an intact block of unweathered granodiorite is

---

Box 4.2. Strength of natural materials

Natural materials exhibit strength. In popular day-to-day understanding there are three types of strength: the strength of a material to withstand a pushing stress or pressure, that is the compressive strength, the strength to withstand a pulling stress, that is tensile strength, and the resistance against shear displacement, that is the shear strength. Mostly not obvious in everyday use is the dependency of strength on the confining pressure, that is the pressure exerted on the sides of the material that are not stressed by the compressive or tensile stress. Therefore, a difference is made in so-called 'unconfined' and 'confined' strength. Unconfined Compressive Strength (*UCS*) and Tensile Strength (*TS*) are the compressive respectively tensile strength of material without confining pressure (Fig. 4.6a respectively b). Shear strength is depicted in Figure 4.6c and triaxial (=stress along three axes) strength is the strength under confining pressure (Fig. 4.6d).

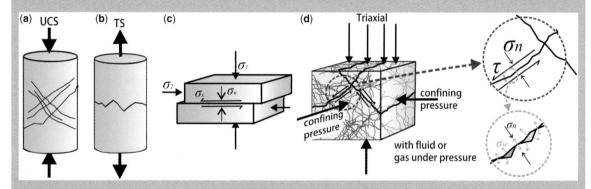

Fig. 4.6. Strength.

Shear strength is one of the most important properties of ground material, both for intact material and for groundmasses. If loaded, many materials fail in shear. Figure 4.6c shows two blocks loaded under shear and normal stress simulating, for example, the contact between two grains in a soil, or a joint or fracture (i.e. a discontinuity) in a rock mass. When the horizontal stress (σ_2) exceeds the shear strength between the two blocks (σ_s), the top block will move (e.g. 'fail'). The shear strength depends on the stress normal to the plane of movement (σ_n); the higher the normal stress, the higher the shear strength. The pressure of fluid or gas in-between the two blocks reduces the normal stress and thus the shear strength. Shear strength of grain contacts or discontinuities is mostly approximated with a relation that depends on the normal stress on the shear plane and two material properties the so-called *cohesion'* (or *Si'*) and 'effective angle of shearing resistance' denoted by φ'. The shear strength increases with higher *cohesion'* and φ'. The apostrophe after the symbol is commonly used to indicate that the properties are in terms of effective stress on the contact plane, that is the stress corrected for possible fluid or gas pressures.

Figure 4.6d shows a mass under confining pressure. Failure occurs along planes where the shear stress (τ) exceeds the shear strength along the plane. Shear failure in a mass occurs due to breaking of crystals, grains, or the contacts between them, movement along grain contacts or by movement along new or pre-existing discontinuity planes. Again, the pressure of fluid or gas (σ_w) present in the pores of the material or in the discontinuities reduces the shear strength. Commonly, the strength of a groundmass is expressed also in terms of cohesion' and φ' (then mostly denoted as 'effective angle of internal friction'). Note that these do not need to be the same as those for the shear strength between grain contacts or discontinuities. Strength may vary with direction due to the orientation of the grains or minerals, small micro-fractures, joints, etc.

Laboratory strength tests are normally done on fairly small samples in the order of 10 cm length and 5 cm diameter. Unconfined compressive and tensile strength of intact rock are measured routinely in the laboratory. Shear strength values of soils and discontinuities in rock masses are measured routinely as well, but very seldom, those of intact rock as the tests require high stress levels for most rocks that demand for expensive test equipment. Tests done in the laboratory on groundmasses are also cumbersome and expensive because the samples and test equipment need to be very large to get representative results. Therefore, most published strength property values for masses are based on classification systems (Section 4.6.1), on back-analyses from slope, foundation, and tunnel projects, or from field tests. Simple (field) tests are often done to get an idea about the strength of intact material and shear strength of discontinuities (e.g. BS 5930 1999 and ISO standards, see Chapter 5).

very high but the same rock type nearby, but weathered, may have a very low strength (Section 4.3.9). Another example is the difference in strength between Palaeozoic limestone in Belgium and the Late Mesozoic limestone in The Netherlands, just 20 km apart (respectively Vinalmont and Sibbe limestones, Table 4.1).

4.3.3. Deformation, stress and strain of soil and rock

If a stress is exerted to a body of ground, the ground will deform. Stress and deformation are coupled properties; there is no deformation without stress and vice versa (Box 4.3). Table 4.2 shows deformation moduli, which are the ratio of stress over the obtained deformation per unit length, for various ground materials and masses. The wide range of possible moduli implies that some grounds show almost no deformation whereas others deform considerably while loaded.

4.3.4. Time effects

A new excavation changes the stress field in the surrounding groundmass. Such changes generate new and different deformation of ground materials, flow of water and gas, development of (micro-) cracks, and (small) movements along new cracks, existing discontinuities, and contacts between crystals and grains. Adjustment to the new stress regime takes time and may lead to failure as new and existing cracks or crystal and grain orientations may line-up to create new failure planes. New stress concentrations may cause ground materials to break while water may flow into previously dry positions and soften for example clay minerals. Originally, stable excavations may thus become unstable after some time.

Apart from adjustment and ageing effects as described above, most ground materials display (very) long-term creep effects over tens to many millions of years (Box 4.3). Most salts and ice display steady-state behaviour, whereas most brittle rocks, such as granite, gneisses, and strong limestones, show increasing deformation rates. Long-term creep is probably responsible for excavation collapses after very long time spans, sometimes up to 2000 years. The mechanisms for long-term creep are still poorly known in detail, but re-crystallization, micro-crack formation under stress and weathering may play a role in the process. Knowledge about long-term creep has significantly increased by extensive research on the long-term behaviour of host rocks for underground nuclear waste repositories (Cui *et al.* 2009; Hudson *et al.* 2009).

4.3.5. Discontinuities

A discontinuity is a plane that marks a change in physical or chemical material characteristics (Figs 4.8 & 4.9). Two types of discontinuities may be discerned: integral and mechanical discontinuities. Mechanical discontinuities are planes of physical weakness, where tensile strength across or shear strength along a discontinuity is less than those of the surrounding ground material. Integral discontinuities are about equally strong as the surrounding material but these may develop into mechanical discontinuities by weathering, stress changes, or chemical reactions altering the mechanical characteristics. A discontinuity can be any plane including bedding planes, joints, fractures, and faults. The shear strength of a discontinuity depends on persistence, material friction of the surrounding material, the roughness and strength of the discontinuity walls, and infill material if present. The orientation of discontinuities together with accompanying shear strength determines options for movement along such planes. The influence of discontinuities on underground excavations has been extensively described by the references given at the start of this chapter and by Anon (1990) and Blue Book (2007).

4.3.6. Groundmass and discontinuities

A groundmass is an assemblage of blocks of intact material with discontinuities in-between. The groundmass may have similar properties for intact material and discontinuities

Table 4.1. Examples of soil and rock mass properties

Name	UCS (MPa)	TS (MPa)	φ' (degrees)	Cohesion' (MPa)	Range of confining pressure (MPa)
Soil					
Sand (rounded particles) (loose to dense)[1](a)			27–35	0	<0.5
Sand (angular particles) (loose to dense)[1](a)			33–45	0	
Sandy gravel (loose to dense)[1](a)			35–50	0	
Residual soil & fill[b c]			25	0	<0.1
Clay (undrained) (very soft to hard)[2](d)			0	0.01–0.3	<0.5
Clay (drained)[2](a c)			20–35	0–0.15	
Peat (drained)[2](c)			20	0.006	<0.15
Frozen dense sand (artificially frozen, T ~ −10 °C) (short/long-term)[3](e f)	7/4	20–50% of UCS	38/22	2/1.4	<1.4
Frozen stiff clay (artificially frozen, T ~ −10 °C) (short/long-term)[3](e f)	6/1.5		1.5/7.5	0.8/0.6	
Ice (natural fresh water ice)[3 4](g h)	1–18	1.3	25–48[i]	0.25[i]	<0.2[i]
Intact rock					
Vindhyan sandstone[j]	102	6.9	37	33	<15
Hawkesbury sandstone[k]	21–60	3.5	47	4.2	<20
Eagle Ford Shale[l]	2.1	0.93	24	0.41	<3
Yucca Mountain-Topopah Spring Tuff[m]					
(TSw2/Tptpmn)	187	11.6	48	40	<15
(TSw3/Tptpv)	16	4.0	47	3.5	<10
Vinalmont limestone[m]	190	7			
Sibbe limestone[m]	3.5	0.38			
Königshain Granite[5](o)					
Slightly weathered	185	18			
Moderately weathered	38	3.5			
Highly weathered	13	1			
Completely weathered			25	0.015	<0.4
Residual soil			35	0.025	

Rock mass[6]				
Sheared flysch[7][p]	4.5		16	0.073	~2.5
Mu-Cha Tunnel Fault (sheared sandstone & shale in clay matrix)[7][q]	2.6		28	0.1	~3.5
Falset Granodiorite[6 8][r]					
Fresh (zone 1)	175		47	0.017	
Slightly weathered (zone 1–2)	110	10.2	46	0.016	
Moderately weathered (zone 3)	80	4.1	38	0.014	
Highly weathered (zone 4)	3	2.7	17	0.008	<0.6
Completely weathered (zone 5)	0.5		6	0.003	
Falset Lower Muschel-kalk Limestone[8 9][r]					
Large blocky	80	8[s]	62	0.027	
Small blocky	70	8[s]	18	0.007	
Sydney-Gunnedah Basin coal[7][t]		0.8	38	1.900	~12.5
Deriner Granodiorite[7][u]	80	16	40	0.350	<1.5

Abbreviations: UCS, Unconfined Compressive Strength and TS, Tensile Strength are the compressive respectively tensile strength of intact material without confining pressure. φ', effective angle of shearing resistance and *cohesion'* are expressions indicating the strength of a material or a mass under confining pressures, with the range of confining pressures as the properties are only validated within the given range.

Notes: (1) 'Loose to dense' refer to the packing of the particles, which influences the φ values. (2) Undrained and drained refer to the dissipation of pore water pressures during loading; generally, undrained applies to fast loading situations and drained for slow. Values reported for undrained strength are 'S_u' (or 'c_u'). (3) Values only indicative; strongly dependent on test conditions, deformation rate, compaction, temperature, and ice structure and quantity. Strength is the highest stress sustained in the test; highest values for fast loading. (4) UCS and TS at temperature -1 to $-16\,°C$; shear properties at $-2\,°C$; shear samples contain rock fragments and air bubbles. (5) Weathering classifiers (indicate an increasing grade of weathering from fresh to slightly weathered, etc.) follow ISO 14689–1, 2003 (the zones follow BS 5930;1999 and not the replacement standard ISO 14689–1 as the replacement is at present considered by some to be inferior to the BS 5930, Price *et al.* 2009). (6) UCS and TS from intact rock. (7) φ' and *cohesion'* determined by rock mass classification and/or back analyses from tunnel construction. (8) φ' and *cohesion'* back analysed from slope engineering. (9) Large blocky implies that most of the blocks in the rock mass are about equi-dimensional with sides between 0.6 and 2 metre, while small blocky implies equi-dimensional with sides between 6 and 20 cm (ISO 14689–1).

Note the large variation between different materials, the wide variation within the same material due to different states of weathering, and the large influence of block size on mass properties while the intact material strengths (UCS and TS) are about the same.

Source: Data from (a) Craig (1978), (b) Chan *et al.* (2005), (c) Bosch & Broere (2009), (d) BS 5930 (1999), (e) Jessberger *et al.* (2003), (f) Zhang *et al.* (2007), (g) Schulson (1999), (h) Gagnon & Gammon (1995), (i) Arenson *et al.* (2003), (j) Dubey (2006), (k) Pells (2004), (l) Hsu & Nelson (2002), (m) Ciancia & Heiken (2006), (n) Swart (1987), (o) Thuro & Scholz (2004), (p) Marinos *et al.* (2009), (q) Yu (1998), (r) Hack (1998), (s) Kouokam (1993), (t) Sainsbury (2009), (u) Cekerevac *et al.* (2009).

Box 4.3. Deformation of natural materials

The deformation per unit length is strain (ε). Under influence of a stress in a particular direction, the material becomes shorter in that direction and wider perpendicular to the stress direction. The amount of shortening in relation to the stress is expressed by the elastic deformation or Young's modulus (E). The amount of widening in one direction related to the shortening in the other direction is determined by the Poisson's ratio (ν). Deformation modulus and Poisson's ratio are anisotropic for most groundmasses, that is the values vary with direction. Figure 4.7 shows various stress–strain curves.

Figure 4.7 (a) shows a linear–elastic deformation; that is deformation increases linearly with stress level; on release of the stress, the sample will return to its original volume and shape. In (b) the material behaves elasto-plastic: the first part is elastic deformation while in the second (plastic) part the deformation increases under constant stress; Figure (c) is similar to (b) but at the boundary between elastic and plastic, the material breaks (brittleness). Figure (d) shows the deformation behaviour of most real ground which is a combination of elastic, plastic and brittle deformation. Intact ground, in particular rock, may deform about elastically for stresses up to 50–80% of the UCS value, however, rock masses very seldom behave elastically but mostly deform permanently with displacements along discontinuities (Box 4.2). Under higher confining pressure or temperature, most materials show a more gradual deformation without brittleness, such materials are said to deform ductile.

Figure 4.7e shows various options for strain v. time for long-term deformation under constant stress. Some materials deform with an increasing deformation rate leading to failure quite rapidly (curve i in Fig. 4.7e). Other materials under stress deform very slowly. Over time, deformation rates may attenuate (curve iv), steady state (curve iii), or re-accelerate after a long steady state period resulting in failure (curve ii). Low-permeability ground shows a time-dependent behaviour because water in pores and discontinuities influences the deformation and makes the deformation time dependent (Fig. 4.7f). The deformation of low-permeability soils is normally denoted consolidation, and characterized by the 'coefficient of consolidation' (c_v – a smaller value indicates that more time is required for consolidation) for short-term deformation, and by the secondary consolidation coefficient for long-term deformation (C_α – a larger value indicates more consolidation in a given time span). The first is mainly related to expelling water from pores, while the second is more related to re-arrangement of grains and changes in material; the latter in particular in organic soils, such as peat.

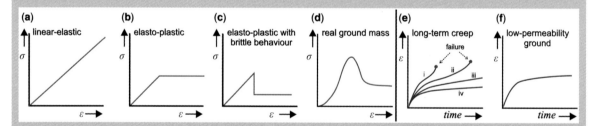

Fig. 4.7. Various stress–strain (deformation) behaviours; σ = stress; ε = strain (Fig. 4.7e: modified from Dusseault & Fordham 1993).

throughout or these may vary causing inhomogeneity. Nearly always, a groundmass is anisotropic because of the presence of discontinuity planes. The overall impact of discontinuities on groundmass behaviour is that they weaken the groundmass by their lower shear and tensile strengths. Infill material in discontinuities may considerably affect mechanical and hydrological characteristics, in particular if consisting of soft materials as clay. A groundmass with discontinuities may be deformed more easily than the same volume of intact material (Goodman 1989). Such deformation will normally occur by relative movement along discontinuities rather than by elastic deformation. Tensile strength of a groundmass with discontinuities is low or equal to zero. Porosity of a discontinuous groundmass is higher than of intact material due to the storage capacity of discontinuities. Its permeability is often much higher due to hydraulic conductivity through discontinuities. Discontinuities lead to anisotropic behaviour of the groundmass and properties, such as deformability and permeability. Discontinuous groundmasses are three-dimensional features, anisotropic in two or more dimensions, and groundmass properties may vary widely making each groundmass unique (Tables 4.1 and 4.2).

4.3.7. (Ground) fluids and gases

Groundwater may be present in porous groundmasses (Fig. 4.10). The porosity is determined by its pore space

Table 4.2. *Deformation properties of various ground materials*

Name	E (GPa)	ν
Soil		
Doha marine loose sand[1)(a)]	0.02	
Sand (Amsterdam)[(b)]	0.035–0.04	0.2
Residual soil & fill[(b,c)]	0.01–0.04	0.15
London Clay (drained)(depending on depth and direction)[(2)(d)]	0.007–0.2	0.125
Aeschertunnel glacial till[(5)(e)]	0.08	0.2
Clay (Amsterdam)[(b)]	0.01	0.15
Peat (Amsterdam)[(b)]	0.002	0.15
Frozen dense sand (artificially frozen, $T \sim -10$ °C) (short/long-term)[(3)(f)]	0.75/0.33	0.3[(g)]
Frozen stiff clay (artificially frozen, $T \sim -10$ °C) (short/long-term)[(3)(f)]	0.3/0.12	0.006–0.13[(h)]
Ice (natural fresh water ice; $T \sim -5$ °C)[(3)(i)]	10	0.33
Intact rock		
Hawkesbury sandstone[(j)]	6–14	0.15
Falset Carboniferous sandstone[(k)]	35–60	0.1–0.2
Vinalmont limestone[(l)]	70	0.31
Sibbe limestone[(l)]	1.2	0.25
Königshain granite[(4)(m)]		
Slightly weathered	50	
Moderately weathered	25	
Highly weathered	15	
Äspö slightly fractured diorite and granite[(n)]	69–79	0.21–0.28
Rock mass		
Hawkesbury sandstone[(6)(j)]	0.05–2.5	
Sheared flysch[(6)(o)]	0.433	
Marly shale (standard zone)[(p)]	1.8	
Mu-Cha Tunnel Fault (sheared sandstone and shale in clay matrix)[(6)(q)]	0.2	0.3
Sydney-Gunnedah Basin coal[(6)(r)]	2.5	0.24
Äspö slightly fractured diorite and granite[(6)(n)]	55	0.26

Abbreviations: 'E', the deformation modulus; 'v', the Poisson's ratio; values for 50% of the failure strength if reported.

Notes: (1) 'Loose' and 'dense' refer to the packing of the particles. (2) Undrained and drained refer to the dissipation of pore water pressures during loading; generally, undrained applies to fast loading situations and drained to slow. (3) Values only indicative; strongly dependent on test conditions, deformation rate, compaction, temperature, and ice structure and quantity; highest values for fast loading. (4) see Table 4.1. (5) Glacier deposit: clayey sand and silt, with gravel and isolated boulders. (6) Properties determined by rock mass classification and/or back analysis from engineering construction. Data from: (a) Chen (2010), (b) Bosch & Broere (2009), (c) Chan & Stone (2005), (d) Karakuş & Fowell (2005), (e) Coulter & Martin (2004), (f) Jessberger *et al.* (2003), (g) Kirsch & Richter (2009), (h) Lee *et al.* (2002), (i) Schulson (1999), (j) Pells (2004), (k) Kouokam (1993), (l) Swart (1987), (m) Thuro & Scholz (2004), (n) Andersson (2010), (o) Marinos *et al.* (2009), (p) Alejano *et al.* (2008), (q) Yu (1998), (r) Sainsbury (2008).

and discontinuities while the storage capacity for fluids or gases in a groundmass is determined by its porosity and permeability. Below the groundwater table, pores and discontinuities are normally filled with water, while water may occur only temporarily above it, for example during and shortly after rainfall. Groundwater tables may vary over time. In urban areas, water may always be present in the ground even well above the groundwater table, as sewage systems always leak and gardens may be sprinkled. Constructions may affect groundwater conditions, for example saturation of the groundmass by impoundment or drying by additional drainage, for example through excavations. For more extensive descriptions of (ground)water in the underground is referred to textbooks, such as Hendriks (2010).

Groundmass permeability may show large differences due to, for example, local cementation or presence of small quantities of clay closing the openings between pores and flow channels in discontinuities. Generally, permeability is high in layers consisting of boulders, gravels and sand, and low in aquitards such as clays. In rock masses, permeability through intact material ('primary' permeability) is low and fluid or gas flow is mainly through discontinuities ('secondary' permeability). In soluble rocks as limestones, karst may have formed causing very large permeability, for example, caves with underground streams or rivers, such as the river Lesse, Belgium (Caves Han-sur-Lesse 2010).

Presence of water influences mechanical characteristics of a groundmass. It adds to its weight and may weaken and soften some materials, for example clay. In particular, in fine-grained groundmasses fluctuations in water presence may cause new mechanical discontinuities to develop that

Fig. 4.8. Discontinuities governing stress and deformation around tunnel and stability of entrance portal slopes (only a few discontinuities are indicated; Bhatwari, Uttarakhand, Himalaya, India; photograph: R. Hack).

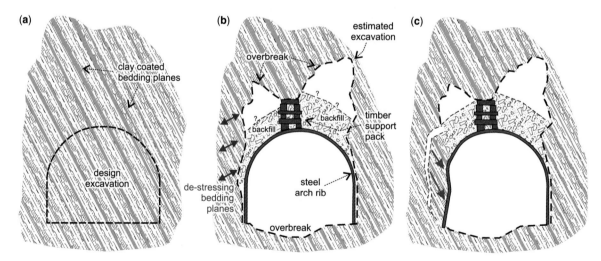

Fig. 4.9. A classic example of the influence of discontinuities on the stability of a tunnel. In (**a**) and (**b**) overbreak during excavation, that is more material was excavated than intended, allows de-stressing of the normal stress perpendicular to the clay-lined bedding planes, causing a reduction of the shear strength along the bedding planes, and allowing the ground-mass to move from left to right bending the steel support arches and leading to subsequent collapse (**c**). (Castaic dam, 40 miles North of Los Angeles, USA) (modified from Arnold *et al.* 1972).

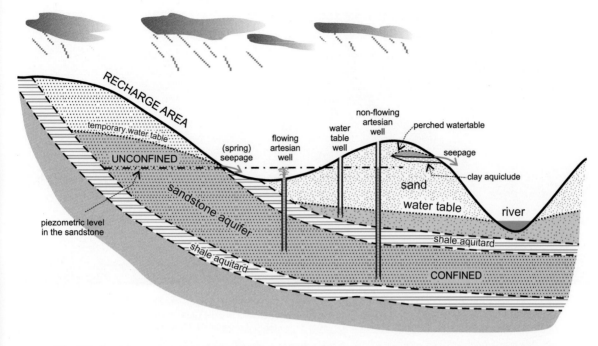

Fig. 4.10. Groundwater presence and pressure in the subsurface depend on the recharge capacity (e.g. from rainwater) of the groundmass, and the geo-(hydro)logical structure and conditions. An aquitard and aquiclude are impermeable layers or bodies of ground that serve as a boundary for groundwater flow while an aquifer is a permeable water-bearing layer. Aquifers may be unconfined or confined. The piezometric level is the level to which confined water will rise in a borehole or excavation (modified from Price *et al.* (2009) with kind permission from Springer Science + Business Media).

reduce the block size and negatively affect its engineering properties (Brattli & Broch 1995).

4.3.8. Influence of water pressure

Water occurrence in pores and discontinuities may reduce the shear strength along grain contacts and discontinuities and thus contribute to groundmass deformation (Box 4.2). However, small volumes of pore water may generate negative pore water pressures increasing the tensile and shear strength (Fig. 4.11). Water is also contributing to the bearing capacity of a groundmass. The ground material together with the water in saturated ground carries the weight of overlying materials and structures. Removal of water by draining or pumping causes a higher stress on and thus compression of the ground material. In turn, this may cause settlement of the overlying layers and of structures at the surface (see also example Amsterdam in Section 4.5.10).

Very problematic and often dangerous conditions arise when pore water pressure causes a sudden large drop of shear strength between particles in soils. Loosely packed, fine-grained granular and groundwater-saturated soils may liquefy when agitated by a relatively small tremor, such as a small earthquake or by excavation vibrations. The loosely packed grain skeleton compacts, but because of the relatively low permeability of the fine-grained soil, the water cannot flow away in time before the next tremor. The repeated pressure built-up in the water becomes larger than the pressure between the soil grains, and consequently the shear strength between the grains becomes zero. At the surface this results in loss of bearing capacity (e.g. tilted and overturned buildings). In the underground, this results in settlement and changes in stress distributions in the ground and uneven stresses on the support structures (Chou *et al.* 2001). Quick sand results from a different mechanism, but with a similar result. Upward water flow reduces the pressure between the grains and thus the shear strength between grains in the soil. Especially during construction of underground excavations, quick sand conditions may happen if water pressures are not carefully controlled (Xu *et al.* 2009). Collapsible soils (Section 4.2) may also liquefy because of collapse of the grain skeleton with excess pore water pressures as result. Quick clay deposits contain clay minerals in a card box structure with very large quantities of water (up to 80%) and are thixotropic, that is the layers become liquefied after agitation (e.g. traffic vibrations, small earthquake), and the shear strength is strongly reduced (Tørum *et al.* 2010).

Fig. 4.11. Small quantities of water in sand create negative pore water pressures and higher tensile and shear strengths in humid sand (left side; dark beige colour) which allows an unsupported tunnel to exist while a tunnel in the dry sand on the right (light beige) collapses immediately (right duck about 8 cm high) (photograph: J. G. A. M. Arnoldus).

4.3.9. Weathering

Physical weathering results in disintegration of intact material into progressively smaller fragments. This is caused by temperature variations resulting in differential expansion and shrinkage of minerals, by water pressures in the spaces between individual solid particles (e.g. in the 'pores'), freezing and thawing of water, or swelling and shrinkage of clays due to water absorption. Chemical weathering is a process that results in decomposition of minerals due to (re-) crystallization pressures, hydration of minerals, and so on. Water and groundwater with dissolved chemical agents are key drivers for reaction with ground material. Biological influences, induced by living organisms (plants, bacteria, worms, etc.), may also cause physical and chemical weathering. Normally, weathering takes place at the surface (see Section 4.2 and Fig. 4.4) where exposure to temperature fluctuations, vegetation, and precipitation is most prominent.

Weathering has a very significant impact on geotechnical properties of rock and soil masses (Hencher & McNicholl 1995). Extensive schemes with descriptions and weathering classification provided the basis for the geo-engineering standards BS 5930 (1999) and ISO 14689-1 (2003). A distinction is made between weathering of intact rock and of rock masses (Table 4.3).

A weathered rock mass surface often consists of a zone of weak and loose material. Intact rock may be present there, but it is often weaker at the surface than the deeper seated, non-weathered, intact rock. The spacing between discontinuities at the surface is smaller and thus the block size, as integral discontinuities have developed into mechanical discontinuities, and the shear strength of discontinuities is reduced (Fig. 4.12). Weathering begins from discontinuities and develops further into the rock material, causing a rather gradual increase in strength in that direction. Depending on the solubility of the material, some or rather substantial quantities of the original rock mass may have dissolved in water and disappeared. The impact of weathering is well demonstrated in the mass properties displayed in Figure 4.13.

Karst is a special case of weathering in limestone, gypsum, or salts where the original rock mass has been dissolved and removed by groundwater (Pesendorfer & Loew 2004; Schmitz & Schroeder 2009). The dissolution process may result in caves. Less common is biochemical weathering by oxidation of pyrite (an iron sulphide, FeS_2). Influenced by bacteria and oxygen-rich water pyrite decomposes into iron oxide and sulphuric acid. Iron oxide is weaker than the original pyrite and reduces mechanical strength parameters in the rock while the acid may be aggressive to the groundmass, engineering structures, and the environment (Brattli & Broch 1995; Oyama & Chigira 2000). Under specific geological conditions, weathering may result in (highly poisonous) arsenic groundwater, for example, in parts of India and Bangladesh (Ghosh & Singh 2009). Arsenic occurs in small quantities in combination with iron minerals and can be set free through desorption from iron hydroxide coatings, the oxidation of sediment locked pyrite (FeS) and the degradation of organic matter resulting in the reduction of iron (Nickson *et al.* 2000). Also see Section 6.2.3. Another type of chemical weathering is oxidation of peat that may disintegrate organic matter causing

Table 4.3. *Weathering description and classification (modified from tables 2 and 13 from (NEN-EN-) ISO 14689-1:2003, with permission from NEN/ISO, Delft, The Netherlands)*

Term	Description	Grades
Intact rock		
Fresh	No visible sign of weathering/alteration of the rock material.	
Discoloured	The colour of the original fresh rock material is changed and is evidence of weathering/alteration. The degree of change from the original colour should be indicated. If the colour change is confined to particular mineral constituents, this should be mentioned.	
Disintegrated	The rock material is broken up by physical weathering, so that bonding between grains is lost and the rock is weathered/altered towards the condition of a soil in which the original material fabric is still intact. The rock material is friable but the mineral grains are not decomposed.	
Decomposed	The rock material is weathered by the chemical alteration of the mineral grains to the condition of a soil in which the original material fabric is still intact; some or all of the mineral grains are decomposed.	
Rock mass		
Fresh	No visible sign of rock material weathering; perhaps slight discoloration on major discontinuity surfaces.	0
Slightly weathered	Discoloration indicates weathering of rock material and discontinuity surfaces.	1
Moderately weathered	Less than half of the rock material is decomposed or disintegrated. Fresh or discoloured rock is present either as a continuous framework or as core stones.	2
Highly weathered	More than half of the rock material is decomposed or disintegrated. Fresh or discoloured rock is present either as a discontinuous framework or as core stones.	3
Completely weathered	All rock material is decomposed and/or disintegrated to soil. The original mass structure is still largely intact.	4
Residual soil	All rock material is converted to soil. The mass structure and material fabric are destroyed. There is a large change in volume, but the soil has not been significantly transported.	5

reduction in volume, and eventually causing land subsidence (Chapter 3).

The rate of weathering is highly variable and strongly depends on local circumstances, as climate, type of groundmass and dissolved chemicals in (ground) water (e.g. fertiliser). Weathering may have a significant impact on an engineering structure within a few years whereas in other conditions no such influence is observed for centuries (Huisman *et al.* 2006). Consequently, the same holds for the weathering depth in the groundmass. In humid and warm environments, *in-situ* weathering may go down to tens and often more than one hundred metres below the surface (Lumb 1983; Shengwen *et al.* 2009). In dry climates, however, the *in-situ* weathered zone is often just a few decimetres or metres deep. The depth of the weathered zone is less where weathered material was removed by the erosion processes (Section 4.3.11).

Materials deep below the Earth's surface may also be subject to active *in-situ* weathering, for example in rock masses surrounding a permeable and water-bearing fault (Katongo 2005). Tectonic and sedimentary processes may have moved weathered material to deeper positions, and they can be encountered almost anywhere in the Earth crust (Fig. 4.4). Moreover, relic weathered material may be found in places where such weathering would not occur today due to changed climatic or environmental conditions (Harris *et al.* 1996; Olesen *et al.* 2007). The same applies to karst features, for example, encountered during construction of the Milwaukee's Deep Tunnels (Day 2004). In this project, bored tunnels for water at depths between 80 and 100 m were constructed in calcareous rocks overlain by much younger, glacial deposits. At the surface, no indications of karst were encountered and evidence from the site investigation was inconclusive. During construction, however, karst features partially predating the glacial deposits caused many problems and partial collapses, resulting in higher costs and delays.

Inside underground excavations, very little, to negligible *in-situ* weathering takes place (this, however, does not apply to degradation of the ground due to stress re-distribution and creep effects). Inside an excavation, environmental conditions are much less variable than at the surface where cyclic variations are the main cause for weathering. There is no direct sunshine with associated heating and cooling, nor wetting and drying cycles due to rainfall and wind in the underground. Therefore, weathering is normally not of great concern for underground excavations except in grounds with unstable or soluble materials or with aggressive groundwater. Rather than weathering, side effects of the use of underground space such as traffic vibrations or radiation of nuclear waste may reduce the lifetime of an underground excavation (Section 4.4).

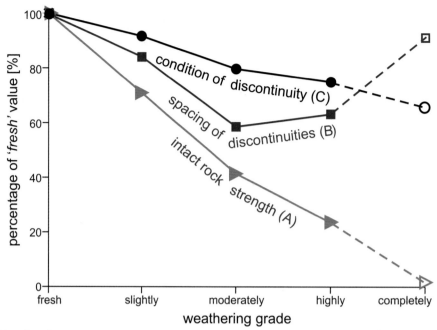

Fig. 4.12. Influence of weathering on intact rock and rock mass properties (from Hack 2012).

4.3.10. Crusts and other 'strong' layers

Normally, weathering makes the groundmass weaker. However, weathering and other groundwater-related processes may lead also to a stronger groundmass by the creation of crust layers. Examples are 'duricrust', 'hardpan', 'calcrete', and 'caprock' in dredging. The use of caprock here, is not to be confused with a '*caprock*' in geomorphology which means a stronger or more resistant rock overlying a weaker or less resistant rock, a '*caprock*' capping a salt body, or a '*caprock*' in the petroleum industry which refers to an impermeable layer capping an oil reservoir. Many more 'crusts' are defined depending on origin and materials (Chesworth 2008). The most important for engineering are described briefly below. Definitions of crusts are not unambiguous and often crusts are simply called a 'cemented horizon' or 'hard layer'. Processes responsible for crust development are precipitation of salts or groundwater-dissolved minerals, chemical reactions between minerals in different groundwater flows or between groundwater minerals and ground materials. Biological processes in the ground, vegetation and influence of humans, for example fertiliser, may also influence the forming of crusts. Some processes typically occur at the surface, such as related to seepage and evaporating groundwater, others at depth, or at or below the sea floor.

A duricrust denotes a strong and often hard layer at or near the surface, normally less than a decimetre thick formed by the accumulation of precipitated minerals from water by capillary action and evaporation. A hardpan may be synonymous to a duricrust, but more commonly refers to an impermeable, strong layer in a soil mass just below the surface, consisting of soil particles bound by silica or iron oxides with calcium carbonate. Hardpans may also be man-made due to compaction by heavy traffic or cementation by pollution chemicals. The strength of duricrusts and hardpans ranges from material that is easy to crumble by hand to strong rock. The term 'caprock' is used in dredging for any relatively thin but stronger cemented layer (with UCS (Unconfined Compressive Strength) from about 10 to 20 MPa) in an otherwise weaker rock or soil mass. In the Arabian Gulf, caprocks are generally layers of conglomerate or breccia (layer of rounded respectively angular rock fragments cemented together), or sand, cemented by calcareous,

Fig. 4.13. Exposure of weathered granodiorite (an intrusive, igneous rock comparable to granite), Falset, Spain. Zones of fresh, hard, and strong granodiorite (bluish-grey coloured) occur between zones with granodiorite decomposed into loose soil material (brownish coloured) (photograph: W. Verwaal).

siliceous, or ferruginous material, normally between 10 and 50 cm thick, and may occur as multiple closely spaced layers at or near the sea floor (Verhoef 1997). Calcretes are comparable to caprocks formed with calcareous cement near the surface, but onshore (Alonso-Zarza & Wright 2010). Calcretes may be metres thick, laterally extend over hundreds of metres and can be very strong with UCS values exceeding 100 MPa.

Crust forming processes depend on many local and regional variables that may differ over short distances both horizontally and vertically (Vervoort & De Wit 1997; Khalifa *et al.* 2009). In arid and tropical environments, crusts and hardpans are quite common but also in Mediterranean climates, highly irregular and often-strong crust (e.g. calcrete) layers may be encountered. In the geological past, crust layers may have developed in areas where they would not develop at present. Often, the occurrence of crust layers is difficult to predict.

4.3.11. Erosion

Erosion is a process of removal of materials by natural agents, for example water, wind, snow, and ice (in some literature, erosion may also include fragmentation of material). The erosion process predominantly occurs at the Earth's surface and is generally less relevant in underground works. However, underground water flows, including water leaking from sewage and water mains, may transport soil or infill materials. This may cause stress relief and displacements in the mass surrounding the underground excavation, which may eventually cause collapse. In contrast to the process itself, erosion products are common in subsurface layers as filled channels created by tidal currents, rivers, wind, or glaciers in the geological past (Wildenborg *et al.* 1990; Liu *et al.* 2009). As the fill material consists often of loose or less competent materials, their incisions in the original, rather consolidated beds may cause rapid lateral changes in rock properties. A classic example of such problems was encountered at the construction of the first Lötschberg Tunnel in Switzerland (Kovári & Fechtig 2000). In this tunnel crossing the Alps, the excavation hit an old, ice-scoured valley subsequently filled by highly permeable sediments in which the water pressure was very high. When the excavation reached the fill boundary, a breakthrough flooded the already excavated tunnel segments with water and loose sediment. A section of the tunnel collapsed and 25 workers were killed. It was impossible to clean-out and restore the tunnel and a new tunnel had to be made. This accident is likely an example of poor management as the fill and its associated risk for tunnelling were identified by a geologist in a second opinion site investigation, but his report was ignored by the management (Beaver 1973).

Almost all excavations in the subsurface intersect erosion horizons but these relatively seldom lead to such consequences as given in this example. Normally, pre-construction site-investigation supplemented with data gathered during construction discloses the presence of changes in material ahead of excavation and appropriate measures can be taken in time.

4.3.12. Swelling

Some materials start swelling when exposed to water or humidity in the atmosphere. The resulting stresses may be extremely high. That may take place in sediments rich in swelling clay minerals such as montmorillonite, for example shales, claystones, mudstones, weathered pyroclastic deposits, and fill material in faults (Brattli & Broch 1995). Anhydrite-bearing rocks exposed to water slowly convert into the more voluminous gypsum, and pyrite-containing shales in which the pyrite oxidizes increase in volume too, with increased pressures as result. Note that stress relief allows volume increase of all ground materials but this is the result of stress reduction. This may be enhanced by the forming of fissures and cracks as, for example, in organic-rich claystone. Rotting of organic material, for example peat exposed to oxygen, causes the opposite, that is a reduction in volume and reduced stresses.

4.3.13. Heat flow and heat insulation

Generally, groundmasses are excellent insulators for heat. Just a few metres below the Earth's surface, the temperature is quite constant even where large temperature fluctuations occur at the surface. This property is used in for example wine cellars. Water or airflows through the groundmass (e.g. through discontinuities) reduce its insulation capacity. In general, this insulation capacity makes groundmasses proper mediums for storage purposes.

Fire is a major concern for underground excavations, not only for the built structures (Duncan & Wilson 1988), but also for the additional heat impact on the groundmass surrounding the excavation. Many cases have been reported where not only the support in the tunnel but also part of the adjacent groundmass disintegrated after a fire. Where the temperatures are high enough, which is not uncommon in the enclosed environment of an underground excavation, also new minerals may be formed that may be weaker than the original minerals (UPTUN 2009; USDOF 2009).

4.3.14. Biological characteristics

Biological activity in the shallow subsurface is generally restricted to mineral-loving bacteria, fungi, and plants. Plants and fungi normally require light although many may live without this source of energy. They may affect stability and integrity of an underground excavation by producing acids or other fluids that may react with the minerals in the ground or with the artificial support systems. Bacteria do not need light to survive and they may live many thousands of metres underground (Pedersen 2000). Animals and plants may migrate from the surface and survive underground, depending on the availability of light and food. For example, large colonies of cockroaches may occur in mines, feeding on timber support.

4.4. Excavation and support

Excavations are made in all types of groundmass. In general, rock masses consisting of not very strong intact rock material, that are not under high or anisotropic stress, that are not subject to swelling, and have few discontinuities with high shear strength, provide the best conditions for excavation. Loose materials, such as sand or poorly consolidated clays, may be more difficult to excavate and require technical solutions to ensure safe underground conditions. There is a wide variety of excavation and support types ('lining'). Selecting the best method for a project depends on many factors. These include the type of soil or rock mass, the time constraints of a project, the shape and size of the required space, the availability of skilled labour, and the economics of a project (Fig. 4.14). For more details one is referred to Hoek & Brown (1990), Thomas (2008), Barton (2000), Maidl *et al.* (2008), Mair (2008), or internet resources (GeotechLinks 2010).

4.4.1. Stress around underground excavation

By excavating a hole in the ground, the stress patterns around the opening change. The excavation allows the groundmass to deform in the direction of the opening. However, closer to the excavation the volume available for the groundmass is smaller in the direction perpendicular to the direction of movement. Hence, the ground pushes in the direction of the excavation but because of the smaller volume available, the stress perpendicular to the direction of movement increases (Fig. 4.15). The result of the deformation is that in unsupported stable excavations, stresses perpendicular to the perimeter of the excavation are zero at the wall of the excavation and the stresses parallel to the perimeter are taken up by the material (Fig. 4.15). This principle is called arching and is the reason why, for example, the curved rooftops in 'geodomes' will not collapse (Section 7.4). The same principle was applied by the Mycenaean and the Romans for their tunnels and bridges (Fig. 4.16). In weaker masses, the stresses parallel to the perimeter cause the material in the perimeter to fail and material will fall into the opening while expanding the excavation size. The groundmass behind the fallen material forms the new perimeter of the excavation and may subsequently fail (spalling failure). In an isotropic linear-elastic ground in which horizontal equals vertical stress, the stresses around excavations are independent of the size of the excavation, whether very small pipelines or major road tunnels.

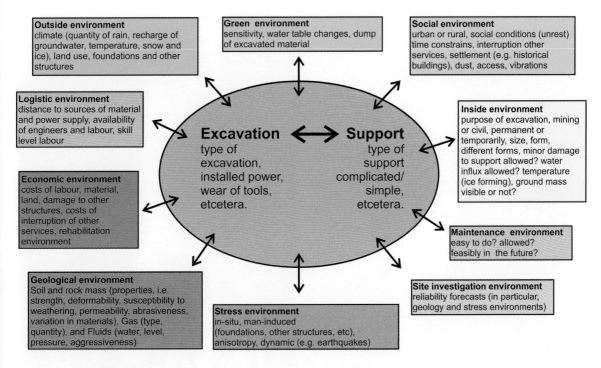

Fig. 4.14. Overview of the main factors that determine selection of excavation and support systems of underground excavations (from Hack 2012).

4.4.2. Failure modes of an excavation

Apart from the failure of material under stress directly adjacent to the perimeter wall as discussed above, underground excavations may fail in a variety of ways by falling of blocks from the perimeter or squeezing, flowing and swelling of the groundmass. Summarized failure mechanisms are as follows:

- Spalling: stresses at the perimeter of the excavation cause the intact material to break (discussed above);
- Gravity failure: material may fall into the excavation due to gravity;
- Squeezing failure: material surrounding the excavation deforms so easily under the influence of the stress field around the opening that it partially or completely fills the excavation and hampers normal operations; this often happens in clays, shales or schists;
- Flowing failure: non-cohesive and non-cemented material may flow into the excavation due to surrounding pressure, under influence of gravity, or is flushed into the excavation by flowing water (loose or weakly cemented sand, residual soil, etc.);
- Swelling failure: materials may start swelling upon exposure to water or air, for example, excavations in anhydrite or swelling clay.

Combinations of failure modes may occur. For example, the clay fill in a fault is squeezed in the opening first. This allows de-stressing of the non-squeezable mass and discontinuities adjoining the fault, which allows gravity failure; or as in Figure 4.9 where overbreak has caused distressing of discontinuities. Failure may be facilitated by re-distribution of stresses in the groundmass upon excavation. Such stress re-distribution may generate new fractures to develop, and displacements along grain contacts and existing discontinuities. This leads to more discontinuities, less shear strength along grain contacts and existing discontinuities, opening of discontinuities, and generally to a weakening of the groundmass (i.e. 'loss of structure') (Read 2004; Jethwa 2009) (see also the ADECO methodology in Section 4.6.5). Strength reduction of the groundmass due to the method of excavation or weathering may further facilitate failure.

4.4.3. Need for support

Support structures are installed in underground excavations to keep the groundmass directly adjacent to the excavation walls, roof, and floor in place. Except for shallow underground excavations where horizontal stresses are low, support structures never counteract the stresses in the virgin subsurface, as arching effects will deflect most of

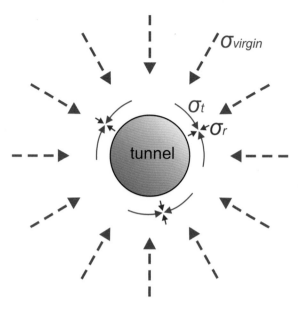

Fig. 4.15. Stress deflection around a circular excavation. If the stresses in every direction in the virgin stress field, that is the stress field before any excavation has been made, are equal – dashed arrows in figure, the stress perpendicular to the wall of the excavation (= σ_r = radial stress) is zero and the stress parallel to the wall (= σ_t = tangential stress) is maximum at the excavation circumference.

the stresses away from the excavation. Deformation of the groundmass after excavation leads to reduction of strength of the groundmass, hence, support needs to be installed as soon as possible after excavation to avoid too much deformation. Rapid action also reduces the required size and volume of the support system as the groundmass itself retains more strength. However, some deformation should be allowed to occur to generate arching effects. Size matters if discontinuities are present as is very often the case. The more such discontinuities intersect the excavation, the higher the possibility of moving or falling blocks (Fig. 4.17).

4.4.4. Stand-up time, time effects, and life times

The timespan between excavation and failure of an unsupported excavation is called 'stand-up time'. Stand-up times allow the excavation to be supported before collapse begins and may range from negligible to thousands of years. Mechanisms controlling stand-up times are only partially known and mathematical expressions are not quite reliable yet. The only rather secure way to determine stand-up times is by collective experience of local miners or construction engineers, assisted by empirical classification systems (Section 4.6). Time-effects may cause serious accidents in underground works. Excavations are often considered safe

Fig. 4.16. 3000-year-old gallery built out of rock blocks. The weight of the overlying blocks is diverted to the sides through 'arching' (Mycenae, Greece; photograph: R. Hack).

if they have been standing up for some time, which may prove to be a fatal error. Experience is required to recognize potential collapse which is often preceded by warnings as small movements in the surrounding groundmass, cracking sounds from the mass or (small) pieces of material falling from roof or walls.

Lifetimes of underground constructions may be affected by inherent technical and external factors. The first concerns the construction and its surrounding rock materials. Lifetimes may be reduced by poor excavation methods damaging groundmasses or by unstable minerals in the groundmass, for example pyrite or gypsum. External factors may include the impact on the stress–strain conditions by other nearby underground constructions, traffic induced vibrations, modified groundwater movements (Delatte *et al.* 2003), and changes in the ground material caused by, for example, heat and radiation from nuclear waste (Hudson *et al.* 2009; Anon 2010).

TECHNICAL CHALLENGES AND ASSETS

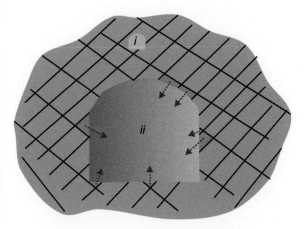

Fig. 4.17. Size matters. The small excavation (*i*) intersects few discontinuities and no options for movement along existing discontinuities exist; the larger excavation (*ii*) intersects many discontinuities and consequently there are many more options for displacement and rock fall or squeezing of blocks (from Hack 2012).

4.4.5. Underground excavation under surface structures

Stress exerted by foundations of structures at the surface adds to the stress field around the excavation. However, the stress–influence sphere (or bulbs of pressure) of foundation pressure is limited. With depth, the sphere extends laterally and the stress reduces equivalently. As a rule of thumb, the angle for the spreading of stress is 45°. However, geological features such as discontinuities may limit spreading or direct the stress in a particular direction (Fig. 4.18).

4.4.6. Excavation and support techniques

Factors determining whether an underground opening may be excavated easily include the strength of intact material, the size of the blocks between discontinuities, and the shear strength between the blocks or grains of intact material. Normally, excavating is faster in weaker rock masses (Fig. 4.19a). Groundwater may generate hydrostatic pressures between the grains or blocks and facilitate transport of materials that require special excavation techniques to allow safe excavation. Time-dependent changes of the mass may also be relevant in some groundmasses. Abrasiveness of the 'ground' material to excavation equipment is often a major factor determining the economics of a project (Fig. 4.19b).

Excavation methods may range from manual to highly mechanized tunnel boring machines (road header, Fig. 4.20; Tunnel Boring Machines (TBM), Fig. 4.21). Regardless of the quality of the groundmass, excavations may be made by any type of method; prisoners dug escape tunnels in rock using just their fingernails and spoons.

Two main types of excavation methods are in use today: mechanical and blasting (Table 4.4). In the past, blasting techniques were used more frequently whereas today mechanical methods are often preferred. Literature often refers to 'cyclic' and 'non-cyclic' or 'continuous' construction operations (ÖNORM B2203-1/2: 2001/2005). In cyclic operations, activities are executed in a successive order, for example drilling holes for explosives, blasting, ventilation to remove the blasting fumes, removing the blasted material, installation of support, followed by drilling new holes for blasting, and so on. In non-cyclic operations, most activities take place simultaneously, in a continuous process, for example by tunnel boring machines.

Mechanical hammering is often used for smaller excavations or for reducing block sizes after blasting. 'Specials' (see Table 4.4) refer to excavation methods by utilizing the

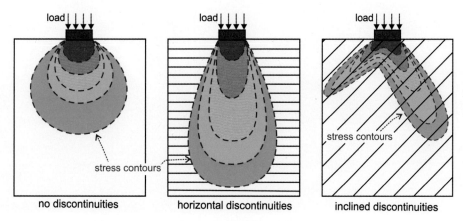

Fig. 4.18. Stress distribution (bulbs of pressure lines of equal major principal stress) in a rock mass due to a vertically oriented uniform load (modified from Gaziev & Erlikhman 1971).

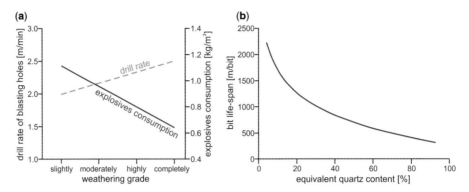

Fig. 4.19. (a) Excavation is faster in weaker rock masses, demonstrated by drill rate of blasting holes increasing and explosives consumption decreasing in granite rock masses with higher weathering grades (modified from Thuro & Scholz 2004); (b) lifetime of drilling bits strongly reduces and thus costs increase with increasing abrasiveness ('equivalent quartz content') (modified from Thuro & Plinninger 2003).

expansion characteristics of wood (if soaked with water) or chemicals, applying high water pressure jetting to erode the rock or soil mass, or using sawing techniques (Section 4.5.5). Hammer and chisel, and fire setting are methods mainly used for very small (often illegal) operations but were used extensively in the past (Section 4.5.5). New developments may be the application of high-energy laser, plasma, and millimetre wave techniques as already investigated for use in drilling of deep boreholes (Woskov & Cohn 2009).

4.4.7. Small underground infrastructure, dredging and boreholes

Boreholes and small infrastructure for cables, sewer pipes, power lines, and so on, constitute a specific type of underground excavation. Such structures are generally cylindrical and less than 2 m in diameter. Mechanical considerations as described above also apply to these 'small' pipes, but smaller diameters allow for other excavation techniques. Trenches for very shallow pipelines and cables up to diameters of about 1 m are made by 'trench cutters' (not to be confused with equipment for digging deep trenches, for example diaphragm walls) (Table 4.4). Trench cutters may install a pipeline or cable and refill the trench with the excavated material. The maximum depth is a few metres (Fig. 4.22).

Trenches for pipelines and tunnels off-shore are often dredged (Herbich 2000). A wide variety of dredgers exist suited for ground masses ranging from soft soil to moderately strong rock. Mostly soft loose soil is sucked with water into a pipe, not much different from a vacuum cleaner, or

Fig. 4.20. Road header excavates the rock by pressing rotating drums equipped with picks of hardened steel to the rock face (right) (photographs: © Sandvik Mining and Construction).

Fig. 4.21. Rock tunnel boring machine – TBM used for drilling one of the 57 km long and 9 m diameter Sankt Gotthard railway tunnels (Gotthard Base Tunnel) in Switzerland. The rock is crushed by the (black) roller disk cutters fitted in the (white) cutter head at the front (photograph: HERRENKNECHT AG 2011).

excavated by buckets mounted on a chain or by a backhoe. The last two also work for cohesive material. Stronger cohesive soil and rock first need to be broken and loosened by cutters resembling the trench or road header cutters described above, or explosives are needed. The maximum depth is normally limited to some 100 to 150 m below the water surface.

Deeper, small-diameter (up to about 20 cm) pipelines are often installed by 'directional drilling': water jets cutting through soft ground or flush borings with pipe pushing. Drilling bits may be attached to the string in stronger groundmasses. The result is a borehole with the hollow drill string left in place serving as the pipe, or the drill string is withdrawn with a cable or other pipe attached. For diameters of up to about 1 m, the original hole is widened with a reamer attached to the drill string, simultaneously pulling in the attached one or more, large diameter pipes. Dependent on the ground conditions and the operator skills, such jetted pipes may reach a length of a few kilometres (Fig. 4.23).

Pipes or small tunnels with diameters from 1 to 2 m are often jacked, a technique also known as 'micro-tunnelling'. The (small) tunnel boring machine with cutting head (the drill bit) together with the support is jacked from the end of the pipe through the ground (Fig. 4.24). Pipe lengths for this technique depend on the shear resistance between the ground and the pipe. For longer stretches, intermediate (booster) jacking stations may be added. The pipe moves through the ground in a comparable way to an earthworm. Bentonite (a material that mainly consists of the clay mineral Bentonite) or other chemicals may be used as a greasing agent to reduce shear resistance along the pipe. Pipe lengths of 950 m have been reported by using this approach. The depth is limited by the depths of the shafts, but the method may be used at any depth between already existing underground spaces.

4.4.8. Water

Groundwater may cause additional problems in underground excavation (see also Sections 4.3.7 and 4.7.4.4). In permeable grounds, groundwater will flow into the excavation often together with ground material thus jeopardizing the stability of the excavation. The easiest solution is dewatering, that is removing the groundwater by draining or pumping, before the excavation starts. If the groundwater

Table 4.4. *Excavation methods (from Hack 2012)*

Mechanical
  Digging
    Man-made/shovel/excavator
  Cutting and grinding
    Borehole
    Road header
    Trench cutter (for shallow trenches for pipelines, cables, etc.)
    Tunnel boring machine (TBM)
    Raise borer (for sub-vertical shafts)
  Hammering
    Jack hammer
    Hydraulic/pneumatic hammer
Blasting
  Pre-splitting[1]
  Smooth wall[1]
  Conventional tunnel blasting
  Conventional large hole blasting
Specials
  Wood expansion[2,3]
  Chemical expansion[3]
  Water (high pressure breaking or jetting)
  Sawing (blade or steel cable), hammer & chisel, fire setting

---

Notes: (1) Pre-splitting and smooth wall blasting are techniques used to reduce the damage to the rock mass and create smoother excavations, (2) wood expands when soaked in water, (3) a borehole is filled with wood or an expanding chemical; when expanding the rock breaks.

cannot be removed, it should be prevented from flowing into the excavation. This may be realized by, for example, installing an impermeable bulkhead (e.g. a shield) in a tunnel boring machine, excavation under pressure, or by making the ground impermeable before the excavation starts, by injecting cement or other chemicals ('grouting'), or temporary by freezing (Pakianathan *et al.* 2002).

4.4.9. Dewatering

Dewatering resolves the problem of water flowing into an excavation and improves the geomechanical characteristics of soils, for example, by eliminating the liquefaction potential. Dewatering by vacuum pumps with vertical or inclined wells lowers the groundwater level with about 6 m at maximum (Fig. 4.25). 'Down-the-hole' pumps, that is pumps at the bottom of the wells, may pump the water from any well depth. The wells may be boreholes but in loose material casing (i.e. steel or PVC pipes) and filters are required to prevent inflow of too much ground material. Free draining of the water is possible by (sub-) horizontal drains from the site of a hill in undulating topography or from other underground excavations. Draining pipes feeding the water to underground pump stations may also be inserted at depth. The drains may be boreholes or pipes inserted by directional drilling. Frequently, dewatering results in settlement of the ground, affecting surrounding structures and buildings, and may distress the condition of surface vegetation. Such impact may be reduced by recharging the water into the adjacent ground (Fig. 4.25). The distance between the well and recharge point may be reduced by installing an impermeable barrier in-between while still achieving the required groundwater lowering. That may be realized, for example, by grouting or installing steel plates (i.e. 'sheet piling') into the ground. Dewatering may cause undesired effects if polluted water flows into the direction of the excavation. Impermeable barriers may be inserted to prevent this.

4.4.10. Ground improvement: freezing, grouting, compaction, pre-loading, heating, chemical and biological treatment

Freezing, grouting, compaction, pre-loading, heating, chemical and biological treatment improve the stability of the groundmass around an excavation (temporarily, in the case of freezing). Freezing and grouting increase the shear and tensile strength of the ground. Freezing may generate swelling of the groundmass as ice has a larger volume than water. This, in turn, may cause displacement of existing structures in the surroundings (cellars, foundations) and may lead to heave if applied near the surface. Freezing functions only if groundwater is present in sufficiently large quantities. The term 'grouting' is normally used for injecting cement with water into the ground and, depending on the application, mixed with aggregate. Chemical substances, such as acrylates, epoxies, and so on, may replace a cement mix but are more expensive. Pressures used for grouting are critical as the grout would not reach all pores and discontinuities with too low pressures while too high pressures may cause breaking of the ground, generate surface heave, displace surface structures, or even cause a blowout. High-pressure grouting may be applied intentionally to compensate for settlement around excavations (i.e. 'compensation grouting') (Woodward 2005). Freezing creates a body of ice in the ground and makes it impermeable. Grouting reduces permeability but seldom makes it watertight, as grout is generally not evenly distributed due to inhomogeneity of the ground. If leaks occur in the grouted zone, these will normally be repaired by secondary grouting.

Compaction techniques are applied to improve stability and geomechanical characteristics of loosely packed soils, including reduction of its liquefaction potential. Various compaction techniques exist for granular soils, also in combination with grouting (Belkaya *et al.* 2008; MENARD 2010). (Pre-) loading a soil is a proven technique for low-permeability soils, such as clay and peat. The load should be maintained for quite some time (i.e. mostly years) to allow the water to be expelled. Its effectiveness is limited to a few metres below the loading surface only as the pressure of the load spreads with increasing depth. Installing drains in the soil accelerates the process, even more so if the drains are kept under vacuum or if electro-osmoses (i.e. increasing water movement by electrically charging the

Fig. 4.22. Rock trenching (Vermeer trencher; photographs courtesy of M. Alvarez Grima).

water molecules) would be applied (Bergado & Patawaran 2000). Clays may be treated chemically, for example, by stabilizing with lime (NORDIC 2010; Tørum et al. 2010). Techniques that improve ground characteristics permanently by changing the minerals or ground texture, as biological (Van Meurs et al. 2006) or heating (i.e. 'vitrification') with microwaves, electric currents (up to temperatures of 1000 °C) or plasma torches (with temperatures up to 7000 °C) (Mayne & Elhakim 2001) are available but still fairly seldom used.

4.4.11. Water pressure

The water pressure of static water (groundwater that does not flow) depends on the water head only. In contrast, water flowing through a groundmass loses pressure due to the resistance of the groundmass; the lower the permeability the more loss of pressure occurs. This affects support systems and methods of excavation. Allowing a slow water flow regime in low permeable ground is often sufficient to reduce the water pressure on the support system to a low or negligible level. Hence, it may be advantageous using a permeable support system in low permeability ground and handling the small quantities of water flowing into the excavation by, for example, an impermeable geotextile. The water may then be collected in a ditch from where it is pumped out of the excavation. That however, is an impractical solution for highly permeable grounds with high groundwater pressures. Moreover, large water flows would flush ground material into the excavation. Hence, in permeable ground, the groundwater pressure should be countered by impermeable support systems and the shield used at the face should be able to withstand the pressure. If the water pressure is high, heavy support is required. In addition, shield bearing TBMs should be able to push the shield forwards against the thrust of the water on the shield. In such cases, it may be more economical to grout or freeze to make the ground impermeable at some distance in the ground away from the support system and ahead of the face of the excavation. The groundwater pressure is thus taken up by the ground itself and the thrust on the shield of the TBM is lower as well. In the case of freezing, the final support should be able to withstand the full water pressures as freezing is likely applied during construction only.

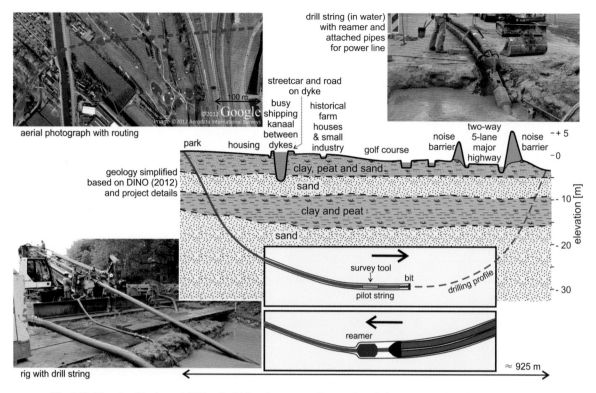

Fig. 4.23. Directional horizontal drilling for high-voltage power lines under existing structures in soft ground (Rijswijk, The Netherlands) (from Hack 2012; photographs left: J. F. Joosse; right: R. G. Bleumink, BT Geoconsult; aerial photograph: Google Earth/Aerodata International Surveys).

4.4.12. Aggressive or polluted water

Surface water and groundwater may be aggressive or polluted and be a threat for the groundmass itself, the support systems, or for the users of the excavation. Water seeping from sewages, landfills, and waste dumps may be aggressive or polluted depending on the type of waste. Acid water seeping out of ore mines or tailings and waste dumps have a bad reputation in this respect (Fig. 4.26). Natural groundwater in some grounds may also be quite aggressive, for example, acid water corroding the tunnels of the London Underground (Raineya & Rosenbaum 1989) or water around the tunnel for the Niagara Tunnel project, Canada (Hughes *et al.* 2007). The latter problem required protection of the concrete lining by layers of impermeable membranes.

4.4.13. Impact of excavation method on groundmass

The quality of the groundmass surrounding the new underground space also depends on the method of excavation. Methods producing large quantities of energy in a short timespan create high stress levels and generate more and deeper fractures in the groundmass. For example, blasting generates more and deeper fractures in a rock mass during excavation than a TBM (Dinis da Gama & Navarro Torres 2002; Hudson *et al.* 2009). Blasting damage may consist of an increased number of mechanical discontinuities and reduction of shear strength due to displacements along grain contacts and existing discontinuities. These result in reduction of the block size, loss of structure of the groundmass, and increased permeability. Such damage depends on the groundmass condition prior to excavation. This implies, for example, that relatively few new discontinuities will be introduced if a groundmass would already contain many of such before blasting (Fig. 4.27) (Martino & Chandler 2004; Jethwa 2009). Excess blasting can destroy originally sound groundmasses. Excavation damage factors for design of underground excavations including factors for poor blasting engineering, demonstrate the influence of the method of excavation quantitatively (Section 4.6.2). In groundmasses with a varying quality on short distance, a very careful planning of the excavation method is required to prevent excess damage in the poorer quality portions of the groundmass.

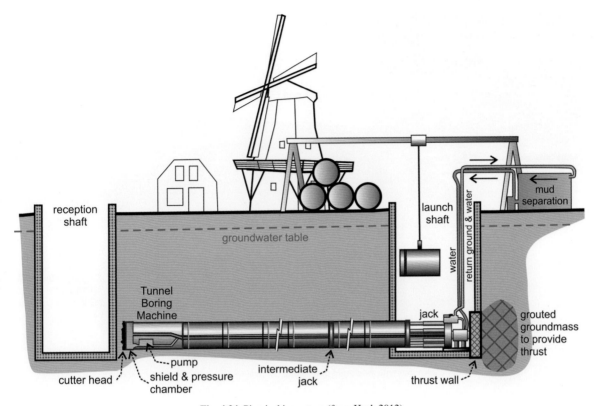

Fig. 4.24. Pipe jacking set-up (from Hack 2012).

4.4.14. Economic viability of excavation methods

As a rule, excavation should be economically viable. Delays in excavation time, extraordinary repair costs, and unexpected failures negatively affect economic viability of excavation projects. If a rock mass consists of larger blocks and the intact rock strength is high, it will take a long time before a mechanical cutting and grinding device has excavated the space. Secondly, the wear on cutters and machinery may be high and consequently the project costs will rise. Likely, in such a project blasting would have been a more economical choice. However, that may be different in the near future. Previously, mechanical cutting and grinding excavation techniques were limited to soil and moderately strong rock masses with an intact rock strength not exceeding 50 MPa. Excavation techniques and equipment have since rapidly improved and mechanical cutting and grinding methods are now becoming more economically viable in stronger rock masses with larger block sizes as well. Even if not economic, mechanical methods such as TBMs may be applied if, for example, blasting is not an option due to the risk of fire or excessive vibrations. Today, TBMs may excavate tunnels up to diameters of around 15 m in one pass.

4.5. Examples of excavation in various groundmasses

This paragraph describes the way excavations are made in various types of groundmasses, what types of support systems are applied, challenges encountered and solutions found, based on a series of examples from around the world.

4.5.1. Rocks with high intact rock strength

The earliest humans mined precious metals from rock masses with very high intact rock strength and with few discontinuities. Apparently, the (past) economic value of the metals made mining in such hard and strong rocks worthwhile (Section 2.1). Normally, ancient excavation was limited to single, small diameter tunnels. Hard rock mining expanded only after the discovery of explosives for rock breaking in the early 17th century and, in particular, when dynamite and mechanized excavation methods became available in the mid-19th century. Often, no support was required and unsupported excavations may stand-up for many hundreds of years, as demonstrated by underground excavations in mines, cellars and water irrigation works

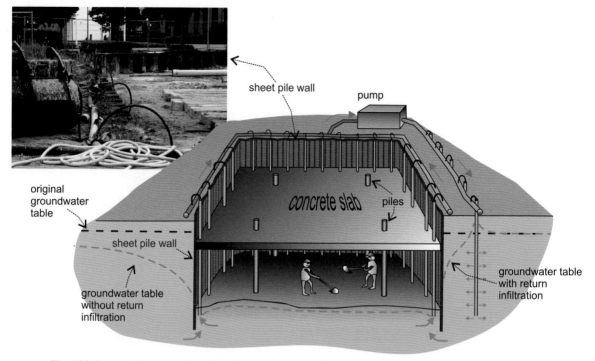

Fig. 4.25. Supported dry excavation with well point dewatering (from Hack 2012; photograph: J. G. A. M. Arnoldus).

still operational today. Some excavations are now historical landmarks, such as the 220 m long Union Canal Tunnel, USA, excavated by hand and gunpowder (i.e. black powder) in metamorphic sedimentary rock (slate with limestone and flint), completed in 1827 (LCHC 2010), and the 192 m long Boolboonda tunnel excavated by mechanical drilling (jack hammers) in granite, completed in 1883 (Fig. 4.28) (State of Queensland 2011).

4.5.2. Flexible and rigid support

Excavations may be supported by flexible support systems even in fairly poor groundmasses (Fig. 4.29). Rock bolts connecting the blocks of rock together are installed to 'reinforce' the rock mass. Shotcrete sprayed onto the rock or soil also binds the blocks of rock or soil grains on the excavation circumference preventing them to tumble down while maintaining the groundmass integrity. By adding wire mesh, tensile strength of the shotcrete cover is improved. If that is insufficient, steel beams may be installed and covered by up to many tens of centimetres of shotcrete. As bolts and shotcrete are flexible to some extent, they allow the rock mass to deform while building up arching effects. If required, a final rigid, reinforced concrete support is installed once arching effects have reached their maximum levels. A membrane sheet of geotextile may be installed in-between the flexible and rigid support preventing influx of seepage water and allowing shear displacements to occur between the two types of support. Geotextile improves the distribution of the stresses from the flexible support onto the rigid support. Often, larger diameter tunnels have been excavated sequential (i.e. in multiple sections) as with the tunnel in Figure 4.29, where the upper section has been excavated and supported first, followed by the bottom section. Smaller sections are often more stable and installation of support is completed more rapidly. This type of excavation and support in which relaxation of the ground is allowed, is often referred to as NATM (New Austrian Tunnelling Method) excavation (Section 4.6.4).

4.5.3. Loose materials and very weak ground

Soft sediments often provide difficult conditions to create and maintain an underground opening. Extraction of the loose material will almost instantaneously result in collapse of the surrounding groundmass. In such cases, stand-up times are very short and extensive measures for temporary support and preventing influx of groundwater are required before a permanent support system may be installed (Fig. 4.30). For TBMs and pipe jacking, complex systems have been developed to maintain the *in-situ* ground and groundwater pressures while minimizing disturbance of the

Fig. 4.26. Acid water causing deterioration of rock mass and steel support in a mine tunnel. The blue and white coloured minerals are sulphates precipitated from seepage water (Rammelsberg, Harz, Germany; photograph: S. Carelsen).

Rock mass type	1	2	3	4	5
Existing discontinuities	None	Few	Some	Many	Crushed
			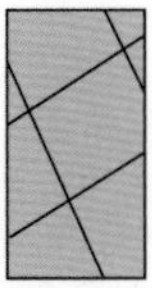		
Rock mass class	Very good	Good	Fair	Poor	Very poor
Damage expected from blasting:					
New discontinuities	Possible			Low possibility	None
Discontinuity strength		Loss of strength	Loss of strength possible		
Dominant damage		Fracturing		Loss of discontinuity strength	

Fig. 4.27. Blast induced damage in various rock mass types. *Source*: Modified from Jethwa (2009).

Fig. 4.28. Unsupported tunnel in granite originally excavated as railway tunnel in 1883, now used as local road tunnel (Boolboonda Railway Tunnel, Qld, Australia; photograph: Department of Environment and Resource Management, Queensland, Australia).

groundmasses surrounding the future tunnel (Fig. 4.31). Caisson-type excavations with pressurized air supporting the surrounding ground and withstanding the groundwater pressures have been applied in some groundmasses (Hewett & Johannesson 1922). Grouting or freezing ahead of the excavation face are other methods to improve ground conditions and making the groundmass impermeable.

The Pipe Roof-Box Culvert Jacking (RBJ) is an alternative excavation method for soft grounds at shallow depth and may be applied if the surface infrastructure should not be interrupted. Impermeable artificial roof, walls, and floor are created first by jacking a series of large sized pipes with a special watertight joint system in-between adjoining pipes (Fig. 4.32). Next, the ground in-between the pipes is excavated in front of pre-fabricated concrete tunnel segments which are jacked into the excavated space (Bai et al. 2006).

4.5.4. Permafrost and permanent snow and ice

Temperatures below 0 °C pose special problems for engineering and underground excavations. Problems can be caused by the ground mass or the air being below 0 °C. The latter is discussed in Section 4.7.4.4. Low ground temperatures occur in so-called 'permafrost' zones where the ground mass remains below 0 °C for more than two years (CRREL 2010). In many permafrost areas, the top layer of the ground thaws during summer or during daytime. Water from rain and melted snow and ice saturates the ground and cannot drain downward as the ground below is permanently frozen. The ground above the permanently frozen layers consists of (over-) saturated soil with low bearing capacity and pools of water. Cyclic freezing and thawing causes displacements in the soil. Where the ground consists of a rock mass, cycles of freezing and thawing also result in physical disintegration of the rock and rock mass (Matsuoka & Murton 2008).

Foundations of structures at the surface situated in permafrost zones built on soil-type material require that the temperature of the foundation layers be maintained permanently below 0 °C to prevent displacement and settlement of the foundation. Foundations into the deeper permanently frozen layers (e.g. on piles or piers) should be able to withstand (or should be protected from) the stresses exerted by the layers of the frost and thaw zone. Frozen material around underground excavations may have to be cooled or isolated from heat sources inside the excavation. During construction, temporarily thawing of ground cannot always

Fig. 4.29. Supporting a highway tunnel in volcanic deposits on Madeira, Portugal (photographs: R. Hack).

be avoided due to the pouring of (relatively) warm concrete and the heat generation of the setting of concrete. This happened, for example, around the Feng Huoshan tunnel in the Qingzang railway to Lhasa on the Tibet Plateau (Zhang et al. 2004).

Permanent ice may be used for foundations or excavations. Ice has a considerable strength under high loading rates (Table 4.1) but exhibits strong plastic and creep deformation characteristics under anisotropic stress. Snow has a low density and is very weak but if compressed, snow becomes stronger and ultimately transforms into ice. Glaciers contain large volumes of ice and the ice may move with velocities of many metres per year. Velocities within glaciers are unevenly spread causing deformations and shearing of the ice and consequently of the constructions on and in the ice. Despite the unfavourable conditions several semi-permanent constructions are built in and on ice, in polar and mountain regions for research stations, military purposes (e.g. Camp Century, Greenland; Clark 1965), cable car towers (Fellin & Lackinger 2007), and as tourist attractions (Fig. 4.33). Regular repairs have to be made to maintain the structures.

Permafrost and permanent snow and ice areas occur in the Polar Regions, Alaska, Greenland and northern Siberia and Canada, but also at high elevations as the Tibet Plateau. Over the last 20 years, increasing temperatures have been observed with consequently thawing of until recently permanently frozen ground together with increased thicknesses of the temporarily thawed top layers. In addition, glaciers become thinner and retreat (IPA 2010) resulting in considerable problems for existing foundations and unstable excavations (Balobayev et al. 2009; PL 2010). Whether a further reduction of frozen ground environments should be anticipated for future constructions, is subject to the current discussions on global warming and climate change.

4.5.5. Mining, existing underground structures, and karst caves

Since millennia, relatively weak rocks were excavated for the production of building stones by sawing or chiselling in 'room and pillar' (or 'pillar and stall') type of mines (Fig. 2.8). This method was applied until maximum stresses around the excavation prevented further expansion. That

Fig. 4.30. Pre-fabricated concrete permanent segment support in a 6.6 m diameter tunnel; New Delhi Metro Phase II, excavated in gravel, sand, and silt (photograph: HERRENKNECHT AG).

determined room and pillar sizes sufficiently stable to stand up for thousands of years without any artificial support. Mostly intact rock strength of the rock is quite low with few discontinuities while the shear strength along most of the discontinuities is relatively high. Around the world, many of such excavations exist in limestone, tuff, and other rock types. The excavation method is still applied locally (Bekendam 2004). Shallow coal and mineral mining is usually done with a shovel, or hammer and chisel from bell-pits, pillar-and-stall workings, or from a small tunnel in a mineral ore vein (Fig. 4.34). 'Fire-setting', that is heating the rock by a fire, is sometimes used to break stronger rock in mineral mining and more recently, gunpowder or dynamite are being used. The entrance from the surface is by a (sub-) horizontal adit in the side of a hill or shallow shafts. The maximum excavated size is reached where the walls, roof, or pillars begin to show cracks. Obviously, mines often collapsed (Zhang & Peng 2005; Bétournay & Lefchik 2008). Timber and brickwork and later steel and concrete support systems are used irregularly in pillar and stall workings and mineral mining, often only in access tunnels and shafts. Apart from mining, extensive underground constructions have also been made for storage or shelter, in particular in urban areas (see Section 2.2).

Access to abandoned mines is often limited as many have been filled by waste or are (partially) collapsed. Usually, support and fill were applied haphazardly, is undocumented, and the stabilizing effect of such support and fill cannot be relied on. Shafts are sometimes closed by wood, iron, or concrete structures and further filled with waste. This often makes these shafts hard to identify. Usually, stability and bearing capacity of the closing structures are unknown and highly unreliable. Open underground spaces that are stable now, may collapse in the future, and spaces that at present are not thought to be a hazard because they are far away from a new construction, may become a hazard because of migration in the direction of the new construction due to collapsing roofs or sidewalls. In 2007, a hole suddenly occurred below the tracks in a tunnel of the Tyne and Wear Metro, UK, due to collapse of an abandoned mining space below (BBC 2007). The tunnel had been in use for years. Failure

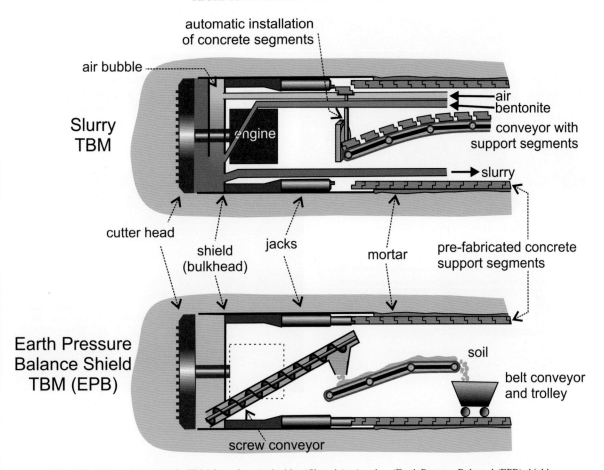

Fig. 4.31. Schematic layout of a TBM for soft ground with a 'Slurry' (top) and an 'Earth Pressure Balance' (EPB) shield (bottom). The TBM cuts the ground with the 'cutter head'. Behind the cutter head an impermeable shield (bulkhead) is mounted which allows control over the pressure in the space between face and shield. In the Slurry TBM, the ground is mixed with bentonite to stabilize the face. The 'air bubble' is used to control pressure in the space between face and shield and to compensate for sudden pressure changes. The bentonite-ground mix is pumped out through a pipe in the bottom of the shield. In the EPB-TBM, the pressure between face and shield is controlled by the 'screw conveyor'. The jacks propel the TBM through the ground. The mortar is injected to fill the void between the excavation circumference and support.

of collapse prone spaces may also be triggered by modifications in the underground environment caused by new underground structures, temperature changes, inflow of oxygen causing renewed rotting of timber, changes in groundwater levels, or vibrations from excavation and traffic.

Shallow underground mining was and is mostly undocumented and quite often executed illegally. For many abandoned underground storage or shelter spaces, the same applies. Therefore, the presence of abandoned underground excavations requires a very detailed and extensive site investigation including research of mining archives and other historical sources (see Section 5.1.13). Controlled filling of the spaces or improving the stability is required if new constructions are to be built or if subsidence poses a threat to existing structures. Fill materials, such as aerated (foamed) concrete or cement grout with fly ash (a by-product of coal-fired power plants), may be pumped into the underground spaces. For example, 600 000 m³ of foamed concrete has been pumped in abandoned mines under the historical heritage sites of the City of Bath, UK (Aldridge 2007). Improving the stability of old excavations while maintaining access is more difficult, expensive, and may be quite dangerous. This may only be worthwhile for spaces with specific economic or historic value, such as the cellars and tunnels in several Central European cities and villages (Meier 2007) (Chapter 2).

In contrast to historical shallow mining, modern mining is normally well documented, and in contrary to the past, many

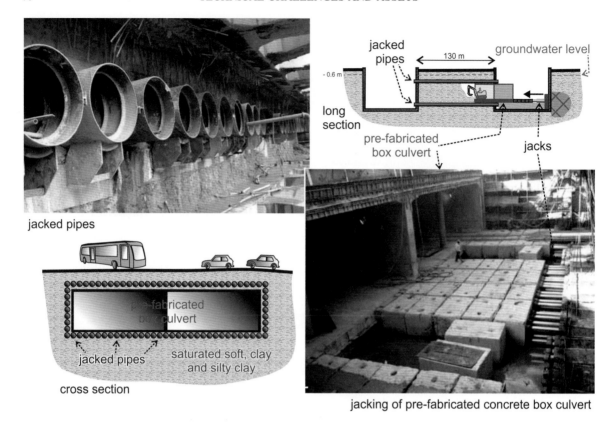

Fig. 4.32. Pipe Roof-Box Culvert Jacking for two-way 8-lane tunnel in Shanghai under saturated soft clay, China. Pipe diameter is 0.97 m, the distance between pipes and street is 4.5 m and the box segments measure 34 by 8 m with lengths between 4 and 18 m (modified from Z. G. Qiao 2006; photographs: Z. G. Qiao).

mines are now backfilled upon termination of the mining. The backfill procedures should comply to a standard well documented methodology and fill properties should be properly recorded and archived. Not backfilled mines, for example, long-wall mining such as used for coal mining (Longwall Mining 1995), may give subsidence and related effects over large areas and for a long time after mining ceases due to time and creep effects in the underground or by changing groundwater levels (Donnelly *et al.* 2008). This may also happen far away from the actual excavation due to horizontal displacement of materials above and in the direction of the excavation. Displacement and deformation in the subsurface may cause loss of integrity, for example, cracks in the groundmass may be formed, existing discontinuities may open, and displacements along discontinuities will cause reduction of shear strength along discontinuities. Freshly opened discontinuities increase the permeability of the groundmass while adding to the technical challenges of building new subsurface structures.

Normally, karst caves are also undocumented and hard to identify. Stability problems of man-made excavations near to or intersecting caves are comparable to those in abandoned mines. An additional problem in karst caves is formed by the often high quantities of water and high flow rates. Filling the karst caves using extensive grouting is normally the only feasible option as was done for the SMART tunnel in Kuala Lumpur (Raju & Yee 2006).

4.5.6. Strong and weak elements in groundmass

Strong and hard elements mixed with or embedded in a weaker groundmass are notorious for excavation problems. They are often not identified during site investigation and, hence, excavation equipment brought to site is not designed to cope with such hard materials. Moreover, they may easily be detached from the groundmass and start rotating when in contact with the equipment or are pushed ahead rather than broken up. That may lead to additional wear of the equipment and retardation or even halting of the excavation operation. Strong elements may then have to be cut and removed manually, which is very time consuming, often dangerous,

Fig. 4.33. Unsupported tunnel in the Jungfrau Glacier, Switzerland (Eispalast 2010; photograph: J. G. A. M. Arnoldus).

and hence expensive. Sudden, large changes in material strength and hardness also generate impact damage on bits of TBMs or road headers (Thuro 2003). In blasted excavations, the strong material may require additional blasting causing excess fracturing in the weaker materials. The weaker ground matrix forming the walls, roof, and floor of the excavation may then also be damaged and may fall out creating an excavation larger than required (so-called 'overbreak') or will require more and heavier support.

Examples of projects encountering strong elements in the subsurface include excavation through calcrete layers by road header for the underground railway in Nuremberg, Germany (Thuro 2003), the excavation of sewage tunnels by TBM in volcanic deposits in Mumbai, India, (Munz & Haridas 2000), dredging the trench for the Oresund tunnel in weak limestones with chert (flint) nodules (De Kok et al. 1997; Brugman et al. 1999), glacial rock boulders in soft clay deposits in the Storebœlt Project (Ovesen 1999),

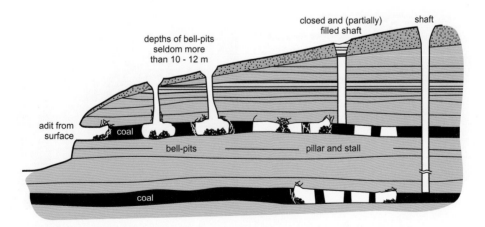

Fig. 4.34. Schematic representation of shallow mining of coal seams. Multiple coal seams may have been mined, the deepest layers most recently (from Hack 2012).

and in excavations by TBM for tunnels for the underground railway in Vancouver (Ciamei & Moccichino 2009). Examples of excavations through mixed grounds are the tunnels for an underground railway made by TBM in Porto, Portugal (Babendererde et al. 2004) (Fig. 4.35) and the excavation made by drilling and blasting for a water pipeline in San Diego, California (Krulc et al. 2007).

4.5.7. Underground excavations in swelling or squeezing material

Squeezing or swelling ground (such as shale or gypsum) may cause a TBM to get stuck because the ground and resulting shear forces on the TBM become too large or cause the TBM to deform if the forces are anisotropic. The latter may cause water and ground to flow into the TBM and in parts of the already excavated tunnel. Both problems are such that it is not uncommon for the planned routing to be abandoned and (part of) the tunnel re-excavated (Shang et al. 2009). These problems do not occur in flexible excavation operations as done by blasting or road headers but then maintaining the required excavation size is often a concern (Fig. 4.36) (Hoek & Marinos 2000; Barla et al. 2008). Box 4.4 shows an example of a tunnel in squeezing ground and the excavation and support methodology. Swelling may be reduced by ground-sealing technologies to isolate the groundmass from the atmosphere and/or water immediately after excavation (Pérez-Romero et al. 2007). That may be done by spraying gunite (cement with water) or shotcrete onto the roof and walls. An excavation may lose its functionality when it will become too small or too irregular in shape. This may be compensated by over-sizing the excavation, or combating squeezing or swelling by applying heavy and expensive support. Even then, frequent repairs of the support may be required, which are dangerous and costly operations.

4.5.8. Excavation in volcanic areas

Volcanic areas may be challenging environments for excavation due to their wide variation of rock and soil masses; very strong basalt rock, weak volcanic tuffs, and weathered soil deposits may rapidly alternate on very short distances, both horizontally as well as vertically. Moreover, big water-filled or empty voids or caves ('lava tubes', Bunnell 2008) may occur in cooled lava flows. Forecasting the exact position of the boundaries between the various rock and soil types is difficult and requires detailed and costly site investigations. In volcanic areas, groundwater often contains aggressive minerals such as sulphides that may corrode steel and concrete of the underground structure (Anon 2004). Temperatures of ground and groundwater may also be high which may give rise to additional problems with excavation equipment and working conditions. Finally, poisonous (e.g. sulphur) gases may be released from discontinuities. Examples include excavation of sewage tunnels in volcanic deposits in Mumbai, India, (Munz & Haridas 2000), highway tunnels in Madeira (Fig. 4.29), and open excavations in Manila, Philippines (Fig. 4.38).

4.5.9. Excavation in man-made materials

Man-made materials may range from landfills to (solid household) waste dumps. Recent landfills for infrastructure, industry, or housing are normally controlled deposits and properties throughout the fill are documented, for example, the land reclamation for the Disney theme park in Hong

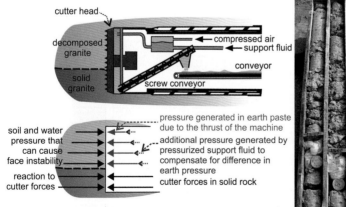

Fig. 4.35. Tunnel excavation with TBM for an underground railway in Porto, Portugal. The figure at the left shows the face support pressures (modified from Babendererde et al./IOS Press). The photographs show the decomposed (left) and solid granite (right) (reprinted from Babendererde et al. 2004, with permission from IOS Press).

Fig. 4.36. Squeezing ground in a fault zone; approximately 1 m of inward displacement is visible in the roof and sidewalls (Nathpa Jhakri Hydro Power Project, India; photograph: E. Hoek).

Box 4.4. Deformable steel rings in squeezing rock mass in Gotthard Base Tunnel

Squeezing ground under high pressures occurred in the Gotthard Base Tunnel (Section 7.3). This tunnel system of two tunnels with a diameter of about 9 m with multifunction crossing stations and shafts is through series of different grounds with and without water. The various geological environments ranging from hard rock to squeezing and soft, soil-like materials, occasionally under very high overburden pressures (up to 2400 m of overburden) and locally with extremely high water pressures, is an example where virtually all available underground excavation and support techniques have been applied (Ehrbar 2008). The length of 57 km invariably complicated the project even more, in particular, the logistics. Part of the Sedrun section (Fig. 4.37a) consists for a large part of a ductile kakiritic (i.e. broken or intensively sheared) rock mass of gneiss, slate, and phyllite (a metamorphosed clay) (Vogelhuber *et al.* 2004) mixed with a brittle, hard rock mass in narrow vertical layers. In about 1.1 km of the section ('Tavetsch Intermediate Massif North') the rock mass is squeezing with an overburden thickness of about 900 m. In this section, the support was successfully made by circular and immediate ring closure after a full-face excavation. The circular excavation was made oversized with varying diameters to accommodate different volumes of squeezing ground (Fig. 4.37b). The ground was reinforced directly after excavation with anchors (up to 18 m in the face and 12 m long radial) and sliding steel inserts consisting of eight segments were installed that allow reduction of the diameter if stresses become too high (Fig. 4.37c). For a length of about 60 m after installing this support, the tunnel was allowed to deform ('deformation phase'), and no further support installed. The schedule included additional anchors and steel rings could be installed where required but these were not necessary. After yielding of the ground was finished (75 m from the face), the steel inserts were concreted in ('resistance phase'). The development of the groundmass and support stresses is shown in Figure 4.37d. The encountered deformations were on average 20 to 30 cm with a maximum of 75 cm. By this support methodology, the groundmass is allowed to deform without excess deformation while reducing the groundmass stress around the already excavated part of the tunnel. The face deformed considerably and the anchoring in the face was necessary for safety and stability.

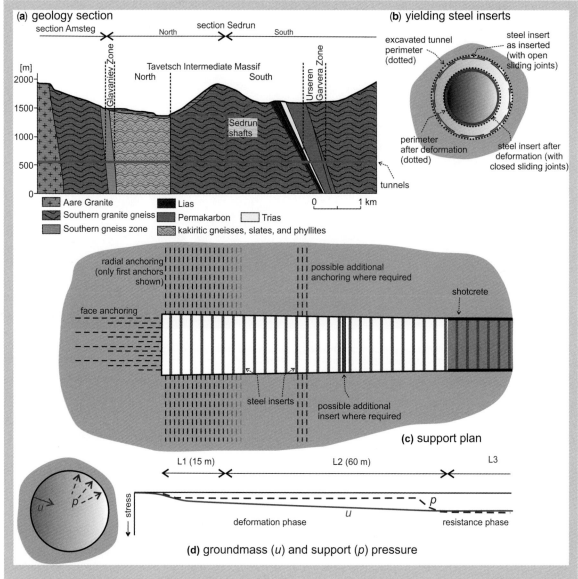

Fig. 4.37. Squeezing rock mass and support measures in the Tavetsch Intermediate Massif North, Gotthard Base Tunnel (modified from Ehrbar 2008).

Kong (Bunschoten 2002). Material mix and properties are undocumented for uncontrolled land-fills or for older fills, which complicates excavation. Often, mechanical characteristics of land-fills should be improved and homogenized by, for example, grouting as was done for the Taskim–Katabaş funicular system in Istanbul (Aykar et al. 2005). Excavation in solid household waste dumps may be even more difficult due to highly irregular compaction. Decaying waste may lead to voids, generate poisonous and explosive gases, and contaminated water (Machado et al. 2008). Such dumpsites may generate extensive displacement and settlement of underground excavations that are hard to predict. Excavation of waste materials also hampers TBMs and road headers in their operations since these cannot cope well with metals, wood, and plastics. Hence, conditions for underground excavation are quite unfavourable for waste dumps and

Fig. 4.38. A 50 m deep open excavation in volcanic tuffs in Manila, Philippines. The walls are unsupported, because the intact ground material is sufficiently strong and contains few discontinuities, none of which is in an unfavourable direction (photograph: G. Gardiel).

preferably avoided. If that is not possible, constructions with foundations or cut-and-cover down to the ground beneath the dumpsite are made. That was done, for example, for the sewage tunnels and shafts in and under a dumpsite of solid household waste in the City of Richmond, USA (Rybak & Brown 2008). *In-situ* transformation of waste materials by heating, for example, with plasma torches might be cost-effective if that would immobilize polluting agents and transform the waste in a 'glass-like' substance (Circeo & Martin 2001).

4.5.10. Cut and cover

Excavation from the surface by 'cut and cover' is often done for structures down to approximately 50 m, a hole is made in the ground ('cut'), the structure built in the hole and the surface restored with previously excavated material ('cover'). Depending on the ground conditions, a hole may be made with freestanding (sub-) vertical walls, if the ground will sustain such slopes. This will normally be the case in rock or cemented soils (Fig. 4.38). In weaker rocks or soil masses, (sub-) vertical slopes are not sustained and artificial walls of steel, concrete, or timber have to be inserted prior to excavation (Fig. 4.25 and Box 4.5). Such walls may become part of the underground structure. Groundwater levels have to be controlled by pumping or by creating an impermeable floor and walls in the excavation. 'Cut and cover' methods are widely used throughout the world.

In 2008 and 2009, several serious failures with 'cut and cover' occurred. During the construction of a railway station for an underground metro line in Amsterdam, The Netherlands, leakage occurred and several historical houses were damaged (Box. 4.5). In Cologne, Germany, the City Archive and adjacent buildings collapsed, probably due to failure of a retaining wall of a nearby underground railway station under construction. That failure allowed a large volume of water with sand and gravel to flow from below the foundations into an already finished part of the underground station. Whether the collapse was due to design or construction errors is still subject of investigation (DW 2010) (see also Section 4.7).

4.5.11. Interaction between foundations and subsurface

Foundations of buildings at the surface are generally deeper in soft soil materials and more shallow in competent materials such as rock. For example, the piles in soft soil for the foundation of the almost 500 m high World Finance Centre in Shanghai go down to 74 m below the surface (Chen 2009) and to 50 m in weak rock for the 818 m high Burj Khalifa Tower in Dubai (Poulos & Bunce 2008). In homogeneous ground, the shear strength of ground materials increases with depth because of the higher overburden pressure (Box 4.2). Often, deeper-seated groundmasses have also a better quality than near surface materials, as they are less affected by surface weathering, and are more compacted and consolidated because they have been buried for a longer time span and under a higher pressure. To economize on the high construction costs of foundations these are often combined with cellars or parking garages.

Surface structures generate stress fields (Section 4.4) and settlement of the ground material below and alongside foundations (Fig. 4.18). New buildings constructed on top of or below existing structures may therefore influence the already existing structures, as, for example, during construction of new Rotterdam Central Railway station (Berkelaar 2009) and the Barcelona underground railway system (Gens et al. 2006). In both cases complicated and expensive structures had to be built to maintain the existing structures, safe and in service. In Barcelona, this included installation of jacks to compensate for settlement of an existing road on top of the underground railway line under construction. Underground occurrences of foundation and structure remains, such as anchors for sheet piling, may further complicate construction (Berkelaar 2009).

Most foundations of structures generate compressive stress in the subsurface. However, this may also be a tensile stress introduced by anchors and so-called 'tension piles'. For example, buoyancy due to high groundwater tables may be controlled by tension piles or anchors. Examples include the permanent tension piles for the underground railway tunnel in Rijswijk, The Netherlands (Zigterman 2009), and the temporary tension piles for the building-pit of the IMAX Theatre in Nuremberg, Germany (Bauer 2010).

4.6. Designing underground structures

In soils and in rather homogeneous rock masses, the required stability and support systems for underground constructions are generally determined by analytical or numerical calculations. However, most rock masses display quite complicated mechanical behaviour. In such cases, designing underground structures solely by geomechanical calculations does not provide satisfactory and conclusive results. Many underground excavations in rock, including tunnels for major highways in Europe and the USA, are designed and constructed largely on the basis of the chief design engineer's experience. An alternative for excavations in rock, in particular for feasibility studies, is designing by empirical 'classification systems', which often are used also for establishing input parameters for analytical and numerical calculations.

4.6.1. Classification systems

Classification systems link rock mass characteristics with recommended excavation and support. Classification systems were developed empirically by establishing relevant parameters, optimizing weighting factors for each parameter, and formulating relations between parameters and stability based on a wide variety of case studies. Underground excavations made in a rock mass with a similar classification rating are assumed to have the same stability appraisal or to require the same support as the ones used for the classification system. Various classification systems have been developed since the mid-1940s and presently most commonly used systems for underground excavations are briefly discussed below.

Box 4.5. Cut and cover leakage

Significant settlement and subsequent damage to adjacent historic houses took place during the construction of an underground railway station in Amsterdam (Fig. 4.39). The houses are founded on wooden piles about 6 m from the building pit. Diaphragm walls had been made to a depth of about 44 m below the surface, and excavation of the ground in-between had started. Where the excavation reached a depth of around 14 m, seepage of groundwater occurred through one of the joints of the diaphragm walls.

The natural groundwater table in the top layers is near to the surface, and both deeper sand layers have separate water regimes with different piezometric levels. The leaking of the diaphragm wall was probably due to installation errors (Fig. 4.40). Water with ground material flowed into the excavation, reducing the water levels and relaxing the ground surrounding the wooden piles. In turn, this resulted in settlement of the layers adjacent to the diaphragm wall, up to 11 cm horizontal displacement of the houses, and the ground relaxation reduced the bearing capacity with a subsequent settlement of some 25 cm. Grouting below and around the buildings was done to stabilize the houses (Bezuijen & Korff 2009). The joints and surrounding groundmass were frozen to allow the underground construction works to continue.

Fig. 4.39. Damaged historical houses due to construction of an underground railway station, Amsterdam, The Netherlands (the figures behind the windows are photographs of paintings of Dutch masters fitted after the inhabitants had to move out because of risk of imminent collapse) (photographs: J. G. A. M. Arnoldus).

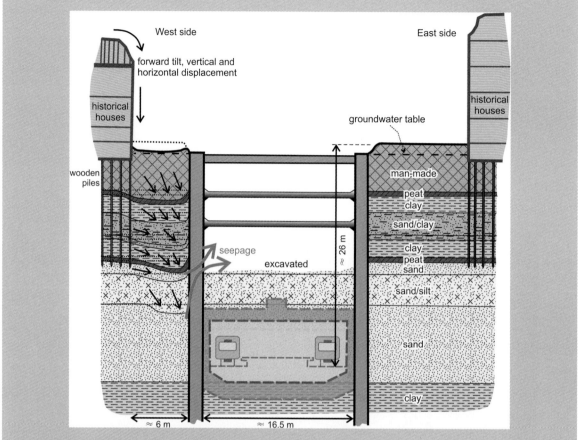

Fig. 4.40. Sketch of the seepage through the (joints of the) diaphragm wall and settlement of the ground and houses during construction by 'cut-and-cover' for an underground railway station in the Amsterdam Metro (modified from Bosch & Broere 2009).

4.6.1.1. Bieniawski's Rock Mass Rating (RMR)

Bieniawski's Rock Mass Rating (RMR) system is one of the oldest still used classification systems (Bieniawski 1989). It was developed in South Africa for underground mining but it is currently widely used in civil engineering as well. The system is based on a combination of five parameters (Fig. 4.41). Each parameter is expressed by a point rating and addition of the points results in the RMR value. RMR ranges between 0 (very poor rock for tunnelling) to 100 (very good rock for tunnelling). Excavation and support is determined by the RMR value and results in five different support classes. The RMR value has been related (empirically) to rock mass cohesion, friction angle of the rock mass, and other rock mass properties (Fig. 4.42). Various adjustment factors and refinements are possible such as compensation for discontinuity orientation and damage due the method of excavation. The merits of the RMR system for underground excavation, a wide variety of modifications, and its application to slope and foundation engineering, is discussed by many authors, including Aksoy (2008) and Hack (2002).

4.6.1.2. Q-system

In the Q-classification system, the 'rock mass quality' is expressed by a factor (the 'Q-factor') which depends on three features of a rock mass: the size of the intact rock blocks in the rock mass, the shear strength along the discontinuity planes, and the stress environment for the intact rock and discontinuities around the underground excavation (Barton *et al.* 1974; Barton 2002). The block size and shape is relevant as excavating and supporting a rock mass of larger, cubical shaped blocks is easier than a rock mass with smaller and irregular blocks. The shear strength is relevant, as blocks do not easily move if the shear strength is high. Finally, the stress environment determines, for example, the possibility of failure by spalling of intact

Parameter		Range of values					
Intact material strength - UCS (MPa)		> 250	100-250	50-100	25-50	5-25	1-5	< 1
Rating:		15	12	7	4	2	1	0
Drill core quality - RQD (%)[1]		90-100	75-90	50-75	25-50			< 25
Rating:		20	17	13	8			3
Discontinuity spacing (cm)		> 200	60-200	20-60	6-20			< 6
Rating:		20	15	10	8			5
Condition of discontinuities		Very rough surfaces, Not continuous, No separation[2], Unweathered wall rock	Slightly rough surfaces, Separation < 1 mm, Slightly weathered walls	Slightly rough surfaces, Separation < 1 mm, Highly weathered walls	Slickensided[3] surfaces, or Gouge[4] < 5 mm thick, or Separation 1-5 mm continuous			Soft gouge > 5 mm thick, or Separation > 5 mm continuous
Rating:		30	25	20	10			0
Groundwater	Inflow per 10 m tunnel length (L/min)	None	< 10	10-25	25-125			> 125
	Ratio of joint water pressure over major principal stress	0	< 0.1	0.1-0.2	0.2-0.5			> 0.5
	General conditions	Completely dry	Damp	Wet	Dripping			Flowing
Rating:		15	10	7	4			0

Notes: [1]RQD expresses the quality of the core obtained from a borehole and depends on the quality of the rock mass; 0%: many discontinuities and weak zones; 100%: sound rock with few discontinuities. [2]Separation is the opening between the two discontinuity walls. [3]Slickensided is a striated smoothly polished surface created by frictional movement between the two sides of a discontinuity. [4]Layer of discontinuity infill material consisting of very fine material (silt or clay), may contain small rock fragments.

Fig. 4.41. RMR parameter ratings. *Source*: Modified from Bieniawski (1990).

rock at the tunnel wall or squeezing of rock. The resulting Q-factor ranges from 0.001 for an exceptionally poor rock mass to 1000 for a rock mass exceptionally good for underground excavation. In general, input parameters may be established by simple measurements or by visual inspection of rock mass exposures or of borehole cores. Table 4.5 describes how one of the input parameters (the number of observed joint sets, J_n) for the block size factor is determined.

The numerical values of the class boundaries for the various rock mass types are subdivisions of the Q range on a logarithmic scale. Figure 4.43 shows the relation between Q rating and tunnel support classes as a function of the span or height of the excavation divided by the ESR (Excavation Support Ratio), an expression for different types of use of an excavation with different safety levels (Fig. 4.44). The tunnel support classes are mainly based on 'reinforcing' types of support, for example bolts that reinforce the rock mass by binding blocks of rock together and shotcrete preventing ravelling of the rock mass and loss of integrity. The use of the modified Qtbm-system for TBM tunnels is under discussion (Palmstrom & Broch 2006).

4.6.1.3. Mining Rock Mass Rating (MRMR)

The Mining Rock Mass Rating (MRMR) classification system (Laubscher 1990; Laubscher & Jakubec 2001) builds on the RMR system as described above, with similar main parameters, but introducing the concept of 'reinforcement potential of a rock mass'. In MRMR, the RMR value is multiplied by adjustment factors depending on future (susceptibility to) weathering, stress, orientation, damage due to the method of excavation and the number of free block faces that facilitate gravity fall. The combination of RMR and MRMR values determines the required

	Rock mass classes				
RMR rating	100-81	80-61	60-41	40-21	< 20
Class number	I	II	III	IV	V
Description	Very good rock	Good rock	Fair rock	Poor rock	Very poor rock
Average stand-up time	20 years for 15 m span[1]	1 year for 10 m span	1 week for 5 m span	10 hours for 2-5 m span	30 min for 1 m span
Rock mass cohesion (kPa)	> 400	300-400	200-300	100-200	< 100
Rock mass angle of internal friction (degrees)	> 45	35-45	25-35	15-25	< 15

Note: [1]Span is the span of the excavation.

Fig. 4.42. Application of RMR parameter ratings. *Source*: Modified from Bieniawski (1990).

Table 4.5. *Number of joint sets (J_n) determination based on visual observed number of different joint sets*

Number of joint sets	J_n (=joint set number)
Massive, no or few joints	0.5–1.0
One joint set	2
One joint set with random joints	3
Two joint sets	4
Two joint sets with random joints	6
Three joint sets	9
Three joint sets with random joints	12
Four or more joint sets, random, heavily jointed, 'sugar cube', etc.	15
Crushed rock, earthlike	20

Source: Modified from Barton 2002 with permission from Elsevier.

support (Fig. 4.45). A rock mass with a high rock mass rating before the adjustment factors are applied has a high reinforcement potential, and can thus be reinforced by, for example, rock bolts, while a low RMR value indicate a rock mass with a low reinforcement potential for which rock bolts will not be suitable.

4.6.1.4. Geological Strength Index (GSI)
The Geological Strength Index (GSI) is derived from a matrix describing the 'structure' and the 'surface condition' of the rock mass (Fig. 4.46). The 'structure' is related to the block size and the interlocking of rock blocks while the 'surface condition' is related to weathering, persistence, and condition of discontinuities. The GSI is one of the constituents of the Hoek-Brown failure criterion (Hoek *et al.* 1992, 2002; Rocscience 2011). The failure criterion does not provide excavation or support recommendations but rather determines rock mass properties (Marinos & Hoek 2000; Marinos *et al.* 2005).

4.6.2. Excavation damage

Classification systems may also evaluate the excavation damage on the groundmass (Fig. 4.47) by the various excavation methods, as described in Section 4.4.13. Normally, natural exposures have not been subject to high stress levels and display less mechanical discontinuities as for example rock masses excavated by blasting. Generally, excavation by manual or mechanical methods (low energy release per time unit) produces less discontinuities than

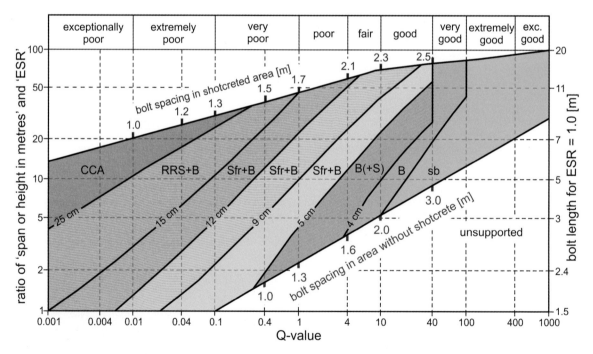

Fig. 4.43. Tunnel support requirements as function of the Q value, the type of use of the excavation (ESR) and the size of the excavation. Support classes range from cast concrete lining (CCA) via steel rib-reinforced-shotcrete arches (RRS) and shotcrete (S) with or without wire mesh (fr) to systematically spaced steel rock bolts (B), and spot bolting (sb). The thickness of the shotcrete layer is marked at the boundaries of the support classes. Bolt spacing is indicated at the top and bottom of the coloured area for bolts in a support with and without shotcrete respectively. Bolt length is given on the right (modified from Barton 2002 with permission from Elsevier).

Drill & blast excavation[a]		Tunnel boring machine[b]	
Excavation category	ESR	**Excavation category**	ESR
Temporary mine openings	3 - 5		
Vertical shafts — Circular cross section	2.5		
Vertical shafts — Rectangular/square cross section	2	Pilot tunnels[1]	2.0
Permanent mine openings, water tunnels for hydropower (excluding high-pressure penstocks), pilot tunnels, drifts, and headings for large excavations	1.6		
Storage caverns, water treatment plants, minor railway and road tunnels, surge chambers, access tunnels	1.3	Water/sewage tunnels[1]	1.5
Power stations, major rail- and highway tunnels, civil defence shelters, portals, intersections	1		
Underground nuclear power stations, railway stations, factories	0.8	Rail- and highway tunnels[1]	0.5 to 1.0[2]

Notes: [1] For temporary openings the ESR is multiplied by 1.5 and the Q value by a factor between 2.5 and 5 depending on the period the excavation has to be open. [2] ESR is 0.5 for long, high-speed rail- or highway tunnels. Data: (a) Barton et al. (1974) (b) Barton (2000).

Fig. 4.44. Summary of ESR (Excavation Support Ratio) values.

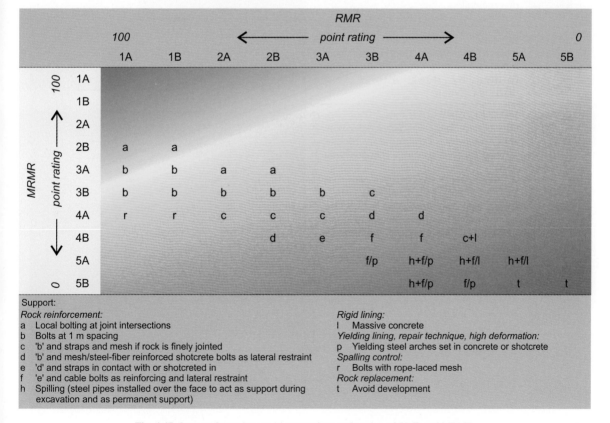

Fig. 4.45. Support for underground excavations as function of RMR and MRMR.

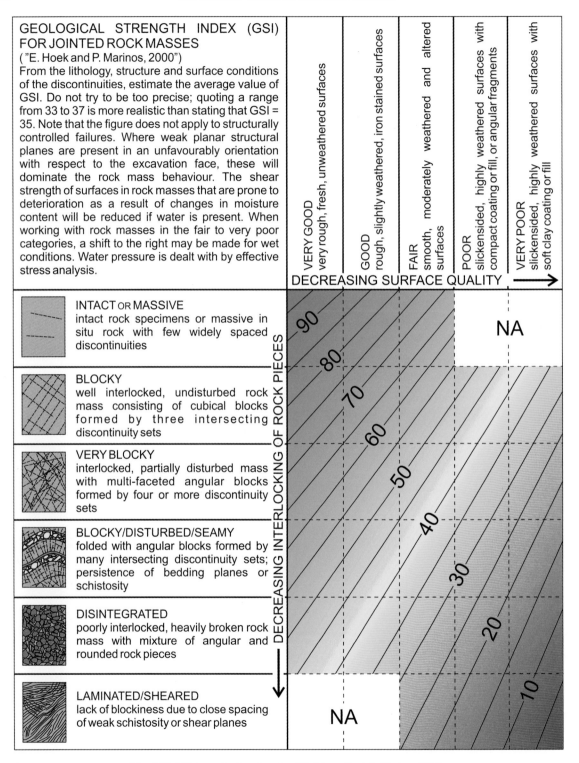

Fig. 4.46. GSI determination chart (modified from Hoek & Marinos 2000).

SSPC[a] (slope)		FRHI[b] (slope)		Hoek - Brown (GSI disturbance factor)[c][d]				MRMR[d] (underground mining)		MBR[e] (underground mining)	
Excavation	Factor	Excavation	Rating	Excavation (slope)	D	Excavation (tunnel)	D	Excavation	Factor	Excavation	Factor
Natural/hand-made[(1)]	1.00	Smooth excavation	-1	In some softer rocks excavation can be carried out by ripping and dozing and the degree of damage to the slope is less (mechanical excavation).	0.7	Mechanical or hand excavation in poor quality rock masses (no blasting) results in minimal disturbance to the surrounding rock mass	0	Boring	1.00	Boring	1.00
		Regular cut	3								
Pneumatic/hydraulic hammer[(2)]	0.76	Manual cut	4								
Controlled blasting	0.99	Controlled blasting	1	Small scale blasting in civil engineering slopes results in modest rock mass damage, particularly if controlled blasting is used. However, stress relief results in some disturbance	0.7	Excellent quality controlled blasting or excavation by Tunnel Boring Machine results in minimal disturbance of the confined rock mass surrounding a tunnel	0	Controlled blasting	0.97	Controlled blasting	0.94-0.97
Blasting with result:	Good	0.77	Regular blasting	5			Where squeezing problems result in significant floor heave, disturbance can be severe unless a temporary invert is placed	With invert	0	Good blasting	0.94	Good blasting	0.90-0.94
	Open discontinuities	0.75										
	Dislodged blocks	0.72			Poor blasting	1.0		No invert	0.5			
	Fractured intact rock	0.67	Poor blasting	8	Very large open pit mine slopes suffer significant disturbance due to heavy production blasting and also due to stress relief from overburden removal (production blasting).	1.0	Very poor quality blasting in a hard rock tunnel results in severe local damage, extending 2 or 3 m, in the surrounding rock mass	0.8	Poor blasting	0.80	Poor blasting	0.90-0.80
	Crushed intact rock	0.62										

Notes: SSPC, MRMR and MBR factors range from 1.00 for negligible damage to 0.62 respectively 0.80 for maximum damage, FRHI is expressed as point rating from -1 for no damage to 8 for maximum damage, GSI disturbance factor ranges from 0 for undisturbed to 1.0 for maxium disturbance. [(1)] Care should be taken that discontinuities due to stress relief are not considered excavation damage. [(2)] This value is based on hammer sizes up to 5 m with a diameter of 0.2 m. [(3)] The description of D is referenced with example photographs of excavation damage. Data from: (a) Hack et al. (2003) (b) Singh (2004) (c) Rocscience (2011) (d) Laubscher & Jakubec (2001) (e) Cummings et al. (1982).

Fig. 4.47. Excavation damage factors for a rock mass. The factors are correction factors to adjust the design parameters of classification systems to account for the method of excavation. As quantified values for excavation damage in underground excavation are rare, this table includes slope damage factors as well.

blasting (high energy release per time unit), except perhaps excavation by hydraulic or pneumatic hammering that also may seriously damage the rock mass.

4.6.3. Monitoring

Monitoring the behaviour of the ground and construction during a project is relevant as it provides an early warning for potential hazards, such as failure or collapse of the structure. Monitoring also gives options to adjust the design during construction (Bock 2001; Kavvadas 2005). Figure 4.48 shows an example of how convergence measurements in a tunnel indicate that, after a long time of stabilization, the excavation begins to deform, pointing to insufficient support on the longer term. Monitoring is standard practice in any project and is often made compulsory by law, insurance companies, or the client. Monitoring should not be limited to the structure that is to be built but neighbouring constructions and the surrounding natural environment should be included as well. Monitoring requires a 'null measurement' as a reference before any construction has started. Without such a measurement, it is virtually impossible to prove that claims from owners from surrounding constructions, or any other party concerned are unjustified. A null measurement may have to be a whole series of measurements for a long time to cover changes that have nothing to do with the new construction. For example, a groundwater table may have to be monitored for years to cover seasonal changes prior to the start of a new project. Finally yet importantly, monitoring may prove very useful in public relations, to generate trust and neutralize any unfounded speculations, rumours, or social unrest.

Monitoring may range from simple, visual observation to sophisticated, automatic measuring systems. Virtually all monitoring is about measuring three features: the relative position of points, stress, and water-level and -pressure. A fourth parameter, temperature, is often required to compensate for temperature-dependent extension and shrinking of the ground and construction. Dependent on their risk of occurrence, special features as the presence of (poisonous) gases or chemicals in water (e.g. acidity) may be measured as well. Until recently, only mechanical and fairly simple analogue electrical measuring devices were available. The digital revolution, the inventions of laser, fibre-optics (Inaudi 2003, 2007, Fig. 4.49), and electro-magnetic, ever smaller measuring devices, and the very significant cost

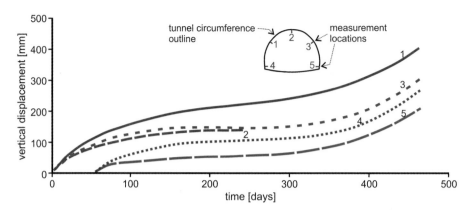

Fig. 4.48. Deformation v. time for a section of the Kallidromo Tunnel, Greece. The deformation of the tunnel wall is measured at various positions along the circumference of the tunnel (insert in graph). After a period (between 170 and 240 days) with almost no increase in deformation, deformation is re-activated due to stress re-distribution initiated by large deformations and failure in adjacent sections of the tunnel. The re-activation may indicate that stability problems in the section in the graph are imminent if no additional support measures are taken, and it may require re-excavation because the size of the tunnel is reduced (modified from Kontogianni & Stiros 2005 with permission from Elsevier).

Fig. 4.49. SOFO sensor installed on concrete reinforcement steel (insert) in buttresses and support to measure deformation and temperature. The fibre optic displacement sensor has a resolution in the micrometre range (South Portal, Bodio, Gotthard Base Tunnel; photographs: SMARTEC SA).

reduction of digital processing equipment, has opened new economically feasible opportunities for monitoring (Lam 2010). Where in the past most measurements had to be done manually at the location of the device; nowadays devices mostly measure on remote request or automatically, and send data wireless or by cable to a remote processing station. Monitoring of surface features from above by (remote-controlled) helicopter and plane or by satellite becomes increasingly useful and is generally very economical. Positioning by satellites (Global Positioning System, GPS) has become a standard tool while other techniques for monitoring surface features with high accuracy are introduced, such as measuring subsidence due to underground excavations with InSAR (Interferometric Synthetic Aperture Radar) (Akcin et al. 2010; Lam 2010).

4.6.4. New Austrian Tunnelling Method (NATM)

The New Austrian Tunnelling Method (NATM) (Von Rabcewicz 1964a, b, 1965; Pacher et al. 1974; Müller 1978) developed in the 1950s, is often referred to as a design and tunnelling method which includes characterization and monitoring of the ground. By this method, the ground surrounding the excavation is used to support itself by allowing small deformations but preventing further loosening of the ground. That is realized through a rather flexible support system of rock bolts or shotcrete in the earliest possible stages of excavation. Another characteristic of NATM is the rapid closure of the bottom part of the support (the invert) thus creating a load-bearing ring. In addition, the method extensively monitors the behaviour of the ground and support. NATM may include legal and contractual parameters that relate the results of classification of the subsurface groundmass to specific tasks to be carried out by project consultants and contractors. Another relevant feature of NATM is that the support should be adjusted to the changing conditions as encountered during excavation avoiding time-consuming re-negotiation with the client and consultants (Field et al. 2005).

NATM is applied widely and often successfully in Central Europe and elsewhere. Successful European examples are the Tauern Tunnel under the Alps in Austria (Ayaydin & Leitner 2009; Weidinger & Lauffer 2009), and the Strenger Tunnel in Germany cutting through highly squeezing ground (John et al. 2005). A less successful example is shown in Figure 4.50. An overview of problems and adjustments made to excavation and support strategies appropriate for the USA is presented by Thapa et al. (2009) for the Caldecott 4th Bore. A successful example of NATM tunnelling in the USA was the Mission Valley Light Rail tunnel, East extension, in San Diego, USA, that opened in 2005. The shallow (less than 20 m of ground cover) twin-track tunnel is excavated in conglomerate and gravel with varying levels of cementation and sand lenses partially below the groundwater table (Ragland et al. 2003). Extensive monitoring of surface settlement and deformation measurements in the tunnel together with an adequate supervision and management structure during construction contributed to the successful operation. Since its launch, the question whether NATM is a genuine, new technology or a collection of existing techniques, and whether the fundamental ideas of the methodology are correct has been the subject of discussion (Kovári 1993; Karakuş & Fowell 2004). That discussion intensified after the collapse of the Heathrow Express Tunnel (Section 4.7.1) (Gudehus 2001; Kolymbas 1998, 2008). Appraisal of the benefits of the complete methodology is often difficult as frequently is claimed that NATM is applied whereas in reality only parts of the NATM methodology are used, for example an excavation supported by shotcrete is claimed to be made following NATM while the other features of NATM are not applied (Romero 2002; Wallis 2010).

4.6.5. Analysis of Controlled Deformations (ADECO) methodology

Since the 1980s, methods were developed to prevent not only generation of excess deformation along the circumference, but also deformation of the groundmass in front of the face. The results were presented in the so-called 'Analysis of COntrolled DEformation' (ADECO) methodology. The formulation of this methodology was initiated by Lunardi (2000, 2008; Lunardi et al. 2008). He observed that any tunnel failure and excess deformation behind the face in the already excavated tunnel, is preceded by failure and deformation ahead of the face. Hence, if deformation in the core (the not yet excavated groundmass) could be reduced, the groundmass surrounding the core will deform less too and that will result in a more stable tunnel. The core may be stabilized by techniques supporting the face, core, and the groundmass surrounding the core ahead of the tunnel face. Full-face excavation should be applied, as sequential excavation will allow more deformation of the groundmass for a longer time span. As an additional benefit, larger and more powerful equipment may be applied in the larger space available reducing the time between excavation and completion of the support. In weaker grounds, the methodology proved to be more successful than other methodologies. For example, the Tasso Tunnel in the high-speed railway line from Rome to Florence, in Italy (Tonon 2010), where sequential excavation failed, but full-face excavation was successful if the core was stabilized with fibreglass tubes (Lunardi 2008). The sediments at the tunnel site consist of sandy and clayey silts and clays, some parts saturated with groundwater.

4.6.6. Numerical modelling

Mathematical programmes for calculating the stability of underground structures emerged since the 1970s. Available programmes include Plaxis (2010), Examine and Phase from Rocscience (2011), UDEC, FLAC, and PFC (Itasca 2011); many others may be found via GeotechLinks (2010). In a numerical model, a computer programme simulates the mechanical behaviour of the groundmass and the engineering structure. Numerical modelling may be done

Fig. 4.50. Collapse of the underground railway at Trudering, Munich, Germany, 1994. The tunnel was planned to be constructed with NATM techniques under a supposedly impermeable water-sealing clay layer with water-bearing layers above. Shortly after commencing the work, the thrust on the face became too large and large quantities of water and ground flushed into the tunnel. A crater formed in the street above so rapidly that a touring car fell backwards into it. Three people drowned and 36 were injured (photograph: AP, Reporters, Frank Augstein).

in one-, two- or three-spatial dimensions and may be either static or dynamic. For modelling ground geometry and properties, two approaches may apply; a *discontinuous* and a *continuum* model. Discontinuous models divide the groundmass into cells; each cell simulates the behaviour of the intact ground material while the boundaries between the cells simulate discontinuities (Fig. 4.51). A continuum model also divides the groundmass into cells but each cell simulates groundmass behaviour, including discontinuities.

Modelling of the excavation and support for the Second Avenue Subway in New York illustrates the use of a continuum model (Carranza-Torres & Nasri 2007, pers. comm.; Nasri *et al.* 2008). The rock mass consists of a mixture of gneisses, calcareous rocks and schists, intruded by veins of intrusive rock, dissected by many discontinuities; it is heavily fractured and degraded in places, folded and weathered. In the continuum modelling, the groundmass properties used are determined based on the Hoek-Brown failure criterion (Section 4.6.1.4). The discontinuities are thus integrated in the groundmass properties via the Geological Strength Index (GSI) and are not modelled as individual features. The continuum modelling shows the overall groundmass and support behaviour, the development of deformation and the suitability of the support design (Fig. 4.52). Discontinuous modelling in the same project was used to study, in particular, the behaviour of individual blocks of rock for features as risk of gravity fall.

A relatively new approach is the so-called '*Particle Flow Code*' (PFC) modelling (Itasca 2011). Materials are modelled with spherical particles that may be shaped in objects of any form by bounding (cementing) particles together. PFC is particularly suited for problems where many objects interact, flow features, and objects that fracture in many segments. Figure 4.53 shows a simple example in which the impact force of a block of rock falling from the roof of a tunnel is modelled as a solid block and as a block that can break apart in many fragments on impact. The latter displays a considerably lower impact force and is probably more realistic.

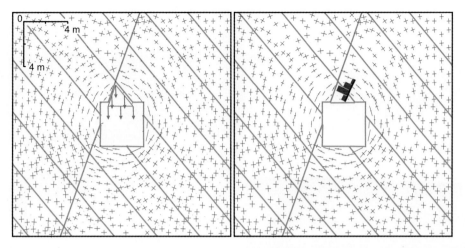

Fig. 4.51. Example of a numerical calculation of tunnel stability in a two-dimensional discontinuous model. Left: block in roof is unstable; right: block is stabilized by a rock bolt. Green continuous lines represent discontinuities; the small crosses (purple) represent the principal stress trajectories. The rock bolt axial load is plotted as a bar graph along the axis of the bolt (courtesy Itasca Consulting Group, Inc. (Itasca 2011) UDEC – Universal Distinct Element Code, Version 5.0. Minneapolis: Itasca).

Although anything may be calculated with a high degree of detail and accuracy, in practice a series of problems occur in using numerical modelling programmes. A basic problem is that the ground behaviour is normally not well known, in particular over longer times. Additionally mostly the number of known or measured quantified property values is small compared to the large number required to correctly model the behaviour of grounds that are more complicated such as discontinuous rock masses. In contrast to analytical and empirical models, these shortcomings are particularly manifest in numerical models as such models are supposed to simulate groundmass behaviour in a correct and detailed manner, while that may not be realistic due to input of inaccurate ground properties and a poorly represented ground behaviour. However, numerical methods may be successfully used for sensitivity analyses that will provide the most likely and worst-case scenarios for a groundmass calculation. The same applies for the various methods of stochastic calculations incorporated in analytical or numerical calculations.

4.7. Operational risks and geohazards

Failures or major problems with underground excavations normally occur if (daily) routine of the engineers and labour are interrupted, or if something unanticipated happens. They very seldom occur from known features or practices underground. Normally, a single failure does not develop into a catastrophe because virtually every project includes safety measures and fall-back scenarios. Disasters nearly always result from a sequence of smaller events that are not disastrous by themselves. Moreover, in normal operations, adequate measures will likely have been taken to interrupt the sequence and prevent the development of a disaster. However, at night or other times when the engineering staff and experts are temporarily unavailable and supervision is limited or absent, a catastrophe may fully develop because adequate actions are not undertaken in time (Wallis 2010).

4.7.1. Routine

Underground excavation works and decision-making are quite distinct from building constructions at the surface because of the greater reliance on personal experience and expertise of labour and engineers underground. Generally, underground excavation works require more teamwork by construction people who often work together for years, in particular in mining but also in civil engineering. Routine is often crucial in such environments and mutual communication frequently takes place through special work slang or pidgin languages, for example, 'Mine Fanagalo' has been used for decades in the mines throughout Southern Africa (Fanagalo 2010). Changing construction methods may impose additional risks to such working environments. If a failure occurs, it may not be recognized and communicated immediately and proper remedial measures may not be taken in time. This has probably been one of the causes of the collapse of the Heathrow Express Tunnel in 1994. That was one of the first tunnels built in the United Kingdom by using NATM. Collapse took place on 21 October 1994 at Heathrow Airport near London, producing a crater at surface and considerable damage to the surface infrastructure. Investigation afterwards identified major organization and communication shortcomings that led to inadequate

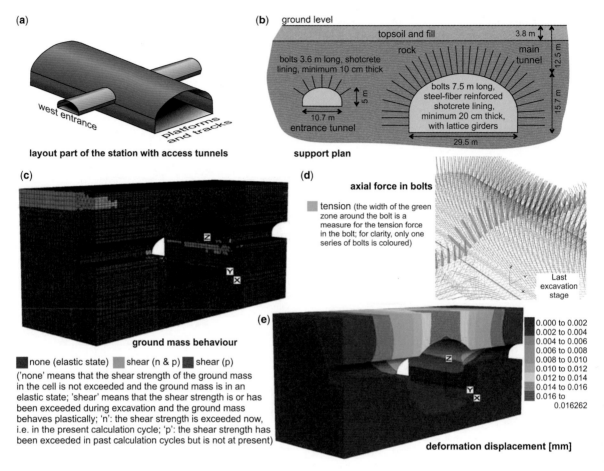

Fig. 4.52. Three-dimensional mechanical modelling in FLAC3D (Itasca 2011) of excavation and support for the intersection of entrances and main gallery for a new underground railway station at 72nd Street/2nd Avenue, New York. The tunnels are excavated sequential in multiple excavation and support rounds. The groundmass behaviour, stress, and deformation shown are those existing after the last excavation and support round (modified from Carranza-Torres & Nasri 2007, pers. comm.; Nasri et al. 2008).

construction and to a less than optimum risk management (HSE 2000; Thomas et al. 2004; SCOSS 2004, see Section 4.6.4).

4.7.2. Unexpected ground conditions, data limitation and use of expert knowledge

In principle, design engineers should ensure that sufficient qualified data about the subsurface is available or will be obtained for a proper design and construction of the project. In practice, however, data is often limited to small-scale geological maps and limited information from other sources, as a few boreholes, penetration tests, and geophysical data. Developing reliable subsurface models from such limited numbers of data requires the input of very significant expert knowledge by an engineering geologist (Fookes 1997; Ozmutlu & Hack 1998, 2003; see also Chapter 5). Expert knowledge is an un-quantified parameter that influences the quality of the interpretation significantly and adds a potential risk to the project. Such risk is low where dealing with fairly simple geological conditions and with abundant subsurface information, but the risk may rapidly increase in more complex geological conditions and if less data is available (Section 5.4). This data interpretation risk is a key reason for adjusting the design and support to the actual ground conditions encountered during excavation on a continuous basis while the project progresses. Such adjustments should often be made instantaneously, as otherwise the collapse may have happened already.

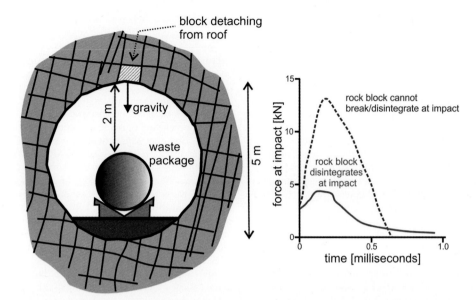

Fig. 4.53. The impact force of a block of rock falling from the roof of a tunnel modelled in PFC to estimate the potential damage to nuclear waste canisters, Yucca Mountain, Nevada, Repository (modified from Hart & Fairhurst 2000).

4.7.3. Risk management

Among many other lessons learned, the collapse of the Heathrow Express Tunnel (Section 4.7.1) triggered a more formal and extensive application of risk management systems (Van Staveren 2006; Fenton & Griffiths 2008) for underground excavations in most of the major projects in the world. Clients, in particular insurance companies (ITIG 2006), have been the driving force for such developments as they demand information on expected risks and on appropriate actions to be taken to minimize the impact and costs of a failure. Risk management includes assessment of the potential risks for a project, including operational, technical, logistical, managerial, and financial risks. All relevant risks are analysed and particular attention is paid to questions as to how these risks may permute through the project. Risk registers with all possible risks to the project, their maximum influence, maximum impact, and possible mitigation options are then made. Examples of risk management formally organized in a tunnel project are the Mission Valley Light Rail, East extension, in San Diego, USA (Ragland et al. 2003) (Section 4.6.4). In this project, monitoring and communication between the various parties in the project were specifically optimized to avoid any damage to surface buildings. An extensive scheme for assessing and calculating risk during construction has been developed by Šejnoha et al. (2009).

Šejnoha et al. (2009) also observe that often the risk analyses are ignored at the time when contractors and consultants are selected, although building codes (e.g. the code of the insurance companies, ITIG 2006) demand the client, contractors and consultants to have available and apply all required expertise necessary for the project. Apparently, the general codes leave ample options for just selecting the cheapest rather than the best but cost-effective options. This significantly reduces the effectiveness of risk management. If properly managed, formal risk management will reduce risks. On the other hand, formal risk management may prove counter-productive as it adds to the bureaucracy, to the costs and may even reduce motivation of the experts and labour forces, which, in turn, adds to the project risks.

4.7.4. Geohazards

4.7.4.1. Earthquakes and volcanoes

In general, geohazards have less impact on underground than on the surface constructions as many geohazards are weather-induced. Also earthquakes have a limited influence (Aydan et al. 2010). Mostly earthquakes do not cause complete collapse of underground structures, although damage can be severe (Li 2011). Obviously, massive damage can be expected if an earthquake occurs at a fault zone intersecting an underground structure, but damage is often remarkably small even if the structure is in the direct vicinity of an active fault (Wang et al. 2009). One of the reasons is that the resonance frequency of most underground structures is not within the range of frequencies common in earthquake waves, in contrast to constructions at the surface. During an earthquake wave, the underground excavation may deform, and compression and extension of the structure may take place. Amplification of stresses during an earthquake at the walls of an excavation does not normally exceed 10–15% of the initial stresses. However, tension cracks may develop in the groundmass and structure. During tension phases of an earthquake wave, the normal stress on discontinuities may reduce, resulting in movements in the surrounding mass. This may lead to a loss of integrity of the mass and support and subsequent collapse. Therefore, support should provide reinforcement to the ground surrounding the excavation in earthquake prone areas and sustain tensile stresses. For example, support types as concrete segments, normally consisting of unconnected separate elements, should be connected by a bolt and nut system preventing loss of support integrity during an earthquake. In contrast to the relatively low impact on the underground excavation, the exits and portals to the surface are quite vulnerable to earthquake damage (Huang et al. 2009; Shen et al. 2009). Although volcanism may be accompanied by earthquakes, the main impacts of this geohazard is related to surface features such as lava flows and landslides that may disrupt underground structures. Rather than being endangered by volcanic hazards, underground excavations and constructions may provide shelter to volcanic hazards.

Forecasting volcanic eruptions and earthquakes is the subject of extensive research. Volcanic eruptions can be forecasted with a reasonable notice time, as most eruptions are preceded by shocks, gas eruptions, or changes in the topography well in advance to the eruption (Brancato et al. 2012). Methodologies for earthquake forecasting are unreliable. Developed methodologies miss either major earthquakes or become useless because of many false-positives. Preparedness for the consequences of an earthquake, strict application of seismic building codes and standards, and general forecasting by probability of earthquakes of a certain magnitude in a given area over a given time (generally years) are considered the best for minimizing the impacts of earthquakes (USGS Earthquake 2011; NEPEC 2012).

4.7.4.2. Gas, explosions and spontaneous combustion
Intrusion of poisonous, explosive, or radioactive gases that may have been evoked by the excavation process, more in particular by gas producing excavation joints, is a quite serious hazard (Edwards et al. 1988; Doyle 2001). Examples include methane entering the underground spaces during construction of the Budapest Metro in Hungary (Greschik 1975), and methane and deadly quantities of hydrogen sulphide likely from bacterial sulphate reduction of anhydrite with methane from coal in the Alborz Service Tunnel in Iran (Wenner & Wannenmacher 2009). Organic waste material present in the subsurface may generate carbon monoxide, methane, and hydrogen sulphide; the latter caused panic during construction of the Amsterdam underground railway (AD 2010). Other examples are radioactive radon gas (Li & Chan 2004; Gendler 2008); in volcanic areas, poisonous (often sulphur) gases, and in areas near gas or oil reservoirs natural gas may seep upwards and intrude in underground excavations. Normally, ventilation of underground excavations and completed constructions solves the problem, as quantities of such gases are generally small (McPherson 2012). However, in non-ventilated spaces these may well reach poisonous or explosive levels, for example poisonous CO-gas build-up or oxygen depletion due to rotting timber support in an abandoned tunnel.

Another hazard is spontaneous combustion or heating due to sudden exposure of organic material or other minerals to oxygen. That may happen with coal seams or coal in waste dumps and stock piles of mines (ICMC 2008), or with particular minerals (e.g. sulphur and sulphides such as pyrite, Wu et al. 2006). Dust control and ventilation may be required to avoid explosions of dust and released gases during excavation. Normally, halting the flow of oxygen to the combustible material immediately after excavation is sufficient to stop the heating, if it is a new excavation in an otherwise undisturbed subsurface. If underground fires already have burned for some time, for example in coal seams, reduction in volume of the burnt material will cause loss of structure and cracking of the groundmass adjacent and above allowing oxygen to access. In such cases, the fire gets beyond control and will require very expensive measures to extinguish (e.g. prohibiting oxygen access to mostly a very large volume of ground by grouting or water injection) with no guarantee that it will not re-ignite after some time when the groundmass further settles due to the loss of structure or temperature changes. Hence, any development in or above already burning subsurface materials should best be avoided (see also Section 3.1.3).

4.7.4.3. Biohazards
A biological geohazard is the release of diseases by underground excavation works. Leaking sewage pipes may cause an unhygienic and hazardous environment with potential threat of infectious diseases. More serious may be dormant (spores of) bacteria and viruses causing such diseases, as the plague, anthrax and hepatitis that may become re-activated when disturbed by excavations in ancient burial sites (Crossrail Bill 2010). The chances of these diseases being activated are supposedly remote, but consequences may be large as modern humans may have no natural defence against the bacteria and viruses of hundreds of years ago. This would allow the disease to spread with many victims as result. The hazard is most severe during excavation but likely air or water seeping in a completed construction could be hazardous too. The risks are regarded seriously enough to take precautions where a suspected area (e.g. a burial site for plague victims) is to be excavated and are likely reduced to acceptable levels by proper hygiene, (biohazard-) protective clothing, and dust control (Healing et al. 1995).

4.7.4.4. Water, snow and ice
Underground excavations or constructions may be flooded by surface water or by groundwater in case of contact with a permeable water-bearing layer in the subsurface (Figs

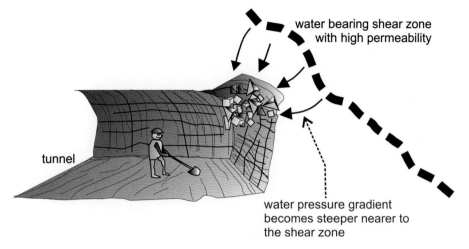

Fig. 4.54. Water from localized high permeability zones may break through in the tunnel before the tunnel reaches the zone (from Hack 2012).

4.50 & 4.54) (Schwarz *et al.* 2003). In cold regions, snow may form a hazard for surface constructions and portals of underground excavations due to the weight of snow coverage or avalanches. If temperatures inside an underground space are regularly well below zero, ice may form. Falling ice sheets during thaw periods and reduction of the clearance of the excavation by ice may hamper normal operation (Frandina & Witt 2002; Hack & Slob 2008). Ice forming may also jeopardize the stability of an excavation. In particular, freezing-thawing cycles cause seepage water in and in-between the groundmass and support to freeze and as ice has a larger volume, the freezing opens discontinuities and pores, displacing ground and support. In the next frost period, more ice may form opening the spaces further, eventually leading to disintegration of the ground and support. This may happen in an excavation near the portals, but seldom well inside an excavation as groundmasses have good heat insulation capacities and small quantities of heat from traffic or other use of the excavation are mostly enough to prevent freezing conditions. However, if this is not enough, problems from ice forming can be avoided by insulating the ground and support, or by additional heating (Frandina & Witt 2002). Natural examples of ground failing from the growth of ice lenses underground are the pingos found as geological phenomena.

4.8. Concluding remarks

During the last decades, underground excavation design and construction underwent a revolutionary change. Drilling and blasting were supplemented by continuous types of excavation with road headers and tunnel boring machines. Over that period, support systems also changed to more flexible types of support, such as bolts and steel sliding inserts with shotcrete. In soft grounds with high water tables, significant progress was made by developing a full suite of TBMs which were further improved to excavate harder and stronger rock.

These improvements were based on science and technology and driven by ever higher economic, environmental and safety demands. Nevertheless, one of the most crucial factors in underground construction remained the experience and expertise by the underground workers. Although all means of excavation and support have become less sensitive to changes in underground environment, unexpected ground conditions are still a main threat to underground excavation projects. Improved investigation methods for the subsurface may contribute to reduce this problem. Another relevant factor in this respect is optimizing and utilizing adequate subsurface data, data processing, and information handling, as will be discussed in Chapter 5.

Chapter 5   Information about the subsurface

The underground is largely a hidden realm. Without special tools, it cannot be directly observed and specific, geological knowledge is necessary to understand its structure and properties. In the past, understanding the behaviour of the underground was based solely on experience gained by trial and error. In the twentieth century, science and (geo-)technology introduced the tools and methods to identify and understand the groundmass properties necessary to develop sustainable, safe and stable underground constructions. Knowledge of the subsurface through sound data and information is a prime requirement for realizing such constructions that do not jeopardize existing structures or interfere with current types of land-use. This chapter describes how to obtain subsurface data and information and how to handle such information to arrive at reliable models to predict behaviour of underground constructions based on engineering geological knowledge and expertise.

5.1. Site investigation

The process of collecting surface and subsurface data, integrating the data and determining their interaction with surface, subsurface, engineering structure and the environment, is normally called 'site investigation'. There are various options to collect such information. The first stage of a site investigation is a 'desk study' in which easily accessible existing data are assembled and interpreted, for example, the data from nearby structures and from government agencies. This leads to a first model of the surface and subsurface and to a first impression about the feasibility and risks of a project. In addition, a desk study provides insight in the number and type of additional data required to arrive at a reliable prediction model for the stability of the underground construction.

The next phase in a site investigation concerns the collection of new data, often in an iterative, cyclic process. New data are interpreted in combination with existing data and integrated in a draft geological and geotechnical model. Ideally, the result is a four-dimensional conceptual site model (3D and time) of the subsurface and its planned construction (Section 5.2). Data collection methods and standards are briefly described below. More extensive descriptions and practices are provided by Bell (2007), Hunt (2005), and Price *et al.* (2009) and in the below mentioned standards and codes.

5.1.1. Site investigation standards and codes

International standards or codes for geotechnical and engineering geological site investigation, including data collection, testing, and reporting, have been developed by the 'International Organization for Standardization'. These include ISO 14688-1/2:2002/2004 (2002/2004) for soil masses and ISO 14689-1:2003 (2003) for rock masses (ISO 2011). The 'European Committee for Standardization' (CEN 2011) produced Eurocode 7. Most countries host an institute for standardization that develops and registers standards and codes, such as 'ASTM International' (formerly 'American Society for Testing and Materials', ASTM 2011) in the USA, the 'British Standard Institution' in the UK (BSI 2011), the 'Deutsches Institut für Normung e.V.' in Germany (DIN 2011), the 'NEderlandse Norm' in The Netherlands (NEN 2011), and the 'Instituto Nacional de Tecnología Industrial' (INTI 2011) in Argentina. In some countries 'de-facto' standards are issued by other organizations as in Hong Kong where the 'Geotechnical Engineering Office' publishes technical guidance documents (GEO 2011). An overview of standards and specifications applicable to tunnelling is provided in the British Tunnelling Society's 'Specification for Tunnelling' (BTS 2010). Many professional societies have also published standards or 'best practices', such as the International Association for Engineering Geology and the Environment (IAEG 2012), the International Society for Soil Mechanics and Geotechnical Engineering (ISSMGE 2012), the International Society for Rock Mechanics (ISRM 2012), and the International Tunnelling and Underground Space Association (ITA-AITES 2012). Issues and the level of detail for standards and codes may vary widely from country to country. Many countries have begun to implement international standards such as ISO and Eurocode, replacing their national codes, supplemented with country-specific amendments and addendums. Not all countries embedded such standards and codes in law and doubts may exist about their legal status. In this chapter, the ISO codes and standards are followed where applicable without discussing national implementation differences. For specific projects, it is advised to check with the national standardization organization on applicable standards.

5.1.2. Maps with surface features only

Onshore topographic, land-use, road, cadastral, historical, and other special purpose maps may be published by public organizations in an analogue (printed on paper) or digital format (Fig. 5.1a). Offshore, public information is often limited to bathymetric maps. Such maps may represent the water depth of river-, lake-, or seabed together with navigation information. Other information on offshore maps may include ownership, situation of protected zones, exploration concessions for natural resources, and locations of offshore facilities, such as drilling platforms. Maps may be available in small-scale (large area – little detail) or large-scale (small area – high detail). For remote areas, relatively small-scale maps might be available only.

More detailed and larger scaled maps may be available, in particular for urbanized areas onshore. Normally, such maps represent information about surface topography, location of rivers, houses, infrastructure, and type of land-use. Implicitly, these may also reveal information about the subsurface, such as valleys in a fault zone or indications of mining sites. Maps with historical geographic information may either have been printed in the past or be based on recent historical research. As a result of massive digitization activities, more and more maps become available in digital format and accessible over the Internet. Open access clauses increasingly apply when developed with tax-payers

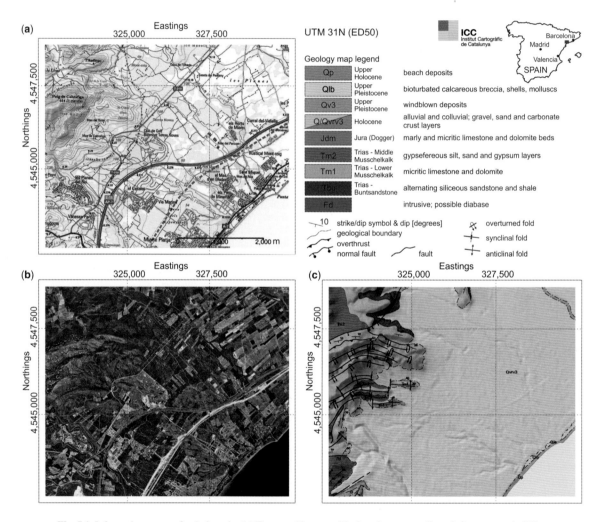

Fig. 5.1. Information sources for desk study: (**a**) Topographic map with elevation contour lines, infrastructure, buildings, land-use, etc. (**b**) Aerial photograph: and (**c**) Geological map; all of the same area and at the same scale. The colours in the geological map indicate different soil and rock formations exposed at surface or present in the subsurface at shallow depths, and the symbols indicate the position of folds and faults. The maps and photograph are digitally stored in the repository of the Institut Cartogràfic de Catalunya (ICC). Shading in (**c**) shows elevation differences.

money leading to large databases freely accessible through the internet (Fig. 5.1; USGS 2012)

5.1.3. Subsurface maps

Geological maps represent information on soil and rock types and their tectonic structure as faults and folds (Fig. 5.1c). Map features are described in a legend and cross-sections may be included to clarify geological interpretation. As geological features are often not exposed, most of the information on geological maps is inferred from scattered observation points in the field and from indirect types of information as described below. Among others, this applies to boundaries between geological units (Fig. 5.2). Pedological and soil maps describe and classify soils based on origin, history, morphology, and agricultural features (WRB 2010). Special purpose or thematic maps may be derived from, or are related to, various applied geological information sources of an area (De Mulder 1986; Culshaw 2003), engineering–geomorphological maps giving information on geomorphology interesting for engineering (Rupke & Cammeraat 2001), and engineering geological maps (KIMAP 2002) with information about properties and features of the ground important for civil engineering projects. Furthermore resource maps displaying rocks suitable for building stone (MineralsUK 2010), groundwater maps with relevant geohydrological properties (GRM 2011), maps showing occurrences of artificial (man-made) ground (Ford et al. 2010) and seismic hazard maps with information about expected intensity and frequency of earthquakes (EHM 2011) are other examples of thematic geological maps.

5.1.4. Remote sensing

Techniques for collecting data of the Earth by space satellites, planes, helicopters or balloons, but also terrestrial, are collectively called remote sensing. These are a major source of information for geotechnical and engineering geological information and may be applied to development and management of the subsurface. Ever since the launch of the first balloons and airplanes aerial photographs were used to produce most geographic and geological maps (Fig. 5.1b). Today, imaging by remote sensing is no longer restricted to the visual light spectrum but digital sensors in nearly every electro-magnetic wavelength, including infrared (temperature) and gamma rays have become available (Section 5.1.12). Apart from passive sensors that just measure the intensity of received electro-magnetic waves, also active remote sensing systems, such as radar or laser emitting and receiving waves reflected from an object, are being applied today. These methods, such as 'Interferometric Synthetic Aperture Radar' (InSAR) and 'Light Detection And Ranging' (LIDAR), allow high accuracy measurement of topographic modifications as land subsidence over wide areas, (see Section 4.6.3 and Fig. 3.1).

Normally, remote sensing techniques are not applied offshore except for very shallow water depths. However, LIDAR ('LIDAR bathymetry') may be applied to profile under water features to depths of some 50 m under optimum conditions (Fugro 2010). For offshore work the main system used is multibeam bathymetry combined with geophysics. Geophysical airborne surveys (Section 5.1.12), often in combination with remote sensing surveys, measure physical properties of the Earth materials and may be applied to identify variations in density and type of groundmasses, in mineral occurrence, and water content and salinity.

Remote sensing imagery is often free or available at marginal fees from space agencies (e.g. ESA 2011; NASA 2011), meteorological, topographic, geological surveys (e.g. JMA 2011; USGS 2011), or commercial organizations (ASTRIUM 2011; Google Earth 2011). Sometimes remote sensed imagery may be cheaply acquired from organizations that are less obvious such as insurance companies that have obtained aerial imagery for investigations of damage after an earthquake. Multi-and hyper-spectral remote sensing techniques have opened new options for obtaining data from the Earth's surface. They measure the intensity of electro-magnetic waves in multiple (up to hundreds) (very) small frequency bands. The combination of intensities measured in different bands gives information about the

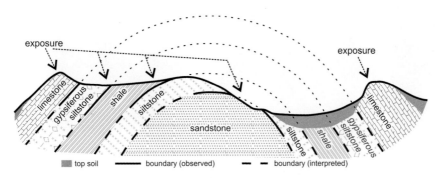

Fig. 5.2. Geological interpretation in a vertical section. In the exposure on the left, a series of layers with different materials is identified, matching with similar layers in the exposure on the right. Knowledge of the general geological structure of the area (broad folds) results in the interpretation of this section.

material reflecting the wave and the medium that the wave has passed through. Often, the type and health condition of vegetation, minerals present in rocks and soils, or occurrence of environmental pollution (Van der Meer & De Jong 2001; Choe *et al.* 2008) may be established with multi- and hyperspectral remote sensing.

Resolution of satellite imagery has increased rapidly over the last decades. Whereas satellite imagery was previously only used for regional mapping these techniques are now applied in site-specific investigations for larger civil engineering projects at the surface. For subsurface projects, remote sensing is used on a routine basis to interpret surface geology features to the subsurface. Satellite imagery of the last few decades may be available for areas where historical airborne images are not. Remote sensing images may be used to investigate temporal changes in the landscape, as landslide and glacier movements, and impacts that may possibly be related to climate change (Kääb 2008). This may be relevant for the stability of entrance portals of underground structures or for predicting future extent of permafrost. InSAR deformation measurements may provide information on stress development and anisotropy of stress fields in the subsurface (Hearn & Fialko 2009). Examples of terrestrial remote sensing include surveys of deformation (Soga *et al.* 2010), size and discontinuities in underground excavations by LIDAR (Fig. 5.3) and monitoring and forecasting damage to tunnels due to moving groundmasses by ground-based radar techniques (Noferini *et al.* 2007).

5.1.5. Field campaigns

On-shore site investigations include field campaigns to prepare a detailed map of the surface conditions together with all other relevant site features, for example geology as mapped in exposures at the surface or underground (Fig. 5.4), evidence of water in- or outflow, of old mining operations and other types of land-use (Dearman 1991; KIMAP 2002; Price *et al.* 2009). During field campaigns, geotechnical properties of soil and rock are determined in exposures while existing maps, remote sensing images, and interpretations are checked and validated against the additional information obtained. Groundmasses in a surface exposure deviate from groundmasses in the subsurface due to different weathering and stress conditions, and therefore, even when the geology is the same, the survey should not be restricted to the surface only. If options exist to survey underground exposures (e.g. in existing tunnels),

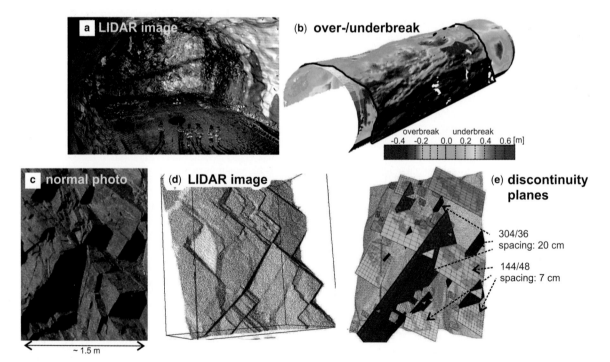

Fig. 5.3. LIDAR image processing. (**a**) and (**b**) mapping over-/underbreak (i.e. too much/insufficiently excavated) (modified after Fekete *et al.* (2010) with permission from Elsevier); (**c**) (**d**) and (**e**) semi-automatic interpretation of discontinuity planes with orientation (dip-direction/dip in degrees) and spacing in-between planes (modified from Slob *et al.* (2005) with permission of the Transportation Research Board; copyright, National Academy of Sciences, Washington, DC, 2005).

Exposure description (following ISO14689-1:2003 (2003) & BS 5930; 1999), and discontinuity description following SSPC (Hack et al. 2003)

Exposure description:
Exposure created during building of tunnel (1884). Rock mass is fresh. No evidence of excavation damage. Probably all damaged rock mass already removed. Visible evidence of some recent and likely still active spalling of the mudstone is visible throughout the exposure, resulting in fractured intact rock and debris on the floor.

Description of rock mass:
Rock mass consists of interbedded mudstone (55%) and siltstone (45%).

Mudstone:
Moderately strong, thickly laminated, very small tabular (flakes), fresh, grey to black, clay-sized grained, MUDSTONE.
Material: clay minerals
Bedding: orientation 244/50, spacing: 0.01 m, persistent along strike and dip, large-scale roughness: straight, small-scale roughness: smooth planar, infill: none.
Joint set 1: spacing: 0.05 m (*1); Joint set 2: spacing: 0.03 m (*1)
(*1: spacing based on block size; other details of joint set could not be obtained).

Siltstone:
Strong, thin bedded, small blocky, fresh, grey, fine sand to coarse silt sized grained, SILTSTONE.
Material: fine sand to coarse silt sized (quartz) grains with probably calcitic cement.
Bedding: orientation 244/50, spacing: 0.15 m, persistent along strike and dip, large-scale roughness: straight, small-scale roughness: smooth planar, infill: none.
Joint set 1: orientation 030/50, spacing: 0.2 m, persistent > 0.1 m along strike and abutting 0.25 m along dip, large-scale roughness: not applicable, small-scale roughness: smooth to rough undulating, infill: none.
Joint set 2: orientation 070/85, spacing: 0.12 m, persistent > 0.03 m along strike and abutting 0.15 m along dip, large-scale roughness: not applicable, small-scale roughness: rough undulating, infill: none.
Joint set 3: orientation 136/85, spacing: 0.12 m, persistent > 1.1 m along strike and abutting 0.24 m along dip, large-scale roughness: wavy, small-scale roughness: rough stepped, infill: none.

Fig. 5.4. Example of a rock mass description in a 'field' survey. In an exposure in a niche of an existing tunnel, siltstone and mudstone layers are identified and described for the design of a new, larger tunnel. (Beskyd Tunnel, Ukraine; photograph: S. Slob; courtesy Witteveen + Bos&ITC).

the opportunity should be taken as it will show the subsurface ground mass *in-situ* in the subsurface and on a metre scale.

5.1.6. Pits, trenches, and inspection holes

Project specific information of underground conditions may also be obtained by digging pits, trenches, or inspection holes to expose soils and rocks just below the surface. Pits and trenches (Fig. 5.5) dug manually or mechanically to a maximum of about 5 m, are a cheap option if no support is needed for the sidewalls. In non-cohesive grounds, support of the walls is required or mandatory by local labour safety regulations. In mechanically drilled, large diameter holes, an expert may go down to investigate the *in-situ* groundmass. Groundmass description is done similarly to that of exposures (Fig. 5.4) and boreholes (Section 5.1.7). The benefit of pits, trenches, and holes is the large volume of groundmass that can be examined *in-situ*, in contrast to the small volumes of samples obtained from standard boreholes.

5.1.7. Boreholes

For deeper sampling and investigation, boreholes are required. A wide variety of drilling equipment, ranging from manually operated drilling tools for sampling to a few metres depth, to large mechanized drilling rigs which may go down to thousands of metres, is available. Drilling may be done onshore or offshore. The latter by divers, remote controlled rigs on the river-, lake-, or seabed or from drilling platforms and vessels. A classification for drilling equipment is based on the energy applied to loosen or cut the groundmass and to remove the cuttings. In 'percussion drilling', a hammer action provides the energy, alternatively a tube can be pushed into the ground. In 'rotary drilling', the energy is provided through rotational grinding (Fig. 5.6). During drilling, a fluid or airflow may be required to cool the cutting device at the end of the drill string (bit) and to flush the soil and rock chippings to the surface. A drilling fluid (or 'mud') may consist of water only or water supplemented with additives (e.g. bentonite) to increase the specific weight of the mud preventing groundwater or gas inflow in the borehole and to stabilize the borehole walls.

Fig. 5.5. Engineering geologists examine the soil in a trench dug near the Lake Isabella auxiliary dam to evaluate the activity along the Kern Canyon Fault, which runs underneath part of Lake Isabella, California, USA (photograph: William Warren Byrd II, U.S. Army Corps of Engineers Sacramento District, USARMY).

In non-cohesive groundmasses, stabilization of the drilling hole might be obtained by inserting a steel or plastic pipe ('casing'). Sampling may be done in a destructive way by fragmenting the groundmass or in a non-destructive way by coring methods. The latter can result in almost undisturbed samples.

Destructive drilling with disturbed samples is cheapest. The soil is loosened and larger objects such as rock blocks are broken or grinded into small fragments ('chippings'). This may be done by dropping a bit with a bucket or shell sampler in a borehole (Fig. 5.6a–d), by a rotating auger (Fig. 5.6e) or by a rotating drill bit mounted on a drill string. The soil and chippings are lifted to the surface by respectively the bucket or shell sampler, forced upwards due to the rotating flights of the auger or are flushed to the surface by a water or airflow. As sampling depths cannot be determined precisely, they provide crude information about mineral composition only. Moreover, samples might be mixed with soil or chippings from deeper or higher layers, which may provide a challenge in areas with complex geological conditions.

Relatively expensive, non-destructive drilling and undisturbed sampling techniques may be required to collect information that is more reliable on groundmass properties, depth, and lithology. Dedicated sampling devices may be pushed or hammered into relatively soft ground. Rotary core drilling (Fig. 5.6g–j) may be applied for rock. The most sophisticated sampling tools deliver almost undisturbed cores, even from loose soils and weak rocks (Fig. 5.7). In 'undisturbed 'samples, the original structure and properties of the ground mass are quite well preserved. Therefore, such samples and cores allow for detailed inspection and assessment of geotechnical properties. However, their relatively small diameters (2.5–25 cm) hamper options for full description of the groundmass. Wide-spaced discontinuities (sub-) parallel to the core might be missed, and properties obtained from testing might not be representative for the large volume of groundmass. Stress relief or changes in water content may create additional fractures. Uncased boreholes might be inspected further by lowering logging tools in the borehole, as a video camera and devices measuring inclination and diameter of the borehole or intersections with discontinuities. Mechanical and geophysical tests may be executed in boreholes as well (Section 5.1.12).

5.1.8. *In-situ* strength and deformation tests

Penetration or 'sounding' tests are widely performed in soils and weak rocks. The Standard Penetration Test (SPT) and the Cone Penetration Test (CPT or 'Dutch Cone Penetration Test' – DCPT) are most frequently used (NCHRP 2007; Schnaid 2009). SPTs are conducted in boreholes at the bottom of which a cone or sampler is hammered into the ground. The number of hammer blows to penetrate a certain (standard) distance is counted as an indication of the strength of the ground. After the test, a standard interval (1 to 1.5 m) is drilled and the test is repeated. SPT samplers deliver small samples as well. Cone Penetration Tests are normally used to determine the strength of the ground in soils (Fig. 5.8). Here, a small cone (mostly with a diameter of 36 mm) is pushed into the ground at a constant velocity (2 cm s^{-1}). The force required for this presents a measure for the strength of the soil. Sophisticated electrical cones may register a wide range of other soil properties, including sleeve friction, pore water pressure, permeability, and electrical resistivity as well. For environmental purposes, CPT-based tools with optical and chemical sensors might be applied (Van Ree & Carlon 2003; NCHRP 2007).

Deformation and strength characteristics of a groundmass may be measured *in-situ* by probes that can be expanded in the ground, such as a pressuremeter (PMT), dilatometer test (DMT) (Schnaid 2009), or flat jack (Figueiredo *et al.* 2010). The probe is pushed into the ground, lowered in a borehole, or placed in a pre-sawn slot in the wall, roof, or floor of an

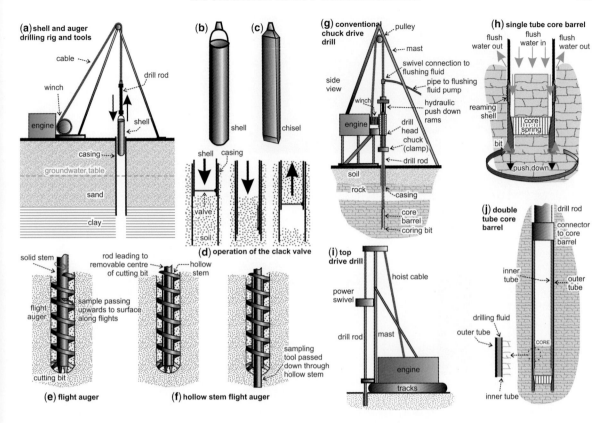

Fig. 5.6. Drilling and sampling methods. (a) (b) (c) and (d) percussion, (e) and (f) auger, and (g) (h) (i) and (j) rotary drilling. (modified from Price *et al.* 2009, with kind permission from Springer Science + Business Media).

underground excavation. The probe may be a type of modified balloon (pressuremeter), a flat steel plate on which a membrane is mounted (flat dilatometer), or a flat jack. The probe is filled with water, oil or gas under pressure and the probe will deform the groundmass surrounding the probe. The expansion of the probe and pressure of the fluid or gas are measured to calculate deformation and strength characteristics of the groundmass. Deformation characteristics may also be determined by installing extensometers in a borehole. They consist of a steel rod in a borehole fixed to the groundmass at the bottom of the borehole. The length of the steel rod extending past a fixed point at the top of the borehole is a measure of how much the bottom of the borehole has moved compared to the top. The extensometer is mainly used as a monitoring tool (see Section 4.6.3), but if loading conditions are known, deformation characteristics of the ground may be determined as well. A similar principle to measure deformation characteristics is by loading the surface while measuring the deformation at the surface simultaneously. As the load is normally applied to a plate, the test is commonly known as a 'plate bearing' test (Fig. 5.9).

Shear deformation and strength characteristics of the groundmass may be measured with a 'vane test' by pushing a steel rod with vanes into the ground which is successively rotated (small version shown in Fig. 5.10) (Schnaid 2009). This test is limited to relatively soft soils only, due to the risk of the vanes to break. Full-scale *in-situ* tests to determine the strength and deformation characteristics of groundmasses include *in-situ* measuring of deformation characteristics of a tunnel or adit. The excavations are sometimes specially made for this purpose. Such tests are rather expensive and therefore conducted in major projects only such as large hydropower schemes.

5.1.9. Laboratory and simple field tests

Besides the *in-situ* tests discussed above, a wide variety of geotechnical properties, including density, grain size, mineral and water content, compressive and shear strength and deformation characteristics, may be measured from samples in the laboratory. These are normally quite small and may not be representative for the larger volume of the *in-situ* groundmass. Also, the process of sampling and transport may have disturbed groundmass properties. A very useful additional source for properties and often good alternatives are simple field tests that may be performed on

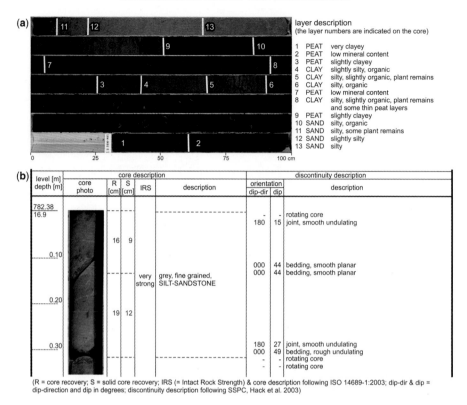

Fig. 5.7. Examples of borehole cores; (**a**) box with cores obtained by a special undisturbed soil sampler of peat, clay and sand, with description (photograph: Deltares). (**b**) core obtained by rotary core drilling in rock with description (photograph: S. Buykx, Witteveen + Bos & ITC). Scale in cm at the bottom in (a) and in the left column in (b).

samples from exposures during a field campaign and on borehole samples. The increased number of simple tests may provide a better estimate of the property and its distribution throughout the subsurface than a limited number of laboratory tests. Examples of such field tests are hand-held penetrometer and vane testers as indicators for the strength of soils (Fig. 5.10). Even simpler means as finger or hand pressure or impact of a geological hammer are for the experienced engineer very useful as indication for strength of soil and rock. Standards have been developed for field tests such as 'how far can a thumbnail be pressed into clay', or 'how easy is it to break a lump of rock by hand or geological hammer' (Hack & Huisman 2002; Price *et al.* 2009). Other simple field tests are 'angle of repose' for estimating the angle of internal friction of granular material, and the 'tilt test' for estimating the shear strength along a discontinuity (Price *et al.* 2009).

5.1.10. *In-situ* stress tests

Stress fields in the subsurface are investigated by stress measurements. In most standard methods, a volume of the groundmass is allowed to expand after unloading. Re-loading until the original volume is attained gives a measure for the original stress condition. Examples of stress measurements are over-coring, flat jack tests, and dilatometer tests. *In-situ* stress measurements are relatively expensive and often have to be repeated many times at the same position to obtain reliable values because the tests are quite sensitive to local variations, such as the presence of discontinuities. *In-situ* stress tests are detailed in the various standards and codes, for example ASTM (2011).

5.1.11. Fluid pressure, permeability and flow

In-situ water pressures are measured by piezometers. Simple types consist of a pipe with a slotted section or porous filter at the bottom installed in a borehole. The water level in the pipe is a measure for the pressure of the fluid at the position of the filter. Electrical piezometers consist of an electrical pressure transducer mounted in a porous filter. These may contain internal dataloggers to store data for longer times before retrieval. With such devices other parameters, as conductivity, temperature, and pH may be measured as well.

INFORMATION ABOUT THE SUBSURFACE

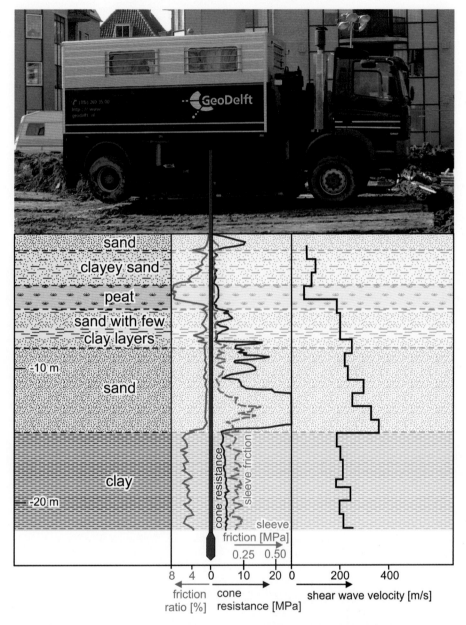

Fig. 5.8. Example of a CPT taken by a truck. The graphs represent the various measurements. On the left-hand side an interpretation into soil types based on the measurements (photograph: Deltares (formerly GeoDelft)).

Electrical water pressure measurement is often mounted in a CPT (Section 5.1.8).

Predicting water influx in underground excavations may best be done by measuring the *in-situ* permeability of the groundmass by filling a borehole with water and registering the water dissipation time. In more sophisticated test equipment 'packers' seal the borehole at specific depths while water is pumped in-between the packer and the bottom of the borehole, or between two packers (i.e. 'packer' or 'Luguon' test, Singhal & Gupta 2010). Registered water pressures and quantity give a measure of the permeability of the groundmass. *In-situ* groundwater flow velocity may

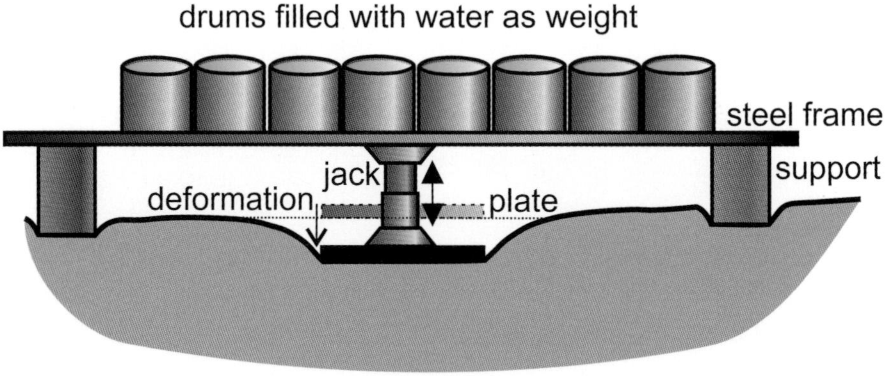

Fig. 5.9. Plate bearing test (from Hack 2012).

be determined by performing a tracer test (e.g. using 'dye') between two boreholes. Groundwater flow velocity and direction may also be determined by temperature differences. The water is heated at a particular location and accurate temperature measurements in the surroundings indicate velocity and direction of the water (Environment Agency 2005).

5.1.12. Geophysics – non-destructive sampling

Geophysical investigations are non-destructive methods to obtain data from the subsurface without physically penetrating and disturbing the ground. With these methods, physical properties of the groundmass and their lateral and vertical variations can be registered including boundaries between layers with different physical properties. Most common properties measured include deformation characteristics (by means of seismic surveys), electro-magnetic, geo-electrical, magnetic and density properties. Spatial distribution of environmental pollution has also been successfully estimated by means of geophysical methods. Some geophysical surveys may be performed from space or air, from the surface onshore, offshore from a vessel, or on a river-, lake-, or seabed. Geophysical techniques may also be applied in or between boreholes to register groundmass properties of the penetrated rock or to infer these for the groundmass in between boreholes. Another application is detection of voids, cavities or abandoned underground mines and shafts (Butler 2008).

In contrast to the point and line type of data obtained from samples and boreholes, geophysical methods may provide (semi-) continuous, two- or three-dimensional information of the subsurface and thus may be used for interpolation of information between point and line data. They may also be used to check lateral extension of features, for example layer boundaries and distributions of properties as weathering (Lee & De Freitas 1990). Interpreting geophysical data requires proper expertise and experience. Normally, geophysical investigations do not provide geotechnical information directly, as for example strength, but rely on correlations, which are of a rather indirect character, risking unreliable interpretation (Box 5.1). Therefore, geophysical methods should be used in combination with ground truth data obtained from boreholes, penetration tests, and field surveys. Some of the geophysical methods most commonly applied in engineering geological or geotechnical engineering are described below. Figure 5.12 provides a tentative overview of their suitability for various purposes. More detailed descriptions of the methods may be found in Abdallatif *et al.* (2009), McDowell *et al.* (2002), Price *et al.* (2009), Telford *et al.* (1990), and the massive amount of information on the websites of various societies for geophysicists (e.g.

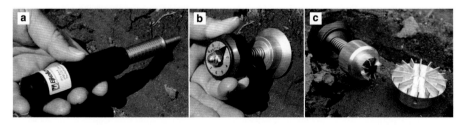

Fig. 5.10. Pocket penetrometer (**a**) and torvane (handheld shear vane tester) (**b**) and (**c**) (photographs: J.G.A.M. Arnoldus).

Box 5.1. Misleading geophysical survey

Geophysical surveys are a valuable tool for determining the subsurface, but may also lead to erroneous interpretations. Figure 5.11 shows empty and filled karst solution holes investigated during an electro-magnetic ground radar (GPR) survey. Significant differences in electro-magnetic properties between air and limestone are well marked in ground radar profiles. Much less significant differences, if any, exist between calcareous silt in a filled karst hole and limestone, running the risk of filled karst holes to be missed on the records.

Interpretation may be even more complicated because the zone with enriched manganese and iron displays a large electromagnetic contrast with the surrounding non-enriched limestone or calcareous silt and will feature prominently in the records. In the filled karst hole, the enriched zone is continuous through the silt because the enrichment has taken place after filling of the karst hole. The continuous presence of the enriched zone through the filled karst hole may erroneously be interpreted that no such karst hole occurs (from Hack 2000).

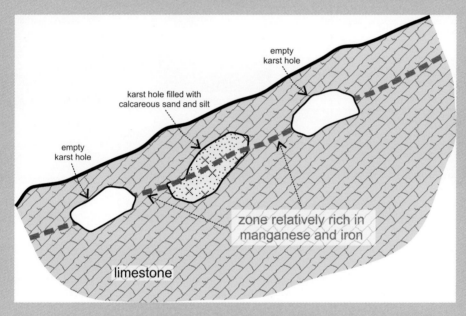

Fig. 5.11. Ground radar investigation.

the digital library of the Society of Exploration Geophysicists, SEG 2010).

5.1.12.1 Seismic methods

Seismic methods operate on the principle of a shock wave transmitted into the ground that is refracted or reflected by an object or boundary in the subsurface (Fig. 5.13). The velocity of the waves and the refraction and reflection depend on the deformation characteristics and densities of the groundmass materials. A seismic signal may be generated by a hammer, explosion, vibration tool or by any other means that creates a deformation wave into the ground, or for some applications may employ natural vibrations, such as created by traffic. The waves are received by geophones (onshore) or hydrophones (offshore) in direct contact with the ground or water. Offshore transmitters and receivers are normally mounted on a vessel or towed through the water. Various types of waves, as pressure or shear waves (wave movement parallel respectively perpendicular to the direction of wave propagation), may be used. All of them are influenced differently by the ground materials. Successful application of seismic methods depends on the contrast in deformation and density properties between different materials. Refractions and reflections are more clearly visible if contrasts are high. Wave velocities may be used to calculate seismic deformation moduli and Poisson's ratios

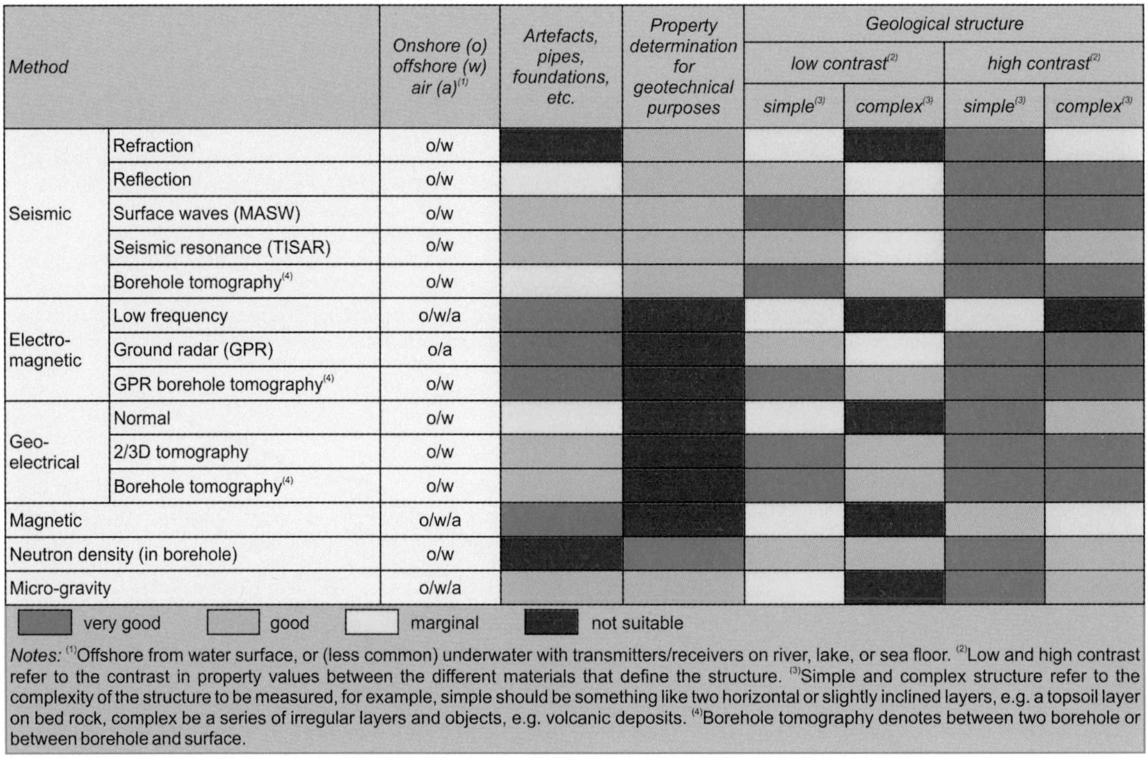

Method		Onshore (o) offshore (w) air (a)[1]	Artefacts, pipes, foundations, etc.	Property determination for geotechnical purposes	Geological structure			
					low contrast[2]		high contrast[2]	
					simple[3]	complex[3]	simple[3]	complex[3]
Seismic	Refraction	o/w						
	Reflection	o/w						
	Surface waves (MASW)	o/w						
	Seismic resonance (TISAR)	o/w						
	Borehole tomography[4]	o/w						
Electro-magnetic	Low frequency	o/w/a						
	Ground radar (GPR)	o/a						
	GPR borehole tomography[4]	o/w						
Geo-electrical	Normal	o/w						
	2/3D tomography	o/w						
	Borehole tomography[4]	o/w						
Magnetic		o/w/a						
Neutron density (in borehole)		o/w						
Micro-gravity		o/w/a						

Legend: very good | good | marginal | not suitable

Notes: [1]Offshore from water surface, or (less common) underwater with transmitters/receivers on river, lake, or sea floor. [2]Low and high contrast refer to the contrast in property values between the different materials that define the structure. [3]Simple and complex structure refer to the complexity of the structure to be measured, for example, simple should be something like two horizontal or slightly inclined layers, e.g. a topsoil layer on bed rock, complex be a series of irregular layers and objects, e.g. volcanic deposits. [4]Borehole tomography denotes between two borehole or between borehole and surface.

Fig. 5.12. Indication of suitability of geophysical methods for various applications. This graph assumes that some basic boundary conditions have been fulfilled; for example, no highly conductive materials occur in topsoil where a GPR is to be done of the underlying groundmass; identifying a pipeline by geo-electrical methods, requires the pipeline to be conductive (e.g. of metal).

(Hack 2000). Relatively new developments are based on analysis of surface waves for shallow (down to some-tens of metres depth) investigations (e.g. Multi-channel Analysis of Surface Waves – MASW) (Park et al. 2007) and resonance imaging (Testing and Investigation using Seismo-Acoustic Resonance –TISAR) (Arsenault & Chouteau 2002). The necessity of direct ground contact for transmitters and receivers and the generally bulky and more complicated equipment make on-shore seismic surveys time-consuming and rather expensive.

5.1.12.2. Electro-magnetic methods

Electro-magnetic methods are based on transmission and reception of electro-magnetic waves. A distinction should be made between Electro-Magnetic (EM) surveys using relatively low frequencies (kHz) and Ground Penetrating Radar (GPR) surveys using high (MHz) frequencies. EM methods transmit an electro-magnetic field that induces a secondary field in the underground. The different electro-magnetic characteristics of the ground materials generate varying differences between primary and secondary fields. EM surveys are generally easy and fast to perform. EM resolution depends on the frequencies used, and height and speed above the surface (Fig. 5.14). The method operates well for determining, for example, the presence of rock boulders in clay or sand (De Graaf & Rupke 1999). Other examples are airborne detection of underground spaces (Patterson & Brescia 2008), landslide hazards for tunnel construction (Pfaffhuber et al. 2010) and permafrost mapping (Mühll et al. 2002; Smith et al. 2010). In an urban environment interactions with existing infrastructure may hamper the survey.

GPR transmits and receives a pulse of high frequency electromagnetic waves, similar to radar. The wave is reflected by underground objects with different electro-magnetic characteristics. The time interval between transmission and reception corresponds with the distance to the object. The higher the frequency of the introduced pulse, the higher is the resolution, but also the more absorption of the wave occurs in the ground. Highly conductive layers reduce penetration depth. In a low-conductivity groundmass, for example sand, penetration depth may be tens of metres under very favourable conditions, but would be 10–15 m or much less in clay and negligible in salt water environments. To reduce energy losses transmitter and receiver

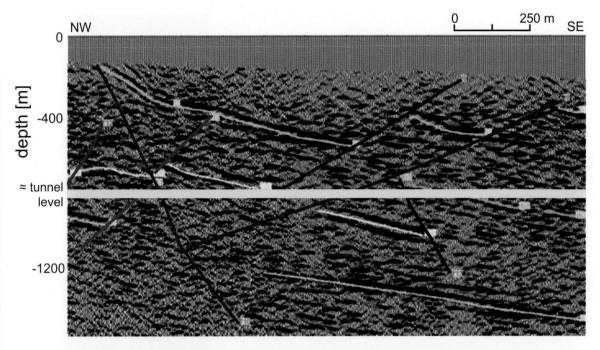

Fig. 5.13. Seismic section and interpretation of the Koralm tunnel between Graz and Klagenfurt (Austria). Grey shaded, wobbling lines are traces of reflected seismic waves. Light-coloured, sub-horizontal lines represent interpreted boundaries between layers with different rock properties and the dark-coloured, steeper dipping lines represent tectonic faults (after Lehmann et al. 2010 with kind permission from Springer Science + Business Media).

should be as close to the ground surface as possible. As in EM, GPR surveys are easy and fast to conduct with a transmitter and receiver mounted on wheels or sledge, or airborne. Again, urban environments may give accessibility restrictions and reflections from existing infrastructure might complicate the survey. Examples of GPR surveys include determining the depth of fractures from tunnels or surface exposures (Kovin & Anderson 2007; Torres Acosta 2008), underground infrastructure, as pipelines and foundations (González-Drigo et al. 2007), shallow layer boundaries (Francke & Utsi 2009) and permafrost mapping (Mühll et al. 2002).

5.1.12.3. Geo-electrical methods

Geo-electrical methods register differences in electrical conductivity or (reciprocal) resistivity of the ground. This is done by introducing an electrical, direct current (DC) in the ground and measuring differences in electrical conductivity/resistivity of geological units. Such methods may also be applied to measure the induced polarization to characterize subsurface groundmasses. Four electrodes are used of which two are used to create a current into the ground and two are used as measuring ('potential') electrodes. All electrodes should be in direct contact with the ground, or water in case of offshore surveys. Apart from the electrodes, the equipment consists of a direct current (DC) source and a measuring device. Computer-operated equipment may apply tens to hundreds of electrodes while pairs of electrodes at different positions are automatically switched on and off. This type of operation might provide a two to three-dimensional resistivity/conductivity image of the subsurface. Water and in particular salt-water strongly influences the resistivity/conductivity of the groundmass. This imaging methodology may provide excellent results to identify boundaries between different groundmasses and to detect underground caves (Casagrande et al. 2005). Geo-electrical methods (tomography) may also be applied in and in-between boreholes (Fig. 5.15). Setting-up the imaging equipment with tens or hundreds of electrodes with associated cables and the measuring time is rather time-consuming.

5.1.12.4. Magnetic, micro-gravity, gamma-ray, and neutron & gamma–gamma density

Magnetic surveys register changes in the Earth magnetic field in boreholes or by airborne tools (Fig. 5.14). Such anomalies may result from subsurface materials with different magnetic properties. Magnetic surveys may be quite useful for identifying buried objects such as metal pipelines. In airborne surveys, the tool is often combined with other methods as EM. Magnetic surveys may be applied on- and offshore. Micro-gravity surveys measure density differences in the ground. Magnetic and micro-gravity surveys may be

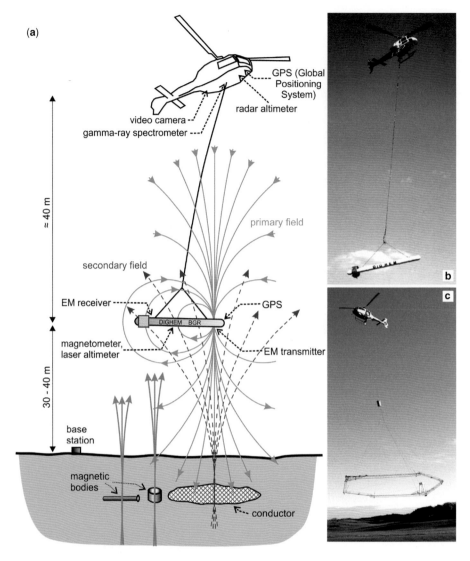

Fig. 5.14. Airborne electro-magnetic survey setup; (**a**) BGR (Bundesanstalt für Geowissenschaften und Rohstoffe) system; (**b**) Fugro DIGHEM; (**c**) SkyTEM (a) modified from Steuer *et al.* 2009 with permission from Elsevier; (b) Fugro Airborne Surveys; (c) SkyTEM Surveys ApS.

used successfully to identify voids and open spaces, such as abandoned shafts or tunnels in mines (Culshaw *et al.* 2004). Gamma-ray registers the emission of gamma rays by ground materials on the surface, airborne or in boreholes.

In contrast, neutron and gamma–gamma density logging tools actively emit neutrons and gamma rays in boreholes. They interact with ground materials and the amount of interaction provides an indication for porosity and density of the material in the direct vicinity (10–30 cm) of the borehole. The tools are used to log boreholes or may be combined with CPTs. These are applied to identify different layers and to test effectiveness of compaction in, for example, an artificial fill (Karthikeyan 2005).

5.1.12.5. Geophysical borehole logging

Several geophysical techniques are used in boreholes to identify differences in properties and structure of the penetrated groundmass. They include seismic, electro-magnetic, geo-electrical, gamma, neutron and gamma–gamma density methods (Fig. 5.16). The results of several methods in combination generate more detailed and reliable descriptions of the borehole profile.

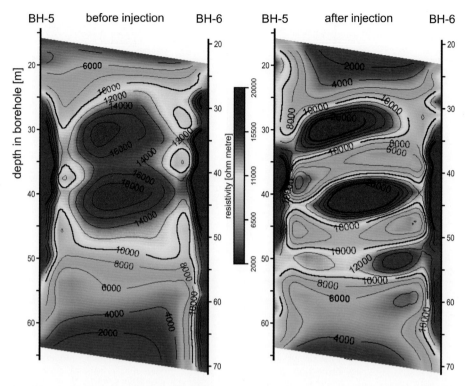

Fig. 5.15. Electrical resistivity borehole tomography between two boreholes for detecting weak and fractured zones in limestone for tunnel construction; left: the electrical resistivity tomogram before injection; right: after injecting a brine (salt water) in BH (borehole) 5. The drop in resistivity values shows the flow paths of the brine through the underground and thus the fractured zones. (Highway tunnels South of Daejon, South Korea) (modified from Ha *et al.* 2010 with kind permission from Springer Science + Business Media).

5.1.13. Site investigation in areas with a historical record (archaeological sites) or caves

Artefacts or signs of past human activities may be present in the subsurface. This may occur in projects for subsurface development in ancient urban areas or at historical mining sites, such as for the metro construction in Athens (Marinos *et al.* 2004; see also Section 4.5.5). Site investigations in these areas will include investigation of archives of mining companies and public (historical) archives. Lupo (2009) and Donnelly (2009) present overviews of site investigation techniques and remedial measures applicable for areas and constructions with historic mining. In areas with natural caves speleologists might keep records and maps with the location of such caves. In areas with past underground activities or caves, site investigation will never uncover all underground spaces unless a metre scale drilling grid is applied (Price *et al.* 2009). The depth of the boreholes should reach well below the level where any spaces are to be expected. Alternatively, the depth can be limited to the level where the influence of the new construction becomes negligible. However, this may be dangerous because of future growth and collapse of existing underground spaces. The grid should also be beyond the limits of the new construction as existing spaces may grow sideways.

5.1.14. Forensic site investigation

Forensic site investigations are another type of site investigation often requiring extensive archive investigations. Forensic site investigations are normally executed after failures resulting in claim cases, serious injuries or mortalities. The purpose is trying to pinpoint the deficiencies in design, building or any other aspect of an underground work that have caused or influenced the failure (Brown 2006; TCFE 2012). The results are used two-fold: trying to identify the cause and organization(s) and people responsible, and the results may be useful for expanding the knowledge on underground construction. Often standards and building codes are adjusted based on forensic investigation results. A forensic investigation differs from a normal site investigation. Structures may have existed for a long time and documents with design and construction details may be lost or have never

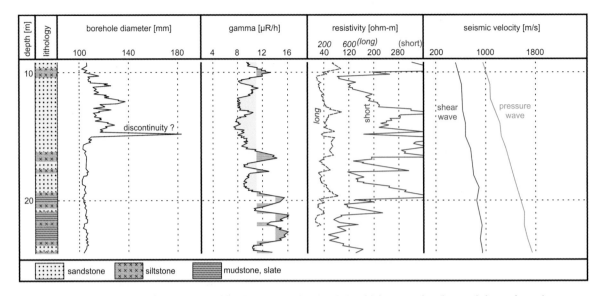

Fig. 5.16. Example of three geophysical logging methods in a borehole with interpretation (long and short refer to the distance between electrodes) (courtesy: Witteveen + Bos & ITC).

been made. Hence, a reconstruction of the design and construction based on present-day visual inspection and evaluation of the site may be the only available information. Properties of ground and man-made materials as derived for the time of failure may have become different from those at the time of construction (Section 4.3). Design and construction methodologies and techniques may also be different. The forensic expert should thus be able to recreate and back analyse the situation in the past. This may require extensive literature reviews of the building practices used at the time of construction, and interviewing engineers and labourers who worked at the site. A specific problem in forensic investigations is posed by the inhomogeneity and resulting uncertainties inherent to natural materials (Section 5.4). Notoriously difficult to answer questions are likely to arise such as 'What is the likelihood that a particular ground property had a certain minimum value at a particular location at the time of construction?' Legal aspects will be important and thus experts are expected to have a fair legal knowledge too. Last but certainly not least, lawyers, public attorneys and others involved will try to interpret the findings of the expert to their own liking, which may be very different from the interpretation intended.

5.2. Geological and geotechnical models

In the past, three-dimensional models were made in wood or plastic for large and costly projects while plan views and sections on paper were constructed for smaller projects. Today, software is available to produce full 3D engineering geological and geotechnical models (Fig. 5.17). Such computer models may consist of a 'boundary model' positioning the boundaries between the defined geotechnical units and of a 'property model' for distribution of geotechnical properties in these units (Fig. 5.18). In general, 3D models are more reliable as all units fit in detail everywhere in the model mathematically, in contrast to individual components in most hand-constructed maps and sections.

5.2.1. Model dimensions

Inherent to the use of geo-information is data dimensionality. In geo-information modelling, ground samples taken from specific points in the subsurface are 'zero-dimensional'. Information to be stored of such a sample consists of a single geographic position (x, y and depth) with one or more properties (Fig. 5.19). Borehole logs and CPTs provide property information along a single line and are one-dimensional. A picture of the Earth's surface (e.g. an aerial photograph) is two-dimensional. Geophysical data of the subsurface may be two- or three-dimensional. Boundaries between geological layers are two-dimensional even if the boundary would be undulating or deformed. Properties might be attached to a boundary. The value of such a property may be equal for the whole boundary but may also vary along that boundary. A model of the underground with boundaries between geotechnical units only is three-dimensional and may be used for modelling the geometry of the units. However, such models are incomplete since no geo-data is given between the boundaries and are therefore often labelled as 2.5 D models. Full, three-dimensional models

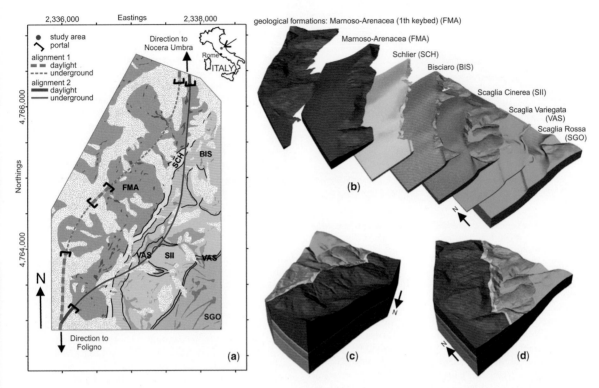

Fig. 5.17. Geotechnical/geological model for a railway through the Apennines, Italy. (**a**) Geological map with two alternative routings, (**b**) expanded model showing every geological formation separately, (**c**) visualization of a 3D model from South-West and (**d**) from North-West (modified from Tonini *et al.* 2008 with permission from Elsevier).

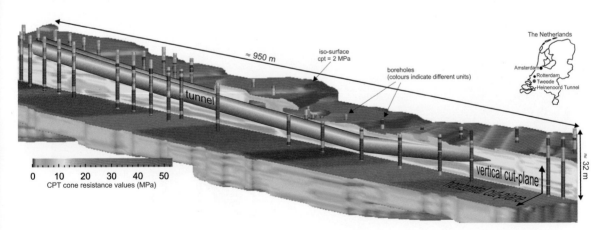

Fig. 5.18. Example of a 3D-GIS visualization of a proposed tunnel alignment. The various colours show the statistically modelled distribution of Cone Penetration Test (CPT) values. The boreholes show the geotechnical units as vertical columns. To show the CPT value distribution around the tunnels more clearly, the top part of the model is partially removed (Tweede Heinenoord Tunnel, The Netherlands; modified from Hack *et al.* 2006 with kind permission from Springer Science + Business Media).

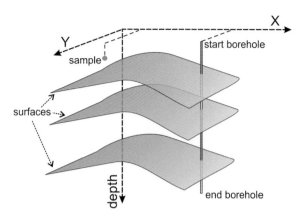

Fig. 5.19. Dimensionality of entities. A sample location has zero-dimensionality, a borehole is one-dimensional while a surface is two-dimensional.

(Fig. 5.18) of the subsurface include property data for all positions in the model.

'Fence diagrams' may serve to visualize the boundaries and property distributions in a full 3D model. The model is presented in a grid of crossing diagrams (or panels). Fence diagrams might also serve to present individual profiles connected where they cross (Fig. 5.20). Full information is available in-between the panels in the first option while the latter option provides no information in-between the diagrams.

5.2.2. Data formats

Digital data may be stored in 'vector' or 'raster' formats. Data stored in a vector format contain information about the geographical position together with the value of the property. The geographical position of data stored in a raster (or pixel) format is determined by cells in a raster (or grid) where each cell has a property value. For example, a line with a particular colour in a vector format consists of the coordinates of the starting and ending point of the vector and the properties of the thickness and colour are assigned to the line. In a raster format, the line consists of a series of grid cells starting at the starting point and ending at the ending point of the line while each cell contains the property of the colour (Fig. 5.21).

A 2D picture or map may be stored in a vector or raster (pixel) format. A 3D model of the subsurface may also be stored in a vector or in a voxel format, which is a 3D raster or grid cell. In a 3D vector model, space is divided by vector representations of surfaces. The part of the space entirely enclosed by surfaces is a volume or a solid. In databases and data handling software, surfaces enclosing a particular volume are marked as enclosing this volume. To such volumes, property values may be assigned. In a 3D voxel model, however, voxels are cells, each with its own, unique set of properties and property values, of which one may be an indicator that the voxel belongs to a particular volume. Subsurface models may also be a combination of a vector and voxel representation. This may be a suitable representation for the (statistical) variation of a property within a geological unit (Kessler *et al.* 2008).

Ground property data in vector format are stored with precise geographical positions, only limited by the original measuring accuracy or by the digits available for digital storage. The geographical position of such data in raster format is given by the pixel cell (in 2D) or voxel (in 3D) and its accuracy is determined by the pixel or voxel size. Computer conversion from vector into raster formats is rather simple but with loss of accuracy if raster dimensions

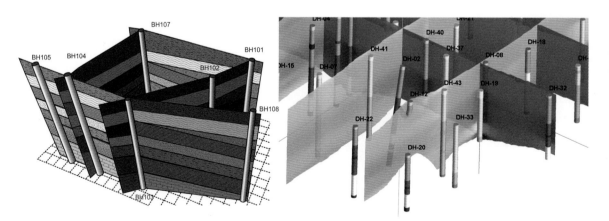

Fig. 5.20. Example of two types of fence diagrams. The left example shows interpretation of geology from borehole to borehole; the right example is created from a full 3D model based on borehole (cylindrical columns) information in and between the panels (modified from: left: StrataExplorer software, courtesy of GAEA Technologies Ltd. 2010; right: RockWorks software, courtesy of: RockWare Inc. 2012).

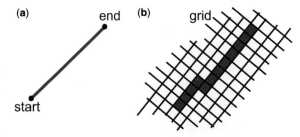

Fig. 5.21. Two formats for the same entity in two-dimensions: (**a**) in vector format, a red line consists of starting and ending point and information that a line is connecting the points. The properties are the red colour and the line thickness. (**b**) In raster (pixel) format the line of (a) is represented by a series of grid cells that have the property 'red'.

are larger than the accuracy of the original measurement. However, conversion from raster into vector might generate significant deviations from the original. For example, a line converted from a raster to vector format will get the property 'thickness', but this thickness is the thickness of the raster and is unrelated to the original measurement and accuracy of the thickness of the line or to the object representing the line.

5.2.3. Data file formats

Digital data, in particular representing visual items as drawings, photos, or videos (e.g. files with extensions as dxf, dwg, cdr, cpt, bmp, jpg, avi), may be stored in a wide variety of formats. Container type files may hold files in different formats as well. Some formats belong to a particular programme, or are proprietary and may have secret specifications, while others are public domain. Translation from one format into another is done by conversion programmes. Such programmes and information about formats and file containers may be obtained through the Internet. However, the quality of conversion programmes is not always secured and data might well be lost during conversion. Loss of information is particularly problematic if not immediately noted. That might apply to relatively small reductions in accuracy, loss of entities in a large and complicated data structure, and loss of associated data such as sources and accuracy of data and loss of the purpose for which the data have been collected and interpreted. This may cause inaccurate or erroneous models in a later stage in the process.

5.2.4. Scale, detail, resolution and data density

The relation between scale and the detail of maps and models originate from paper representations of maps and was determined by pen thickness, paper size and considered area. Enlarging or reducing map sizes affects the scale but not the level of detail on maps (Fig. 5.22). However, digital geo-information stored in a database has no scale, detail may be infinitely high or low, and the area covered is practically unlimited. Despite disconnection from its original meaning, scale is often still applied to illustrate the level of detail (Fig. 5.22). Resolution refers to the number of information points per unit length in raster formats, for example the number of pixels per unit length of a photograph or the number of voxels per unit length in a volume of a digital 3D geological model. Here, resolution might not be similar in all directions and might vary in the same direction. In regional 3D geological models, for example, X and Y directions are often modelled in a lower resolution than the depth, resulting in tabular shaped voxels. This is done, as normally information density in depth is much higher than horizontally. In groundmass mechanical models, voxels remote from the area of interest (e.g. a tunnel) are dimensioned larger as that considerably saves calculation time and barely affects model results. Resolution is not a term commonly used in vector models but is equivalent to the accuracy expressed in the number of digits used for X, Y, and Z representation, which is unrelated to the original data accuracy. Hence, scale and detail have lost their traditional meanings in digital modelling of the subsurface.

In digital subsurface geo-information models, nowadays resolution is determined by data density and the required level of detail that, in turn, is related to accuracy of subsurface models (Section 5.4). Generally, models covering a larger volume are less detailed and contain less data than a smaller volume model (Fig. 5.23). In advanced stages of project development, more data have become available allowing for increasingly detailed and finer resolution subsurface models. Hazard and risk acceptance also puts boundary conditions to the minimum level of detail required. Although feeding models with more precise data normally contributes to the quality of model output, this does not necessarily reduce model uncertainty. It is rather the combination of local geological complexity, economic relevance of the construction, risk on jeopardizing human life, and environmental considerations that determines acceptable detail and uncertainty levels.

5.2.5. Temporal data

Data produced for the design of a construction ('as designed') are not necessarily the same as the data after completion of the construction ('as built'). Small but sometimes quite significant deviations from the design may develop during construction. Such deviations may not always be documented in the 'as-built' data sets normally stored in cadastres or land registration offices. This may particularly apply to underground structures as these are generally not visible and deviations from the reported 'as-built' data might be overlooked rather easily by cadastre employees at inspection. Subsurface data may also be dislocated over time. In soft deposits, cables and pipelines may 'walk' through the subsurface due to varying stress or temperature conditions (Hededal & Strandgaard 2008). Even larger

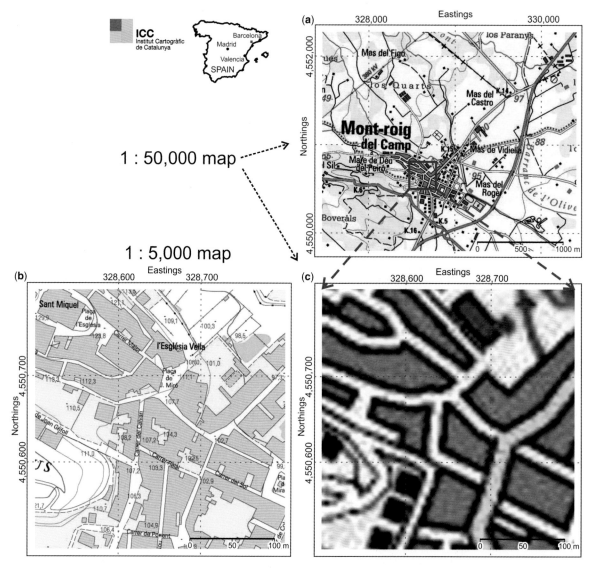

Fig. 5.22. Scale and detail on digital maps; (**a**) an area of a 1:50 000 scale map within (**c**) a blow-up of that map to a scale of 1:5000; (**b**) represents the same area as in (**c**) but derived from a 1:5000 scaled map. The level of detail in the blow-up of the 1:50 000 map is much less than that derived from the 1:5000 map (source: Institut Cartogràfic de Catalunya, ICC).

structures may be displaced due to unforeseen settlement and consolidation. Virtually never, such 'as-is' information is documented in data repositories.

5.2.6. Computer aided modelling

Until quite recently, most common modelling programmes were two-dimensional Computer Aided Design (CAD) or Geographical Information System (GIS) programmes extended in three-dimensions with limited functionality for subsurface modelling. Also many programmes were purpose-made for modelling a specific process in the subsurface, such as modelling groundwater and contaminant flow or mechanical (e.g. stress – strain) relations. Full 3D computer programmes for modelling the subsurface originate from mining and the oil and gas industry. The mining industry focused on spatial modelling of geology and ore bodies interpreted from a dense network of borehole data. The oil

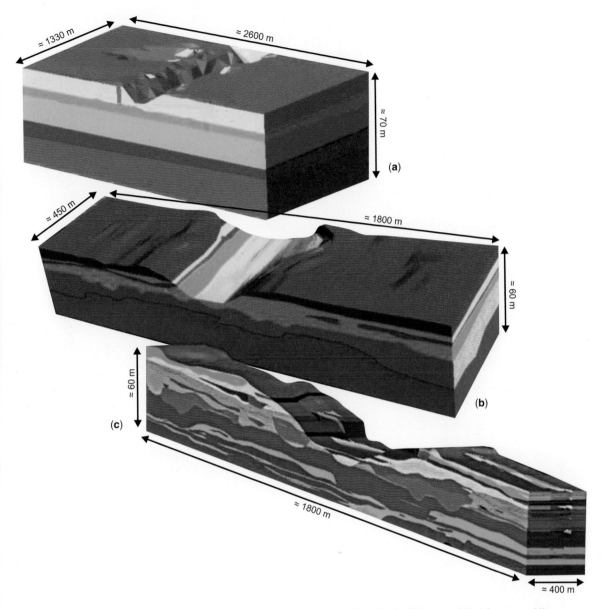

Fig. 5.23. Subsurface model for the Tweede Heinenoord Tunnel, The Netherlands; Volume modelled decreases while detail increases from (**a**) via (**b**) to (**c**) (modified from Hack *et al.* 2006).

industry developed geological modelling tools based on rather limited numbers of borehole data but on extensive and detailed seismic exploration data (Hack & Sides 1994). 4D (i.e. space plus time) programmes with a fully 4D database structure are currently being developed and storage of sequential 3D data sets, analysis and visualization in 4D are now possible (Hack 2010). A further development are five-dimensional analyse tools and database structures in which the 5th dimension may be used for a variety of parameters, such as costs, thermal behaviour, uncertainty and accuracy (Caumon 2009, 2010; Snijders 2009; Table 5.1 – GrADS), or level of detail (Van Oosterom & Meijers 2011).

Today (2012), a wide variety of 3D computer programmes for modelling the subsurface is available (Table 5.1). Generally, GIS-based systems are easier to operate than

Table 5.1. Examples of programmes and tools for 3D subsurface modelling (after Hack 2012)

Program	Suitability for all-purpose modelling of underground excavations	Remarks	Developer/supplier/ consortium leader	Web-site
2/3/4D Move	+	Family of 2D–4D programs for oil & gas, mining, geology	Midland Valley Exploration Ltd., UK	http://www.mve.com/
Amine	+	AutoCAD based geological modelling & mining system. Used for geotechnical, environmental & geochemistry modelling	Flairbase, Canada	http://www.flairbase.com/WEBSITE/EN/amine.html
ArcGIS 3D Analyst	+	3D extension of the popular ArcGIS cartographic software	ESRI, USA	http://www.esri.com/
AutoCAD	−/+	CAD system, not a 3D modelling program, but with add-ons used for subsurface modelling	Autodesk, USA	http://usa.autodesk.com/
Cadsmine	+	CAD based geology, mining & geotechnical modelling	Gijima, South Africa	http://graphicmining.com/node/178
CoViz 4D	(+)	4D visualisation	Dynamic Graphics, Inc., USA	http://www.dgi.com/coviz/cvmain.html
Datamine	+	Mining, geology, mine planning, geotechnical & environmental modelling	Datamine International Ltd., UK	http://www.datamine.co.uk/
Deep Exploration	(−)	Conversion & integration software	Right Hemisphere, USA	http://www.righthemisphere.com/products/
EarthModel	−	Oil & gas, geology & seismic modelling	Fugro Jason	http://www.fugro-jason.com/overview.htm
EarthVision	−	Oil & gas, geology & seismic modelling	Dynamic Graphics Inc., USA	http://www.dgi.com/earthvision/evmain.html
EVS/MVS	++	Mining, geology, geotechnical & environmental (ESRI program environment)	C Tech Development Corporation, USA	http://www.ctech.com/
FEFLOW	(−)	3D Fluid flow (groundwater) & geothermal modelling	DHI-WASY GmbH, Germany	http://www.feflow.info/
FEHM	(−)	3D Fluid flow (groundwater) modelling	Los Alamos National Laboratory (LANL), USA	http://fehm.lanl.gov/
FLAC3D/3DEC/ PFC3D	(−)	3/4D mechanical modelling programs (only some basic geology modelling)	Itasca, USA	http://www.itascacg.com/home.php
FracSIS	++	Mining, geological, geophysical, geochemical, environmental & geotechnical modelling	FracSIS, Australia	http://www.fractaltechnologies.com/home.36.html

Name	Rating	Description	Company	URL
Geoblock	+	Mining, geology, geotechnical, environmental (open source)	Contact: Getos Ltd., distributer: Sourceforce	http://geoblock.sourceforge.net/geoblock.htm & http://getos.chat.ru/#Presentation
Geocap GIM and OlI & Gas	−	Oil & gas, geology & seismic modelling	Geocap AS, Norway	http://www.geocap.no/
GEOL_DH & KAI-2001	−/+	Mining, geology (AutoCad tools)	Alexei Deridovich	http://geol-dh.narod.ru/en_index.html
Geomodeller3D	++	Oil & gas, mining, geology, geophysical, geothermal, carbon capture & sequestration (CSS), hydrogeology, geotechnical & environmental modelling	Basins Consortium/Intrepid Geophysics, GeoIntrepid, Australia & BRGM (Bureau de Recherches Géologiques et Minières), France	http://www.geomodeller.com/geo/index.php?lang=EN&menu=homepage & BRGM: http://www.brgm.fr/index.jsp
Geosoft (Oasis montaj)	−	Mining, borehole interpretation, geophysical & geochemistry modelling. Linked to the ARCGIS software of ESRI	Geosoft, Canada	http://www.geosoft.com
GoCad	+	Geology, oil & gas, mining & geophysical modelling	GoCad, France	http://www.gocad.org
GrADS	(+)	4D visualisation & 5D analysis tool	COLA, USA	http://grads.iges.org/grads/head.html
GRASS GIS	++	Oil & gas, mining, geology, geotechnical & environmental (open source)	Open Source Geospatial Foundation (OSGeo)	http://grass.osgeo.org/
GSI3D	++	Oil & gas, mining, geology, geophysical, geothermal, hydrogeology, geotechnical, archaeological, flood risk management & environmental modelling	British Geological Survey, UK & INSIGHT GmbH, Germany (formerly also: Geological Survey of Lower Saxony, Germany)	http://www.gsi3d.org.uk/
Jewel Suite	−	Oil & gas, geology & seismic modelling	JOA Group	http://www.jewelsuite.com/
Leapfrog mining	++	Mining, geology, geophysical, geochemical & geotechnical modelling	ARANZ Geo Limited, New Zealand	http://www.leapfrog3d.com/mining solutions
Lynx	++	Mining, geology, environmental & geotechnical modelling	Lynx Geosystems S.A. (Pty) Ltd., South Africa	http://www.lynxgeo.com
MathLab	(−/+)	Mathematical program, many tools from third party developers for geological data handling	MathWorks, USA	http://www.mathworks.com/products/
Micromine	++	Mining, geology & geotechnical modelling	Micromine, Australia	http://www.micromine.com/
Micromodel	+	Mining, geology, geotechnical & environmental	RKM Mining Software, USA	http://www.rkmminingsoftware.com/
Microstation	−/+	CAD system, not a 3D modelling program, but with add-ons used for subsurface modelling	Bentley, USA	http://www.bentley.com/en-US/Products/MicroStation
MineGeo	++	Mining, geology & geotechnical modelling (directed to mining)	Cad Cam Solutions Australia Pty. Ltd., Australia	http://www.minegeo.com/index.html
MineSight	++	Mining, geology & geotechnical modelling (directed to mining)	MineSight, USA	http://www.minesight.com/index.php

(Continued)

Table 5.1. Continued

Program	Suitability for all-purpose modelling of underground excavations	Remarks	Developer/supplier/ consortium leader	Web-site
MODFLOW	(−)	3D Fluid flow (groundwater), with many add-ons	USGS	http://water.usgs.gov/nrp/gwsoftware/modflow.html
Petrel	−	Oil & gas, geology & geophysical modelling	Schlumberger	http://www.slb.com/services/software/geo/petrel.aspx
Plaxis	(−)	3/4D mechanical modelling program (only some basic geology modelling)	Plaxis	http://www.plaxis.nl
Promine	+	CAD based geology, mining & geotechnical modelling	Promine, Canada	http://www.promine.com
RockWorks	++	Software tools for geology & related subjects, including 3D modelling program. Extensively used in oil & gas, mining, environmental, geotechnical & groundwater modelling	RockWare, USA	http://www.rockware.com/
Roxar IRAP RMS Suite	−	Oil & gas, geology & geophysical modelling	Emerson Process Management	http://www2.emersonprocess.com/en-US/brands/roxar/Pages/Roxar.aspx
StrataExplorer	−	Interpretation tool for boreholes & strata/unit definition (not a full 3D modelling program)	GAEA Technologies Ltd., Canada	http://www.gaeatech.com/index.html
Surpac	++	Mining, geology, geotechnical & environmental modelling	Gemcom, Australia	http://www.gemcomsoftware.com/products/surpac
Techbase	++	Geology, environmental, geotechnical, groundwater, oil & gas & mining modelling	Techbase, USA	http://www.techbase.com/index.html
Vulcan	++	Mining, geology, geotechnical & environmental modelling	Maptek, Australia	http://www.maptek.com/products/vulcan/index.html
WinFence	−	Tool for fence diagram interpretation (not a full 3D modelling program)	GAEA Technologies Ltd., Canada	http://www.gaeatech.com/index.html
ZOOMQ3D	(−)	3D Fluid flow (groundwater)	OO models, University of Birmingham, Environment Agency, British Geological Survey (BGS)	http://www.oomodels.info/pmwiki/pmwiki.php & http://www.bgs.ac.uk/science/3Dmodelling/zoom.html

− = less suitable; + = more suitable; ++ = extensively used; in-between brackets: uncertain. Suitability for all-purpose modelling of underground excavations is based on the experience of the author, supplemented by public information from the developers and suppliers, and literature references. Suitability may change as programs are regularly updated and new features may make programs less suitable in 2012 to become very suitable in the future. The table is not conclusive as software development is very versatile (e.g. new programs, changes of name or ownership, programs are split in different or combined to new packages, same program may be available under different names). All programs (if not mentioned otherwise) are or claim to be 3D (or 4D) programs; however, some may not possess a full 3D (or 4D) database structure. Geophysical, fluid (groundwater) flow, and mechanical programs have only been included if options for geological modelling are included. Most names of software packages, companies, and consortia are registered trademarks®. Only the leader is included for products developed by a consortium.

CAD-based systems. However, the differences have become less prominent over the years and are expected to reduce further with on-going development. Not all 3D programmes contain a full 3D database structure; some only simulate 3D by software routines. This may lead to problems in modelling, in particular for geologically more complex areas. For more detailed information on modelling and software packages refer to, for example, Kelk (1992), Turner & Gable (2007), Yanbing et al. (2007), Baojun et al. (2009); IUGS (2012); Rengers et al. (2002); and the web addresses provided in Table 5.1.

5.3. Data processing and information

A wide variety of data and information is required to design and build safe underground constructions, including geological data and properties of subsurface materials.

5.3.1. Data harmonization and standardization

Normally, data from different sources are not registered and stored in the same way. Digital data occur in many different formats and following different standards. In a project, such data should be combined which requires standardization and harmonization. That may be done manually or by digital conversion of data (Toll 2007; Hack 2010; Schaminée & Klapwijk 2010). The lack of standardization of data is a major obstacle in geological modelling and is the subject of research (D'Agnese & O'Brien 2003; Chang & Park 2004; Choi et al. 2009). Parallel to the research are many ongoing initiatives to develop standards for geotechnical data storage and exchange. New initiatives arise regularly while others cease. Some of the initiatives that seem to be promising in 2012 are AGS (2012), DIGGS (2012), GEF (2012), GIMCIW (2012) and TUNCONSTRUCT (2012). The Internet infrastructure is often used for exchange of data, discussed in the following sections in more detail.

5.3.2. Data and the internet

The Internet triggered and accelerated data harmonization and standardization significantly, in particular standardization in formats suitable for internet protocols. Standard internet languages as HTML are less appropriate for data standardization and storage of data for specific application fields. However, meta-language XML (eXtensible Markup Language) allows easy extension and adoption of the XML language (Houlding 2001; W3C 2012). XML has been implemented by various programmes (e.g. Geomodeller3D in Table 5.1) and is widely used for model coupling (Liu et al. 2010). Currently, data interoperability is a major research topic (Tegtmeier et al. 2009; Turner & D'Agnese 2009; Yang & Raskin 2009; OneGeology 2012) and some cases most relevant to subsurface data handling are discussed below.

5.3.2.1. Geography Markup Language (GML)
An important implementation of XML is the Geography Markup Language (GML) as developed by the Open Geospatial Consortium Inc. (OGC 2012). The definition allowed standardized geospatial and location-based data to be stored and communicated worldwide. GML serves as a basis for CityGML (2012, see below), LandXML (2012) for land-use planning purposes, eXploration and Mining Markup Language (XMML 2012), further developed in the GeoScience Markup Language (GeoSciML 2012) and the meta-standard for geotechnical data, the Geotechnical Markup Language (Geotech-XML 2012). The latter is a version of XML specifically adopted for geotechnical data and is already applied in soil slope case histories (SlopeSML) and retaining walls. The Joint Technical Committee of the Federation of the International Geo-Engineering Societies (JTC2 2012) is developing standards for digital description and storage of geotechnical data (GeotechML).

5.3.2.2. CityGML
CityGML (2012) is developed for the representation of 3D urban objects by defining geometrical, topological, semantic and appearance properties. Thematic information goes beyond graphic exchange formats and allows deploying virtual 3D city models for analysis tasks in different application domains, as simulations, urban data mining, facility management and thematic inquiries. CityGML is well integrated with Google Earth (2011) and has been extended to incorporate subsurface features, including Geo Building Information Modelling (GeoBIM) (Zobl & Marschallinger 2008; Fig. 5.24). GeoBIM is an extension of BIM (Building Information Modelling) which is a subset of CityGML, integrating CAD/CAM data into the GML environment.

5.3.3. Data sources

A brief overview of subsurface data and information sources for various settings is given below.

5.3.3.1. Small infrastructure
In some countries, the cadastre (Chapter 6) holds information on the location of small infrastructure (cables, pipes, etc.) in the subsurface. Elsewhere, such information might be obtained from other public institutions. Where no public data storage facilities exist, data should be collected from a wide variety of potential sources dealing with construction or maintenance of small infrastructure, for example from gas and power supply and telecommunication companies. Normally, the data is only available in hardcopy but more and more cadastral data, including small infrastructure are being stored in digital databases. Information consists of 2D data describing the location of the infrastructure in relation to the Earth's surface with the depth below surface of small infrastructure entities as property. Often, data about small infrastructure is stored as photocopies or scanned documents that may be distorted, and thus be inaccurate. For small infrastructure in particular, normally 'as-designed' data are

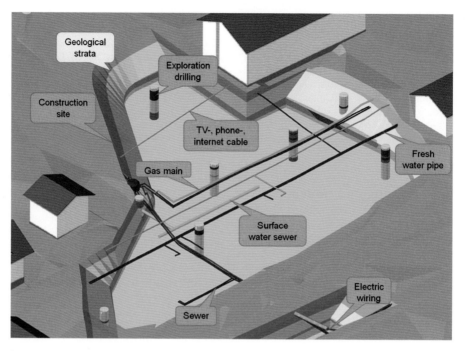

Fig. 5.24. Example of a typical GeoBIM framework combining natural and technical/man-made above- and subsurface objects. Building pit excavation in an area covered with buildings. Small infrastructure, geological features and boreholes are indicated (from Zobl & Marschallinger 2008).

stored and not 'as-built' nor 'as-is' data. Today, several initiatives are being undertaken to integrate small infrastructure in 3D GIS together with surface and subsurface objects (Fig. 5.24).

5.3.3.2. Military
Locations of military facilities might be relevant when dealing with underground installations. This may apply to confidential information concerning underground pipelines for gas and oil supply, telecommunication lines, command centres in bunkers or underground launching facilities for missiles. Not only geographic data may be relevant but also details of construction and the results of site investigations. Such technical data may be difficult to collect and their coordinates might be fabricated. Normally, military reconnaissance data (e.g. remote sensing data) is not accessible. Confidentiality may be released after a certain time span. In some countries, some military data concerning the subsurface are supplied to the public domain (USACE 2010).

5.3.3.3. Topography
National topographic survey organizations are rapidly converting from map production solely based on aerial photographs and geodetic measurements, to storing in hardcopy or digital format the products of a wide range of data sources, such as satellite and aerial photographs, elevation data from laser altimetry (LIDAR), and vector and raster special purpose maps of the surface. Normally, topographic surveys do not store or provide subsurface data.

5.3.3.4. Public works
Governmental organizations involved in land-use planning may be extensive users of CAD/CAM systems, in compliance with consultants and contractors working for the public sector. Data of relatively new structures are generally available digitally in vector or raster format. Data of older structures may be present in hardcopy drawings only. Such organizations often hold data of underground constructions for roads and railways, which may also be stored by cadastres in a less detailed format.

5.3.3.5. Archaeology
Due to the Malta Treaty (Chapter 6), archaeological investigations should precede major construction works to prevent disturbing archaeologically valuable objects in the subsurface. Data obtained from such surveys are normally stored in national or municipal archaeological surveys and normally comprise hardcopy maps at various depth ranges. Results of recent archaeological investigations may have been stored digitally in CAD/CAM systems.

5.3.3.6. Geology

Geological data are stored by national geological surveys. Traditionally, geological surveys produced geological maps with cross-sections displaying the geological conditions in depth and often they made special purpose maps for the shallow subsurface, occasionally for foundation engineering purposes (De Mulder 1988). Since the 1990s, many Geological Surveys began to store their subsurface data and information in digital databases, including borehole data, rock and soil properties and geophysical data, either in vector or raster formats. Digitalization and integration of such data into digital models made at least part of the data accessible to the public domain (Box 5.2) but some data might be kept confidential, at least for a certain time span.

5.3.3.7. Geotechnology

Geotechnical data of the subsurface are generally not stored in centralized and publicly accessible repositories. Normally, such data is held by individual consultants and contractor companies. In a few countries, geotechnical subsurface data are, either voluntary or compulsory by law, stored in national repositories (in the UK; BGS-NGDC 2012; in The Netherlands: Box 5.2 and Geodatabank 2012). The European Union has formulated the INSPIRE directive which regulates (geo-)spatial data and access to data in particular for environmental issues (INSPIRE 2012).

5.3.4. Geo-databases

Geological conditions of most parts of the world are represented in 2D paper maps and stored in archives together with paper records of field observations, borehole logs and other data. Developments in geo-information sciences and increasing availability of Geographic Information Systems (GIS) stimulated digital data storage and production of layer or digital 3D geological models (Ford et al. 2008; Dino 2012; GSI 2012; NSW 2012; BGS 2012). Conversion of paper archives into digital formats has been an unprecedented effort. Sometimes, such conversion activities required collection of additional information to support correct digital storage of underground data, hampering a rapid conversion into digital formats. In many countries, this conversion was completed by the beginning of the 21st century.

Since 2007, more than 115 national geological surveys have undertaken a joint effort to make all geological map information available in digital format, mutually accessible and interoperable. This project, 'OneGeology' (2012) has been a joint contribution by all geological surveys to the International Year of Planet Earth (2007–2009) (IYPE 2010) and aims to produce a digital geological map of the world, eventually at a scale of 1:1M. The project is inclusive thus ensuring that all countries may participate. Depending on their capability and capacity, nations will provide access to their geological map data in various ways. For some

---

Box 5.2. DINO, the internet-based geo-data and information system in The Netherlands

Through DINO, the Geological Survey of The Netherlands (TNO) manages its information task by making subsurface data accessible on the web. DINO is a public data information centre financed by a consortium of four Dutch Ministries and was formally established in 1999.

DINO consists of three main portals. DINOData is the actual database of the Dutch subsurface; DINOMap holds information about the interpretations of data/maps in the DINOdatabase, and DINOServices provides access to the services through the DINO database (DINO 2012).

The database (DINOData) contains over 400 000 shallow to very deep, quality labelled borehole descriptions, 100 000 Cone Penetration Test records, many hundreds of kilometres of geophysical bore hole logs, more than one million soil and rock samples and several hundreds of digital maps. Moreover, the archive contains more than 16 million groundwater data; 12 000 vertical electrical soundings and many other geo-electrical measurements; the results of geological, geochemical and geomechanical sample analyses; as well as seismic data. In addition, DINOData contains public digital and analogue results of exploratory drillings for oil & gas extraction, and salt mining.

DINOMap gives access to geological information on the Dutch subsurface through maps using four types of data models: (1) an atlas of digital map depths of nine important stratigraphic horizons for the deep subsurface; (2) the Digital Geological Model of the Netherlands provides the depths and thicknesses of the Late Tertiary and Quaternary beds to a depths of roughly 1 km; (3) the Regional Geohydrological Information System (REGIS) with detailed information of the hydrogeological structure and groundwater flow systems in the subsurface; and (4) GeoTop is a detailed three-dimensional model of the topmost 30 to 50 metres of the Dutch subsurface. It provides users with a cell-based description of the spatial variability of geological, physical and chemical parameters in the subsurface.

DINOServices provides custom-made and specialized services to active users and clients. These services may be used for monitoring purposes and for pro-active use including early warning systems for floods and droughts. Electronic data delivery is free of charge, or at marginal costs. A next step in this process is to insert subsurface geodata into an overall and nationwide system of data registration (Basisregistratie Ondergrond, BRO 2012), enforced by law.

surveys, coverage will be as raster images at first. Others, with more developed systems, will dynamically 'serve' geological map data for their territories. For the more sophisticated attributed vector data, the project works in cooperation with the IUGS Commission for the Management and Application of Geoscience Information (CGI 2012) and uses GeoSciML (Section 5.3.2.1; De Mulder & Jackson 2007).

Box 5.2 shows an example of a database for geoinformation, and in the following section an outline is given for an information infrastructure and management system for civil engineering projects. Another relevant project also incorporating information management is the European research project on innovation in underground construction (Beer 2010; Chmelina 2010; TUNCONSTRUCT 2012).

5.3.4.1. Information management for civil engineering
In large civil engineering project various surface and subsurface data are used throughout a project from planning, via design and construction to maintenance and demolition, by different professionals, such as planners, architects, and civil and geotechnical engineers. At present, the different professionals, at the same or different stages during the lifetime of a project, use different software with different data sets. Transfer of information between the professionals is often only by hardcopy, or if digital, by some final interpreted model or image, for example, a jpeg or tiff file with a cross-section of the subsurface. Such transfers lead to loss of information as the information transferred is mostly without any underlying information such as the raw data, the reason for which data have been collected, and the reasoning used for interpretation. Information is often also lost over the years during the lifetime of a project and data are re-collected during various stages of a project. An environment providing all data to the designers, builders, etc. in one single digital environment for the entire lifetime of a project would be easier and is expected to result in a serious cost reduction (GIMCIW 2012). The concept is that a central (project) database management system keeps track of every piece of data collected over the full lifetime of a project.

The system requires the following: (1) surface and subsurface data form an integrated model of surface and subsurface. Surface data may be existing and newly designed buildings and infrastructure, and subsurface data may be, for example, the position of existing cables and underground infrastructure, and geological and geotechnical data; (2) the database management system should be able to cope with all types of information (raw data, interpreted models, calculations, etc.); (3) all raw data should be stored and be accessible; (4) the interpreted and modelled data with a full background and record on motivation and methodologies used for the interpretation of the model should be kept available; (5) information on the accuracy of data and the likelihood of models should be available; (6) all users remain owners of their own data and outsource data management only while keeping access to their data. Upon completion of a project all organizations and companies involved are required to deliver copies of the data for future storage; and (7) data formats are not reformatted. Enforced standardization was not seen as a viable option for the project, since that would inevitably lead to higher costs. The GIMCIW project has produced a set of tools to translate metadata from one format into another for mutual access for all data to all stakeholders, among others to allow the data to be integrated into CityGML (2012). Application of this system should structurally reduce data redundancy and should save time and costs. Additional cost savings are expected from reduction of errors due to miscommunication and misinterpretation of data. In 2012, the research was completed and the results are submitted to various international bodies, such as the Open Geospatial Consortium (OGC 2012) and JTC2 (2012).

5.4. Models and uncertainty

Ideally, three-dimensional models containing changes over time ('4D') would be the best possible storage of data for underground constructions, but models may not be correct. The accuracy and likelihood of models are often varying throughout the model and depend on underlying data. This is inherent to the methodology used for making the model and how data have been collected. In addition, forecasting behaviour of geological and geotechnical materials is hampered by an often incomplete knowledge about time effects.

5.4.1. Geological–geotechnical subsurface models

A geotechnical model for a particular application, such as a tunnel design, consists of geotechnical units (Fig. 5.25) separated by boundaries. The units have been defined by the expert making the model. Internally, such units are often not homogeneous and rock or soil properties may vary quite considerably within each unit. The amount of variation depends on the degree of variability of the properties within a groundmass and the context in or the purpose for which the geotechnical unit is used. A groundmass with a wide variation over small distances inevitably results in geotechnical units with wider variations in properties. The more limited the allowed variability in a geotechnical unit is, the more accurate the geotechnical calculations can be. However, limiting the variability can only be based on detailed information about the unit in all its dimensions. This implies availability of sufficient data and often in collecting additional data which is expensive. Therefore, the benefit of a higher accuracy obtained for a calculation based on geotechnical units with smaller variations in properties should be balanced against the economic value of the planned engineering work and against the potential risk of failure of the structure to the environment and human life. Acceptable property variations within a geotechnical unit for the design of a highly sensitive engineering structure, such as

Fig. 5.25. Exposure with geotechnical units of different consistency: dolomites (a calcium-magnesium carbonate rock) develop steep slopes while maximum sustained slopes in shale are shallow. Slope dips indicated by the black solid line (photograph: R. Hack).

a high-speed railway tunnel will be less than for a geotechnical unit in the calculation for a wine cellar under a private house.

There are no standard rules for the division of a groundmass into geotechnical units. Such division depends on experience and the 'engineering judgment' of experts (Fookes 1997; Hack 1998, 2009; Culshaw 2005; Einstein 2007; Caumon 2009). This implies that the structure of the subsurface model also mainly depends on expert judgement. As expertise varies from expert to expert, the reliability of a geotechnical model is very difficult or impossible to assess.

5.4.2. Model uncertainty

In geo-engineering practice, estimation of potential errors in geotechnical properties of the subsurface is generally based on their potential impact on the planned structure. This may be denoted as a hazard and risk analysis. Various methodologies, such as 'geotechnical base-line methods' (Van Staveren 2006), probability studies, and Monte Carlo simulations are applied to quantify such errors in the design of an engineering structure due to uncertainty in subsurface properties (Fig. 5.26) (Wassing et al. 2003; Wycisk et al. 2009; Caumon 2009; Caers 2011). However, these analyses do not or only partially address two quite relevant issues: the quality of the geological and geotechnical expertise needed to develop the subsurface model and the quality of the expertise required for the division of the groundmass into geotechnical units.

In data interpretation geoscientists apply 'a priori' knowledge of the geological environment in which respective units are supposed to be formed, supplemented with their knowledge of the geological history of the site (Fig. 5.27). The quality of this interpretation and of the resulting model therefore depends on the quality of the expertise of the geoscientist(s). This also holds for determining geotechnical units, assessing the unit boundaries and the range of property variation allowed for each unit (see before). Today, reports dealing with the subsurface normally contain analyses of uncertainties in measurable properties. However, this does not apply for the reliability of interpretations and for the allowed variation in various units as these cannot be assessed reliably at present (Davison et al. 2003; Evans 2003; Caumon 2009).

This problem was partly resolved by Clarke (2004) through his 'generic confidence evaluation scheme'. In this

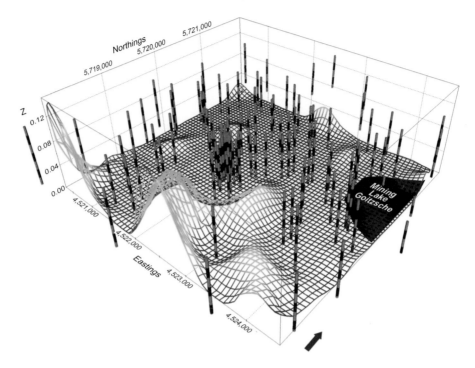

Fig. 5.26. Uncertainties for the top surface of a unit indicated by a mesh with different colours (red mesh – high uncertainty, dark blue mesh – low uncertainty), based on the borehole data (black-grey columns) and on the spatial variability of the z values. The units of the z-axis are uncertainties, varying between 0 and 1 (from Wycisk *et al.* 2009 with permission from Elsevier).

scheme each data point has a confidence value, ranging from 1 to 10, assigned to it. A boundary defined by using this data may then be given a confidence value based on the interpolation of the confidence values of the data points (Fig. 5.28). This methodology gives a fair indication of confidence for a geological model where geological conditions are quite well known and relatively large numbers of data are available.

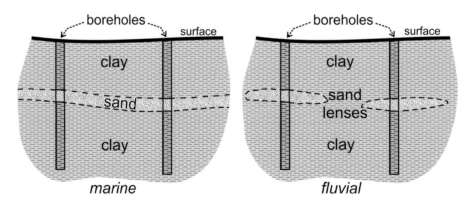

Fig. 5.27. Two cases of geoscientific interpretation. Sand is found in two boreholes. For the sand occurrences in the left case, deposition in a marine environment is assumed. In marine environments, sand layers often have a wide lateral extension. For the case on the right, a fluvial (river) environment is assumed. In such environments, sand layers generally show very limited lateral extension.

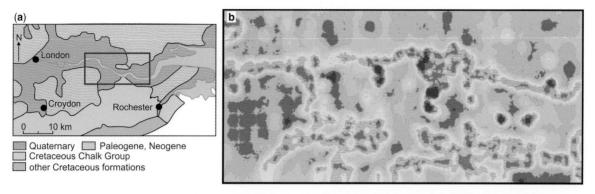

Fig. 5.28. Grid displaying confidence data in the Thames Gateway region between Woolwich and Gravesend for the upper boundary of the Chalk Group. Colour variation from blue to red indicates a change from low to high confidence in figure (**b**). The red rectangle in (**a**) gives the approximate location (modified from Royse *et al.* 2009; British Geological Survey © NERC. All rights reserved. IPR/140-16CT).

5.4.3. Time

Geological processes generate change of rock or sediment properties over time, for example weathering of minerals may impact geotechnical properties of a rock mass. This may jeopardize safety of engineering structures positioned in or on the rock if such degradation would not have been anticipated in the design. Such changes over time may be monitored at specific times and locations and relating changes between monitored locations might be achieved by statistical interpolation. Forecasting behaviour for longer time spans, however, is only possible by using trends in geological processes for which proper quantitative knowledge is often still incomplete.

Fig. 5.29. Real-time augmented visualization (AR). Left photos: user sees on his hand-held display the real world augmented with a semantic 3D model of underground infrastructure while walking on the surface; right photo: a display mounted in the helmet of the engineer shows the geology projected at the correct position on the tunnel face (photographs courtesy of left: G. Schall & D. Schmalstieg (2008); right: G. Beer (2009) with permission of CRC Press/Balkema).

5.5. Visualization

Spurred by the computer gaming and movie industry, three-dimensional visualization of geological and geotechnical models developed rapidly in the late 1990s. Today, 3D models of the subsurface in a two- or three-dimensional projection are common practice for geoscientists, even on low-budget computers. 4D visualization is more expensive but technically feasible. Either goggles with standard TV screens or sophisticated 'visualization rooms' can be used. Most recent techniques do not require goggles anymore but only a modified TV screen, or use projection in space by laser.

A recent development are goggles or handheld viewers coupled to a positioning system, for example on the surface a compass with a Global Position System, or with a camera or laser positioning system in the underground. Such 'augmented visualization' (AR) systems allow walking through the site and viewing the subsurface or any other information in real-time in the same direction as the viewer is pointing or looking, projected over the real environment (Schall *et al.* 2011) (Fig. 5.29).

5.6. Concluding remarks

This chapter advocates making 3D registration a standard for subsurface data interpretation. At present, such models are still an exception. Adding a fourth (time) dimension to subsurface models is seen as a further improvement to develop sustainable underground structures as subsurface conditions may change over time affecting stability and the environmental setting. Uncertainty in data and subsurface models may generate legal complications in projects. Many claims in civil engineering deal with problems originating from uncertainty of subsurface data. A more intense use of the subsurface may lead to more and new types of conflicts relating to competing functions. This requires proper data and information collection and storage as well as a suitable legislative framework for the development and management of the subsurface as described in Chapter 6.

Chapter 6 Legal aspects, policy and management

As the underground is inevitably becoming a more and more serious option for development, the complexities should be dealt with in terms of legislation, policies and management issues. This chapter describes how the subsurface is managed and regulated. It discusses three strongly related and interlinked issues: legal aspects, policy, and management. *Legal aspects* (Section 6.1) deals with (written) laws and regulations concerning the subsurface. Such laws are normally produced by a government, approved by a house of representatives and enforced by the public sector at national, provincial, or municipal levels. Legislation is rooted in history and culture and this may explain similarities among (groups of) nations, also when it concerns the subsurface. *Policy* (Section 6.2) deals with ambitions, plans, or actions relevant to the subsurface as stated by public or private organizations. These provide guiding principles for decision-making and actions, which may eventually result in laws and regulations. Policy may also include less formal agreements, such as covenants between relevant stakeholders, for example between government and industry. Finally, *Management* (Section 6.3) concerns organizational processes including strategic planning, setting objectives, distribution of resources, deploying assets needed to achieve objectives, monitoring changes, and measuring results. Management also includes permitting procedures for activities, as well as data registration and storage and information for (later) use.

In this chapter less attention is given to historical backgrounds of legislation, policy development, or management. This book discusses subsurface issues at relatively shallow depths (down to 250 m) and is directed to civil engineering more than to oil and gas exploitation and mining. However, extraction of natural resources such as construction materials and groundwater are included as these activities are mostly in the shallow subsurface. Definitions of terminology are important in legislation, policy and management. However, with regard to natural resources and the subsurface, definitions vary between countries and can also differ between different laws or regulations within one country. Definitions, if given at all, are often not very clear, and language differences may add different connotations. For example, the term 'soil' can have completely different meanings. Sometimes soil is used for just the top layer of the Earth's crust consisting of loose material a couple of metres thick, whereas in another law or country it is in use to describe all ground below surface (mineral particles, organic matter, water, air and living organisms), or it is used in a more technical sense as in Section 4.2. Describing all these specific differences in terminology between and within different countries is not within the scope of this book, if it would be possible to do so at all. Therefore, the definition or meaning of the terminology used in this chapter should follow from the context in which it is used.

6.1. Legal aspects

This section discusses how individuals and groups (societies) formally deal with their subsurface through legislation. Legal aspects of subsurface development are dealt with in diverse ways and at different levels. Here, it is focused on general legal aspects and approaches for subsurface development exemplified by particular cases and situations.

At a global level, four concurrent systems deal with legal rights, obligations and with resolution of disputes. *Civil law* is the most widespread system and has a prime focus on using a comprehensive set of written rules (or codes) for a given area of law (e.g. the Code Napoleon). These written rules constitute the primary basis for court decisions and sentencing, pronounced by a judge. Among countries that apply civil law significant differences exist based on historical roots, for example the French, German, Polish, and Scandinavian civil law subsystems. Comparative law as to how and why legislation differs as well as its consequences is a field of research in itself and distinction in subsystems is not clear cut. The French and German systems are based in Roman law, however the first evolved from the Napoleonic Code and the second from the German Civil Code in 1900. Polish law is distinguished because it has been influenced strongly by socialist law. The Scandinavian civil law subsystem is more of a mix between civil and common law, is based to a lesser extent on Roman law, and evolved more independently in the 17th and 18th centuries (Mattei 1997; Siems 2006).

A second major legal system is *Common law*, mostly applied in the United Kingdom and in some of its former colonies, including the United States of America, Canada, Australia and India. In common law emphasis is placed on customs and case law. Statutes may supplement case law. Parties may argue their case in court by using facts, customs and previous court decisions (case law), often before a jury that arrives at a verdict based on these proceedings. The process is managed by a judge who applies the law and passes sentence.

The third major system is *Islamic law* (Sharia and its expansion with jurisprudence: Fiqh), which is applied in Arabic countries such as Morocco, Libya, Sudan, Saudi Arabia and Yemen. In this system, the Islamic principles of jurisprudence are applied. The Islamic law has some similarities with common law.

Figure 6.1 shows the distribution of legal systems across the world. Two countries (Republic of South Africa and Namibia) and one State (Quebec, Canada) apply a mix of civil and common law due to their history. These are former colonies of a country with common law (United Kingdom) and a country with civil law (The Netherlands and Germany, and France respectively). A fourth, minor, system is also identified (*Customary law*) which in a number of cases is mixed in with the other legal systems, impacting the way legal decisions are taken in some countries.

As to legal aspects in subsurface development, *land* (i.e. surface area) should be distinguished from its (potential) *contents* or assets in the subsurface such as minerals, oil and natural gas, groundwater, space, etc. Although this book will not pay particular attention to the surface or to its related soils (in a pedological sense), (surface) land and subsurface cannot always be distinguished in a legal sense.

Landownership and control over assets in the subsurface do not necessarily coincide and may be regulated separately. The environment is another issue that has evoked a significant body of legislation concerning the subsurface.

This leads to a description of legal aspects along three lines:

- Landownership and spatial planning (Section 6.1.1)
- Natural resources and subsurface construction (Section 6.1.2)
- The environment (Section 6.1.3)

6.1.1. Landownership and spatial planning

Landownership includes the right to use the land, for example for farming, cultivation, living, or for transferring the rights by rent or lease. By law, the landowner's permission is required to use the subsurface. To access the subsurface, land may be bought from the landowner. Selling rights are an essential component of full landownership. Depending on who legally owns the land, public, private and customary land holdings may be distinguished. There is extensive literature about landownership, land information systems, and reforms in such systems that occurred in the last two centuries (Ezigbalike *et al.* 1995; Williamson 2001; Abdulai 2006).

In most Western democracies, land rights are designated to individual parties (legal entities). These may be private or public. Public land refers to land occupied, used, or acquired by the Government or other public authorities. Private land is owned, held, used, or occupied under a freehold title, a leasehold title, or a certificate of claim that is registered as private land. Many countries also exercise customary land holdings. In the customary system, land belongs to the community but individuals of that community have the right to cultivate and use the land as if owned. These include Indian lands (USA, Canada, South America), and Australian Aboriginal land. For land in countries such as Malawi, Ghana, Uganda, Papua New Guinea, Solomon Islands, customary rights are an important element of land use. Customary rights differ significantly from ownership rights because such lands cannot be sold (the right of alienation, Ezigbalike *et al.* 1995; Williamson 2001).

Today, landownership is often quite fragmented in contrast to the past when landlords owned large tracts of land. In communist and socialist countries as China, Cuba, Vietnam and North Korea, most of the land is owned by the State. The same has been true in the former German Democratic Republic but lands were privatized after reunification in 1989. While moving towards a market economy, China and Vietnam developed a system where individuals may hold the rights for use (only) for periods up to 50–70 years and benefit from what the land brings by private exploitation, excluding the right of sale. Private ownership of land was (re-)introduced in China by law in October 2007. This law gave privately owned land the same status as state owned property, including automatic renewal of residential land use rights. This also provided the basis for developing a cadastral system documenting ownership- and lease-rights (Wang 2007).

Historical, cultural, and political factors may thus play a significant legal role in landownership and in land transfer (Kaza 2004). In constitutional democracies, landowners' rights are generally registered and asserted by Land registration Acts. Separate governmental or administrative bodies keep records of such rights. Registration of ownership or property rights is needed to avoid conflicts. As for the land surface, land title registration is done through a cadastre (see also 6.3.2). Cadastral systems were first developed in Europe to support land taxation as a major source of government revenue. Today's land administration systems and resulting 'best practices' often date back to Napoleonic times when 'modern' land administration systems were introduced by colonizing powers such as England, France, The Netherlands, Germany, Portugal, and Spain (Williamson 1983). Cadastral systems were also used for demographic purposes. Examples are the Doomsday Book of William I in England and the modern cadastre of Napoleon I in France and large parts of Europe (Ezigbalike *et al.* 1995). Cadastres for legal registration of land are in place in most European countries. In some countries or regions (e.g. Austria, Denmark, Finland, France–Alsace, Germany, Greece, and Poland) this task belongs to the local court.

Landowners should respect the rights of owners of neighbouring lands which may be regulated by other legislation, such as environmental and mining legislation.

Landowners may exert rights to control the land. Beside straight ownership, there are other legal options to regulate use of the subsurface and multiple use of space amongst parties, for example to grant access or allow mineral exploration and exploitation. The main relevant legal options in

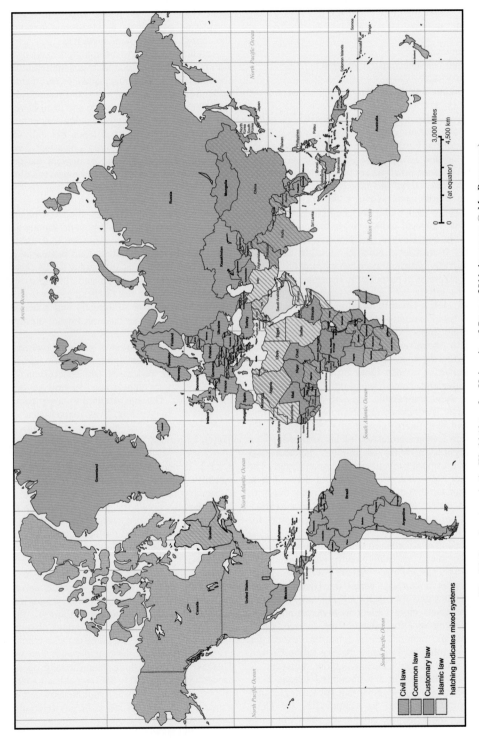

Fig. 6.1. Legal systems in the World (data after University of Ottawa 2011; basemap: © MapResources).

developing and managing the subsurface are described briefly below. Sandberg (2003) and COB/RAVI (2000) distinguish four options:

- (full) ownership
- condominium law and cooperative model
- lease and easement
- building and planting rights

6.1.1.1. (Full) ownership
Theoretically and traditionally, a landowner owns a cone-shaped portion of the Earth down to the very centre of the planet. This tradition is described in the Talmud and in Roman Law. The same principle applies to the English and American Law since these are largely based on Roman Law. In The Netherlands, Civil Code landownership includes 'all space above the surface, all Earth layers below, all groundwater and all fixtures'. In the Czech Republic, Denmark, Finland, Hungary and Norway, this principle does not apply from a legal point of view. In Denmark, Finland and Norway, private ownership stops at 6 m below the surface for all practical purposes. In Hungary, ownership is limited to the surface but includes underground parts of buildings (ITA 1991). Often multiple pieces of legislation determine the actual extent of ownership. For example in Finland no depth limits exist based on the Land Register (Johnson 2010), however the Building code puts the normal utilization depth to two stories or 6 m below the surface (Rönka *et al.* 1998). Laws on landownership in Germany, Switzerland, United Kingdom, The Netherlands and Belgium, show comparable legislation on the use of the underground. Distinctions mainly deal with the landowner's current and future interest in their land (COB/RAVI 2000). In Europe, protection of landownership is a fundamental right based on the first article of the Additional Protocol to the European Convention of Human Rights. The extent of ownership is not only a matter of depth, but in specific situations and countries restrictions may apply to ownership in terms of the contents of the volume owned. These come from other legislation in force, and may for example be related to mineral rights.

Often, ownership of mineral resources is split between the State and the landowner, depending on the type of resource. Mostly, oil, natural gas and black coal are government property. An exception is Norway, where the landowner also owns the minerals rights for black coal. For other minerals and subsurface assets more variety in ownership occurs. In the United Kingdom, for example, the surface owner owns all metals and minerals in the subsurface except for gold, silver and uranium. The latter belong to the Crown. Underground water in the United Kingdom, however, is not owned unless flowing in a well-defined stream. Pore space is another subsurface property that attracts increasing interest, for example related to technologies for CO_2-sequestration. In the USA, two situations ('rules') in common law co-exist: the 'American rule', giving ownership of pore space to the surface rights owner, and the 'English rule', where the mineral rights owner also owns the pore space (Duncan *et al.* 2009). This is relevant for granting permission as well as in identifying liability concerning the consequences of pore space use.

Another exception to the comprehensive ownership rights occurs when public interest is considered to exceed that of the landowner. This may apply to the rights to excavate underground materials (see also Section 6.1.2) or to use underground space. In practice and in an increasing number of countries, this means that landowners do not have exclusive rights to an unlimited depth, but only related to their actual interests and possibilities of usage. They cannot, for example, claim such rights to prevent the construction of a tunnel or of another underground structure below their property if that would not interfere with normal use of the land, at or, above the surface.

Underground activities are illegal without proper consultation by and agreements with the landowner. For expropriation, most European countries (e.g. Belgium, France, Italy, or Germany) require a court procedure that addresses the legitimate rights of the landowner as much as possible and safeguards appropriate and (financial) compensation. In Greece, normally full financial compensation is provided for the construction of tunnels and other infrastructure, but only in case these have an impact on normal land-use (Stamatiou 2002).

Extreme fragmentation of privately owned (urban) lands related to inheritance rights, occurs in Japan and hampers the permitting processes for underground construction. Therefore, most tunnels are constructed below publicly owned lands. This has resulted in sharp corners and steep inclinations in some railway tunnels and serious cost-inefficiencies due to technical difficulties (Shin 2000; Tongji University 2006). See also Box 6.1.

6.1.1.2. Condominium law and cooperative model (multiple ownership for multiple use)
Ownership of 'all space above the surface' or 'all Earth layers below' already triggered questions in the past about how to deal with, for example, the use of air space overlying a land property and multiple ownership in multi-storey buildings. In legal terms, current approaches to designating ownership to space above the land surface may be applied in analogy to the subsurface. For use of multiple purpose space, ownership may be split into separate, three-dimensional units. In the USA, a long tradition exists in dealing with multi-level and multi-purpose buildings, often combined with transport and traffic routes (thus also mixing private and public property). To address these complex situations in a proper legal way, the *Condominium Law* was developed. This law recognizes three-dimensional property units above ground (Sandberg 2003). The owner has a share in the whole property with the exclusive right to use a certain section and regulations are used to define the interrelationships between individual owners of shares.

As to the underground, the Israeli Supreme Court recognized the option of separating ownership in the subsurface

Box 6.1. Special Measures Act in Japan, 2000

Ownership of land in Japan is very fragmented, particularly in urban areas. Another characteristic in land development is that reaching consensus is crucial in the decision-making process in Japan. This effects arriving at formal agreements among the many landowners, in particular in case of constructing tunnels and other public works underground. For this reason, most of Japan's underground infrastructure is situated under public roads, which are under the jurisdiction of national, prefectural, or municipal agencies and can be used after acquiring permission from the relevant agency (Shin 2000). But roads occupy only 7.8% of the total area of Tokyo. As in all other megacities, in Tokyo underground space under roads is intensively used already, requiring very significant input of time and money to construct new underground facilities (ITA 2000). Therefore, new tunnels should be built under existing underground facilities. As a result, the average level of underground constructions gets deeper year by year. Today, new subway lines and high-voltage electrical conduits are often built at depths of 50 m or more below the surface. At such depths one may wonder if there would be any real need to limit such constructions to areas under roads (Karstens *et al.* 2003).

The Special Measures Act was promulgated based on this perception. That Act aims to facilitate construction of public facilities in the deeper underground of privately owned land. In 2000, the Japanese Parliament adopted the Special Measures Act for Public Use of Deep Underground Space. The new law applies to 'deep underground' zones normally not in any use by landowners; that is below the foundation piles supporting skyscrapers. Ordinances to be passed based on the law are expected to define 'deep underground' as being at least 40 m below the surface. This has been the World's first piece of legislation to establish special measures designed to promote the public use of deep underground areas. Deep underground spaces under privately owned, but largely unused land may now be readily utilized as space for urban infrastructure. Most importantly, optimal routing for infrastructure elements are not hampered any longer by incidental occurrences of public lands and systematic development of urban infrastructure is facilitated by strongly reduced negotiation times with landowners. As financial compensation is no longer required to the owners of such land, project costs have significantly dropped (Shin 2000).

---

from that over other parts of the property in the Akunas Case, which dealt with expropriation of the subsurface for the construction of tunnels. However, that Case did not address the required registration method or the adequacy of the Title Registry for documentation of separate ownership (Sandberg 2003). In the *Cooperative model* ownership of a piece of land is shared with other persons. Assignment of the exclusive use of certain parts (below, on, or above the surface) of the shared property can be to one party only (COB/RAVI 2000).

6.1.1.3. Lease and easement, building and planting rights

Leasing subsurface space only concerns the use of such space through a written agreement by a property owner allowing a tenant to use the property for a specified period of time and rent. Such use can encompass a wide range of activities, including the right to explore and exploit minerals or oil & gas (Section 6.1.2). If used for creating an underground excavation it is impossible to return the leased underground space into its original conditions by the end of the lease contract period and the value of the property is changed. Therefore, the agreement should describe the state of the property upon termination of the lease. Through a *building lease*, the owner may grant the right to use the land and the right to build to a lessee. Building leases may also stipulate the property owner to become the owner of the newly constructed buildings on his land; or *vice versa*, that the right holder becomes the owner of the construction, cables and pipes. Building and planting rights allow separation between the ownership of the land and constructions on the surface or growing crops. In line with this right, the ownership of a subsurface construction can be separated from the right to use the land surface. By *easement*, the property owner allows third parties access to the property under certain conditions and for specific purposes. Easement is frequently used for underground passage of land for utility companies. The right of way across a piece of land is a specific type of easement.

6.1.1.4. Spatial planning

An area of legislation that is different from individual ownership, but is relevant to land and land use, is spatial planning. Spatial (physical or land use) planning is a tool for governing bodies to (re-)distribute lands and human activities across a territory, for example to combat urban sprawl. Environmental sustainability, economic competitiveness, and social cohesion are important criteria in spatial planning. Spatial planning provides governments tools (e.g. through Zoning Plans) to interfere in the rights of private landowners by assessing the use of the land by law. Land use zoning may also extend to the subsurface from a legal point of view, but not in Venezuela and the USA. In Denmark, the Danish Subsurface Law regulates planning of the subsurface. In The Netherlands, since 2007 local authorities must register any governmental or public law inducing restriction on land use and to make such information available to the public through one single source kept by the Dutch Cadastre. Registration does not only include rights on individual plots but also rights related to governmental land use plans. In 2008, a new Spatial Planning Act came into force that requires incorporating spatial consequences at the surface of subsurface infrastructure. It includes a package of spatial planning standards to allow spatial plans to be made available in digital format.

On a supra-national level, the European Union has addressed spatial planning issues through the European Spatial Development Perspective (ESDP) since 1999. This Perspective provides EU countries with a model for spatial development, with special emphasis on urban areas (RPD 1999). However, no issues dealing with the subsurface were included. Following an inquiry at the European Commission (COB/RAVI 2000), it became apparent that at European level:

- no legislation exists on building in the subsurface or on transport through the subsurface
- spatial planning regulations belong to the specific responsibilities of the individual countries.
- no plans exist for legislation of subsurface activities yet, except for ownership of the subsurface.

In the USA, the Federal Land Policy and Management Act passed in 1976. For administrative and protective purposes, this law placed federal lands and their resources under the jurisdiction of the Bureau of Land Management (BLM). The land was to be retained and managed for multiple use, unless the transfer of certain rights will serve a national interest (BLM 2001). This means that over time the land resources (exploitation potential for minerals, forestry, nature reserves, etc.), should be kept intact. Only if for a given area the opportunity arises, rights may be transferred depending on third party interests and BLM policies. The Bureau manages the Federal lands, which cover approximately one third of the USA as a result of historical development since the Revolutionary War. Large tracts of land were at that time purchased from other nations, including Alaska that was bought from Russia in 1867.

Basically, spatial planning is still a two-dimensional exercise. As the subsurface is becoming an increasingly relevant land-use issue, integration of the subsurface in spatial planning and developing three-dimensional (3D) land use legislation will soon be inevitable. Some countries and cities (The Netherlands, Montreal in Canada) have already developed policies to incorporate the subsurface in their spatial planning (see Chapter 3 and Section 6.2).

Table 6.1 contains an overview of the legal system and the type of landownership including possible depth restrictions for countries where legislation relating to these aspects has been identified. Although it is not claimed that Table 6.1 is complete, the table shows that relatively few countries have included depth limitations and/or functional restrictions in legislation.

6.1.2. Natural resources, subsurface construction

This section discusses legislation concerning natural resources and (subsurface) construction. In legislation, most countries make a clear distinction between subsurface functions of significant economic value (e.g. minerals, oil and natural gas) and more socially oriented functions (e.g. storage and spatial planning). Due to its significance to national economies, development of natural resources is normally organized at national/federal levels. Mining legislation developed simultaneously in many countries in the first decades of the 20th century, but in some nations such legislation dates even back to Napoleonic times (B&A Groep 1997a, b). Three main systems can be identified with respect to the ownership of natural resources (Johnson 2010). These are the land ownership system, the concession system and the claim system. In the concession system a national authority grants exploration and exploitation rights. In a claim system the first discoverer can obtain exploitation permits. Many countries updated their mining legislation between 1990 and 2000 to comply with new environmental legislation.

6.1.2.1. Mining law, extraction of construction materials
Principle 21 of the Stockholm Declaration (UNEP 1972) reads:

> States have, in accordance with the Charter of the United Nations and the principles of international law, the sovereign right to exploit their own resources pursuant to their own environmental policies, and the responsibility to ensure that activities within their jurisdiction or control do not cause damage to the environment of other States or of areas beyond the limits of national jurisdiction.

No comprehensive international law on mining exists but most countries do have several components in common. The starting point mostly is the mining code (or law) with its basic principles and policies to mineral exploration and exploitation. But laws on labour, safety, land, water, environment and finances are also part of regulatory systems relevant to mining, influencing mining operations, determining processing of mine tailings and sharing profits with the government (Otto 2000). Concerning mineral extraction and control, differences in mining legislation often relate to local geological conditions, land use patterns, population density, topographic and climatic conditions. Some of the main issues and differences are highlighted hereafter.

Non-metallic minerals, including construction and building materials such as stone, sand, gravel, and clay, are often privately-owned. In contrast, metallic and other mineral resources (including oil, gas and coal) are normally State-owned. There are very few exceptions, as for example in Finland where the Mining Law has no provisions for exploration and exploitation of oil, gas and coal. A Finnish citizen or company has the right to exploit natural resources even under another property (with restrictions). The concession, however, has to be approved and compensation must be paid to the landowners. In case of State-owned resources,

Table 6.1. *Overview legal landownership systems inclusive depth restrictions*

Country	Legal system	Landownership
		Private–Public
Europe		
Austria	Civil	Private, theoretically down to centre earth (unlimited)
Belgium	Civil	Private, theoretically down to centre earth (unlimited)
Czech Republic	Civil	Private, only indispensable space above and below the surface
Denmark	Civil	Underground below −6 m is state owned. Use by concession right
England & Wales	Common	Private (fee simple)[1], theoretically down to centre earth (unlimited)
Scotland	Common+civil	Private, theoretically down to centre earth (unlimited)
Finland	Civil	Private until −6 m.
France	Civil	Private, theoretically down to centre earth (unlimited)
Germany	Civil	Private, theoretically down to centre earth (unlimited)
Hungary	Civil	Private limited to the surface of ground, but incl. underground parts of building
Italy	Civil	Private, theoretically down to centre earth (unlimited). In practice down to the depth of actual interest in usage. Outside this zone state owned
Ireland	Common	Private (fee simple), theoretically down to centre earth (unlimited)
Norway	Civil	Spatial zoning plans extend into subsurface as far as usage is allowed. Private, no surface developments deeper than 6 m without concession
The Netherlands	Civil	Private, theoretically down to centre earth. Others may use the subsurface at large depth, unless interests of owner are impacted
Sweden	Civil	Private, theoretically down to centre earth (unlimited), local authorities may impose restrictions
Switzerland	Civil	Private, theoretically down to centre earth (unlimited), but as far as plausible interests apply
Asia		
Australia	Common	Public
		Private (fee simple)
		Customary (aboriginal)
China	Civil	State owned, since 1990 land leasing
		Private land use rights
New Zealand	Common	Customary (aboriginal)
		Private (fee simple), theoretically down to centre earth (unlimited)
North Korea	Civil	State owned
Mongolia	Customary, since 1990 Romano-Germanistic legal system	Law on land, freehold ownership
Papua New Guinea		Customary
Russian Federation and Central Asian Republics	Civil	Subsurface owned by state
Vietnam	Civil	State owned
		Private exploitation rights
Africa		
Republic of Congo	Civil	Private. Land Law July 1973
		President has jurisdiction over granting surface rights
Ghana	Common	Customary
Kenya	Common	Private property ownership is protected by the constitution
Malawi	Common	Customary
South Africa	Common+civil	Private, theoretically down to centre earth (unlimited)
Uganda	Common	Customary
Latin America		
Cuba	Civil	State owned
Venezuela	Civil	Private, theoretically down to centre earth (unlimited), except when natural resources are discovered

(Continued)

Table 6.1. *Continued*

Country	Legal system	Landownership
		Private–Public
North America		
Canada	Common	Public
		Private (fee simple)
		Customary (Indian)
...Quebec	Common+civil	Public
		Private (fee simple)
		Customary (Indian)
United States of America	Common	Public
		Private (fee simple), theoretically down to centre earth (unlimited)
		Customary (Indian)

Note: [1] The greatest possible estate in land, wherein the owner has the right to use it, exclusively possess it, commit waste upon it, dispose of it by deed or will, and take its fruits. A '*fee simple*' represents absolute ownership of land, and therefore the owner may do whatever he or she chooses with the land. If an owner of a fee simple dies intestate, the land will descent to the heirs (West 2008).

the State may grant exploitation to private parties in return for a fee. In Hungary, development of oil, natural gas, black coal, metals and other mineral resources is designated to State-owned organizations only. The necessity to attract foreign capital investments for exploration and exploitation in China resulted in the Mineral Resources Law in 1986 (amended in 1996). Regardless of a State Council Decree in 2000, allowing fully privately-owned mining companies to explore and exploit mineral resources. The lack of clarity in the legal arrangements guaranteeing these mining rights kept foreign investors away. At the other end of the spectrum of state-owned v. private mining enterprises, are the USA, South Africa and Iran, where 90% or more of the exploitation rights are held privately.

In developing natural resources three stages may be distinguished: exploration, exploitation and closure/decommissioning. In mining legislation, these stages are often treated differently from country to country. Exploitation of natural resources, however, is usually only allowed through governmental permits or licenses. In some countries exploration rights are granted for an indefinite period of time (e.g. Peru and Bolivia), in others for periods from 12 months up to 3 years (e.g. Chile, Ghana, Japan). In a number of Asian countries the average maximum exploration time ranges from 6 to 8 years (Naito *et al.* 1998). Exploration rights for a given area are mostly granted to a single party at one time. In Thailand, more than one entity may simultaneously obtain prospecting rights for a given area.

For exploration to continue into mining rights, strict procedures often have to be followed by submitting information on the mining itself as well as on its potential environmental impact. The latter has become increasingly restrictive on mining in many areas. In general (as is the case in many Asian countries (Naito *et al.* 1998), the party holding exploration rights has priority over others in applying for a mining permit. In exploration, more and more socio-economical aspects relating to local communities have to be considered in addition to compensation to landowners and land users for damage caused by the mining operations.

In the Philippines, it is compulsory by law to include local people in mining operations. In Australia, Aboriginal owners have been granted overriding rights on mining within their reserves, although the Australian Government maintains control over mining in National Parks and World Heritage Areas.

Exploitation rights for the actual mining are granted in most countries for specific periods, ranging from 25 years (often with possibilities for renewal) to indefinite (Naito *et al.* 1998; Bastida 2002). EU Member States have a long tradition of centralized control over mining and quarrying activities, normally managed by their Departments of Trade, Industry or Economic Affairs.

In the USA, the Bureau of Land Management (Department of the Interior) controls the Mineral Rights in the extensive areas of federal public land (approximately 1/8 of the total USA land area). Here different laws deal with different types of minerals:

- The General Mining Law of 1872, for 'locatable' minerals like precious metals and gemstones.
- The Mineral Leasing Act of 1920, for black coal, and industrial minerals as phosphate and potassium, and the Federal Coal Leasing Amendments Act of 1977 for leasing black coal exploitation rights. Mining claims do not cover leasing of all minerals and there are geographic differences as well. This type of exploitation of public land is available in 19 States, and exclude National Parks and Indian Reserves. As to the type of minerals, one may, for example, lease sulphur exploitation rights only in Louisiana and New Mexico. Leases are normally issued by competitive bidding.
- The Materials Act of 1947, for sand, gravel and other construction materials. One may obtain the right to mine these materials through permits. Government agencies, non-profit organizations, churches, and scouting

organizations may get free permits for non-commercial purposes.

Normally, lease cost consists of three components: an initial bonus, rent over the lease period, and a royalty fee related to the value of the mined material. Permits for the production of construction materials, including regulations for site closure, are normally granted at local governmental levels.

A special case in natural resources deals with the continent of Antarctica. In 1959, the Antarctic Treaty was signed as a legal framework suspending territorial claims for 30 years. Seven countries claimed mineral rights for the Antarctic. They signed an agreement in 1988 on the Regulation of Antarctic Mineral Resources Activities for a controlled approach to mineral exploration activities with the 26 Antarctic Treaty nations. However, that agreement became obsolete by the successive Madrid Antarctica Treaty, signed in 1991, putting a 50-years ban on minerals and, oil and natural gas exploration and mining in Antarctica. In October 2007, the United Kingdom submitted a territorial claim to the United Nations, followed by Chile and Argentina. Other countries with claims are Russia, Brazil, Australia, Ireland, New Zealand, France, Spain and Norway. Table 6.2 shows an overview on legislation related to natural resources, specifically relating to mining law and extraction of construction materials as has been identified in countries around the world. Due to their properties, mining of some minerals generated very specific legislation. Health risk evidence forced Japan to stop asbestos mining in 2002; in Australia, asbestos mining was outlawed in 2004.

Table 6.2. *Overview on legislation for natural resources, specifically relating to mining law and construction materials*

Country	Natural resources
	Mining Law, Construction Materials
Europe	
Belgium	Oil/gas/coal/metal/mineral resources exploitation according to specific laws
Czech Republic	Mineral resources state owned
Denmark	Oil/gas/coal state owned, state permissions required for exploration and exploitation
	Permission required from State to explore, extract, utilize sand & gravel to protect existing resources and environment
England & Wales	Land owner has common right of support
	Mineral rights can be severed from surface ownership
	Oil/gas property of the Crown based on 1934 Petroleum (product) Act. All coal and mines of coal state property based on 1938 Coal Act. Extraction may be licensed
	Surface owner owns other metals and minerals except gold, silvers and uranium which belong to the Crown
Scotland	Land owner has common right of support
	Mineral rights can be severed from surface ownership
	Oil/gas property of the Crown
Finland	Finnish Mining Law. No provisions for exploration/exploitation oil, gas & coal.
	Citizen/company has the right to exploit metal/mineral resources even under another property (with restrictions). Concession to be approved and compensation to be paid to land owner
France	Oil/gas/coal/metal/mineral resources exploration/exploitation need authorization and concession
	Exploitation of construction materials and minerals only require authorization for extraction, not a concession
Germany	Federal Mining Law. Oil/gas/coal not property landowner
	Landowner can apply at any time for permit to exploit metal resources
Hungary	Development oil/gas/coal/metal/mineral resources only by state owned organizations
	Surface owner may develop building material resources for own use
Italy	Mineral deposits state owned
	Quarries belong to land owner, exploitation must be authorized by regional government
Norway	Subsurface oil/gas state owned property
	Coal is property of landowner

(Continued)

Table 6.2. *Continued*

Country	Natural resources
	Mining Law, Construction Materials
The Netherlands	All citizens may exploit metals. Minerals are property landowner. Alluvial gold is property of the State
	Mining Act 2002 applies down to depths >100 m below ground level for mineral resources and >500 m for geothermal energy. Mineral resources are owned by the State
Sweden	Minerals Act 1992. Mining law regulates claims and concessions for metal exploitation. State share can be required depending upon cost/income. For minerals permission for exploitation required, must obey environmental laws. Parliamentary sanction needed for oil/gas/coal exploitation
	Peat Act (energy)
Switzerland	Concession to be obtained from Canton for exploitation of oil/gas/coal/metal/mineral resources
Asia Pacific	
Australia	Oil/gas/coal/metal/mineral resources control exploitation in Common Wealth Government
	Aboriginal owners have been granted overriding rights on mining within their reserves. Australian Government controls mining in National Parks/World Heritage Areas
	Natural resources management (financial assistance) Act 1992
	Mining Act 1978
Burma (Myanmar)	Myanmar Mines Law 1994
Cambodia	Mines and Mineral law (under consideration)
China	Mineral Resources Law Jan. 1st 1997
	Resources state owned
	Random excavating and mining of mineral resources prohibited by Environmental protection law
India	National Mineral Policy, 1993
	Distinction of minor minerals
Indonesia	Mining Law No. 11, 1967
Japan	Mining law applies to minerals or resources of economic value. Surface owner should accept development of mining rights, but has priority on surface and shallow subsurface minerals without economic value are at disposition of owner land property rights
Laos	Mining Law 1997
Malasiya	State Mineral Enactment (under consideration)
Mongolia	All natural resources state owned. Law on underground resources. Law on mineral resources (mining)
	Automatic progression from exploration to mining right
New Zealand	Minerals Act 1991, Resource Management Act 1991
	Mining Regulations 1981
	Mining Act 1971
Pakistan	Except oil, gas and nuclear, minerals regulated at federal level, Minerals are a provincial subject, under the constitution of Islamic Republic of Pakistan
	National Mineral Policy in 1995
Philippines	Philippines Mining Act 1995
	Indigenous Peoples Rights Act 1997 requires legal involvement of indigenous local communities in mining projects
Russian Federation and Central Asian Republics	Mining laws from the 90's in some cases Republics have priority right to purchase raw minerals. Granting rights through tenders and auctions (exc. Turkmenistan) Exploration, development based on licensing (expl. 2–6 years; Development time based on depletion time of reserve)
	Kazakhstan on contracts
Kazakstan	Underground Resources Law and Amendments, 1996
Kyrgyzstan	Law on Entrails of the Earth, 1997
Uzbekistan	Law on Mineral Resources 'the Subsoil code', 1994
Tajikistan	Mining Code (under consideration)

(Continued)

Table 6.2. *Continued*

Country	Natural resources
	Mining Law, Construction Materials
Thailand	Minerals Law, 1967. More than one entity can obtain prospecting rights over an area
Turkey	Mineral resources exclusively owned by the State. Mining law 1985, amended 2004. Natural gas and crude oil regulated by Petroleum Law 1954. New law 2007 not yet approved by President
Vietnam	Mineral Law Sept 1 1996
Middle East	
Iran	Mining Regulation
Saudi Arabia	In 1997, in order to coordinate projects and promote efficiency in the mining industry, the Saudi Arabian Mining Company (Maadin) responsible for regulating mineral exploration and overseeing its progress was created
Africa	
Angola	Exploration rights with government, especially oil and diamonds
Cameroon	Mining Code 164 following French law
Central African Republic	All mineral resources in the ground or at the surface – including groundwater and geothermal waters – are the property of the State
Chad	No separate mining legislation
Republic of Congo	Civil war strongly impacted situation. Mining Code 2002, president has jurisdiction over enactment. Exploration and exploitation rights standard construction materials granted at provincial level
Egypt	Egypt's mineral resource development is co-ordinated by the Egyptian Geological Mining and Survey Authority
Kenya	Mining and Minerals Act
	The minerals belong to the country
Madagascar	1990 Mining Code made provisions for three types of mining permits. Exploration and exploitation type 1 permits were granted only to individuals or groups of Malagasy nationality and were valid for 2 years. Types 2 and 3 permits, valid for 3 and 5 years, respectively, are designed for small to large mining companies incorporated under Malagasy law
Mali	Mainly gold mining. National directorate for geology and mines.
Morocco	Morocco's mineral resource development is controlled by several state owned organizations
Mauretania	1999 Decret portant sur les Titres Miniers Mainly iron ore
Namibia	All mineral rights are vested in the state. The Minerals (Prospecting and Mining) Act of 1992
Niger	Uranium and gold
	Prospecting, exploration and mining permits issued by government with time limits
Nigeria	The Mines Manual – containing the Mineral Act (Cap. 121 of 1946). The Quarries Decree No 26 of 1969 and Quarry Regulations 1969. Diamond Trading Decree – Decree No 55 of 1971
	New Mining Law will replace these once approved by the government. All solid minerals are owned by the Federal Government
Rwanda	Civil unrest impacts activities
Zambia	Mines and Minerals Act (1995)
	Government policy is not to participate in exploration or other mining activities or any shareholding other than regulatory and promotional role
Zimbabwe	The Mines and Minerals Act 1961 and a number of amendments have been made since then. All minerals are vested in the President and one requires rights to work mineral deposits through an application to the Mining Commissioners. Mining activity is open to both local and foreign individuals and companies
Latin & North America	
Canada	Public Lands Mineral Regulations, 1995 ('mineral' means any naturally occurring homogeneous constituent of the lithosphere, but does not include petroleum, natural gas or related hydrocarbons, soil, peat, groundwater or non-hydrocarbon gases; (minéral))

(Continued)

Table 6.2. *Continued*

Country	Natural resources
	Mining Law, Construction Materials
United States of America	Department of Natural Resources Act, 1994
	Resource Conservation and Recovery Act
	Federal Land Policy and Management Act
	Oil/gas/coal/metal/mineral resources private companies may exploit resources under government controls after agreement has been reached with landowner
State of Louisiana	Mainly private exploitation rights

6.1.2.2. Groundwater

Groundwater is a vital resource for drinking water, agriculture, industry and as a source for surface water. Under Roman Law, applied as a model for most Western European, African and Latin American countries, landowners had the exclusive right to use the groundwater below their property. They could thus 'mine' it to the extent they wished. Where judges believed that groundwater flowed like a river, they applied 'Riparian' rights as well (riparian means land on a riverbank, thus neighbouring land). Landowners bordering such groundwater streams could take whatever volumes they wanted, but they had to respect the needs of downstream users. In the USA, landowners are entitled to use groundwater for their lands only, but have to avoid unnecessary spillage and are not allowed to use it beyond their land areas (Burchi 1999).

Since access to water supply is so vital to societies, disputes on groundwater rights are quite common, in particular in arid and semi-arid areas where water resources are scarce and can be easily over-utilized. A special case is Israel and its neighbouring countries, where groundwater was specifically addressed in the Peace Treaty (1994) and in the Oslo II Agreement (1995) (Shamir 1998). In these treaties, transborder water systems and common projects are involved. Not only water quantity, but also environmental and water quality issues are addressed. Under these agreements one country (Israel) was even allowed access to the territory of another country (Jordan) for abstraction of groundwater. Israel was also permitted to increase pumping rates with the stipulation that existing Jordan groundwater uses should not be negatively impacted. A permanent Joint Water Committee was set up to implement the agreement. In Israel, water resources are public property subject to control by the State, based on the Water Law of 1959. The same applies for Spain (since 1985) and Italy (since 1994). When State ownership of groundwater was challenged, courts in the USA and Spain almost consistently ruled in favour of the common good (Burchi 1999). In contrast to the original Roman Law principles, in many countries today's ownership of land does not include the exclusive right to use groundwater any longer (Zhou 2006). In some Muslim countries, private ownership of groundwater is not allowed as groundwater is considered a 'gift of God'.

Early groundwater legislation addressed quantitative aspects only, but significant attention was paid to groundwater quality and ecosystem issues in the past few decades. That, in turn, resulted in a considerable increase in environmental legislation.

In the European Union, five (ground-) water-related Directives affect national legislation in the 27 EU Member States (Fig. 6.2):

(1) Groundwater Directive (89/68/EEC) for the protection of all groundwater from listed pollutants (1979). This Directive will be abandoned in 2013 due to the Water Framework Directive (see under 3);

(2) Council Directive (98/83/EC) for the quality of water for human consumption;

(3) Water Framework Directive (2000/60/EC) for the protection of all waters in an overarching system at river basin scale, and maintaining or aiming at a good quantitative and qualitative status in water bodies (including groundwater) within a set time frame (2013). Its main goal is to reach a good quality ecological status. For groundwater the quality status is solely expressed in chemical terms by means of listing chemical compounds and maximum acceptable concentrations instead of ecological functioning as in surface water;

(4) Groundwater Daughter Directive (2006/118/EC) for the prevention of pollution and focused on the chemical status of groundwater bodies. It requires monitoring and assessment of the groundwater quality and aims at the reversal of persistent positive pollution trends. This Directive is the first that comprehensively addresses groundwater bodies shared by two or more countries. Outside Europe, cross-border problems in groundwater are normally still neglected;

(5) Nitrate Directive (91/676/EEC) for reducing water pollution caused, or induced by nitrates from agricultural sources and aimed at combatting the extension of pollution (1991). One of the issues in this Directive is the identification of zones prone to nitrate pollution. The entire territories of The Netherlands, Denmark and Germany belong to that zone, in contrast to the United Kingdom where only a limited area is considered vulnerable to nitrate pollution.

Fig. 6.2. European Parliament hemicycle (© European Communities).

Requirements for these EU Directives overrule earlier, national legislation and must be implemented in the national legislation in all EU Member States. Multiple legal requirements may jeopardize such implementation.

In the European Union, legislation may also require the national or local authorities to undertake an Environmental Impact Assessment as part of the permitting procedure. In France, Detailed Water Plans ('Schéma Directeur d'Aménagement et de Gestion des Eaux', SDAGE) are agreed for specific aquifers or sub-basins. These plans even have a retrospective effect on existing groundwater extraction permits. The Detailed Water Plans may have overriding effects in which legal action always results in withdrawal of the permit even if predating the Plan. Moreover, nations may have more generic or more specific types of groundwater legislation, vis-à-vis the importance of groundwater for their drinking water supply and water resources management in general. Examples of national legislation on groundwater are the Austrian Water Act 1959 (amended 2003 to incorporate the EU Directives), France's Water Act (1992), Spain's 1985 Water Act, and the Dutch Water Act (2009) that integrates the management of groundwater and surface water in a single Act.

Groundwater legislation varies significantly from one country to another and, for example in the USA and India, on State level as well. In the USA, at least nine Federal Acts exist with implications for groundwater use and protection, including the Safe Drinking Water Act of 1974. Amended in 1986 and in 1996, this Act is meant to protect public health by regulating drinking water supply from source to tap. Privately owned wells, serving less than 25 individuals or with less than 15 service connections, are excluded from this legislation that authorizes the Environmental Protection Agency (EPA) to set water quality standards. In the USA, the EPA has delegated supervision of water systems in almost every State. As in France, legislation passed in Texas in 1997, resulted in water plans which also have binding effects on permitting that did not exist before.

Examples beyond the European Union and the USA include the Victoria Water Act of 1989 (Australia), Jamaica's 1995 Water Act and the South African National Water Act (1998) (Burchi 1999). Other countries, for example Nepal, are in the process of adapting legislation to incorporate groundwater resources more properly (Timilsina 2004).

In China, the Water Law (1988) specifically mentions in article 26 the need for remedial measures and compensation for economic losses resulting from lowering of groundwater levels due to mining and underground construction projects leading to subsidence or loss of productivity.

In general government permits for groundwater withdrawal, mining activities, or for the installation of groundwater wells are often required. Normally, such permits are provided by the national or local water authorities, unless only small quantities (generally less than 10 to 40 m^3 day^{-1}) or during short time intervals are to be extracted. Exceptions may apply for areas prone to settlement and construction damage or where archaeological remains can be jeopardized by degradation caused by oxidation through falling groundwater levels. There, permits are also required for such minor withdrawals.

Worldwide management of groundwater resources requires more attention, in particular as availability of clean (and drinkable) water is linked to human health and quality of life. The European Commission presumes that in 2025, 3 billion out of 8 billion people will have inadequate access to clean water (EC 2009); many of them living in megacities, in particular in Asia.

Groundwater extraction may impact the environment in many ways, for example, by lowering the groundwater table causing drying out of the natural environment on the surface, or causing subsidence that results in damage to constructions. Therefore sustainable development of water resources with proper legislation is required to manage conflicting interests, not only with nature but also within society where different user groups may have different interests, for example, drinking water supply companies v. house owners.

6.1.2.3. Subsurface structures and infrastructure

As in the case of mining, the subsurface is physically penetrated in case of the construction of subsurface structures, including tunnels, underground parking garages and storage facilities. Infrastructure also includes small infrastructure, such as cables, pipe lines and sewerage-systems. In the planning- and construction-phase and during operation, clarity is needed on rights and responsibilities. For example: what are the rights and obligations of a landowner regarding subsurface structures, even if he has no direct access as in the case of traffic tunnels? Should these rights and obligations be split; and if so, how? Should separate ownership entities in three-dimensional units in the subsurface be created? Who is responsible if an incident in a subsurface property impacts the surface or adjoining landowners or vice versa, that is, an incident on surface impacts the subsurface structures?

Unclear or lack of proper legislation may hamper underground development as it has done in the past. Individual parties involved may not be obliged to cooperate and/or willing to accept responsibilities or risks in realizing new developments. Individual rights compared to common interests by society can be at stake which may be dealt with in expropriation legislation.

In the European Union, legislation for specific issues exists for tunnels of the Trans European Road Network (EC 2010). In 2004, a European Union Directive was adopted on minimum safety requirements for longer (>500 m) tunnels (EC 2004a). By 2002, some 500 such tunnels were identified in the wider European area, most of these in Italy, and probably expanding to 600 to 700 in 2010 (EC 2002). Amongst others, safety requirements are related to infrastructure elements, as single or twin tubes, escape routes and emergency exits, drainage and ventilation.

For building underground constructions often many authorities must be consulted for collecting the required permits, some of which are overlapping or even conflicting (ITA 2000). Attempts to coordinate the permitting procedure in an urban setting have been more or less successful in some countries, for example:

- In Helsinki (Finland), multi-purpose utility tunnels carrying water, electric power, heating and communication lines were constructed between 30 and 80 m under the city, with a total length of 40 km. The tunnel system is 5–7 m wide and accessible to vehicles for inspection, services and repair (Chow 2002). A comparable, but shallower solution was applied in a development area in Amsterdam, The Netherlands (Fig. 6.3).
- In The Netherlands, many occurrences of cables and pipes are registered (see also Section 5.3.3.1.). Following some serious disruptions, legislation was developed to ensure registration and information exchange on underground networks. The Cables and Lines Information Centre (KLIC 2012) was established to register all activities related to cables and pipes at a national level. KLIC has legal status and is part of the Dutch Cadastre as a result of the Underground Cables and Pipelines Information Exchange Act (Wet Information-uitwisseling Ondergrondse Netten (WION)) that came into force in 2008. The Spatial Planning Act of 2008 also incorporates subsurface infrastructure, including safety zones for pipelines in spatial and zoning plans. It is estimated that the total length of pipelines in The Netherlands is 300 000 km, 18 000 km of which is used for transport of dangerous substances.
- In Hong Kong, China, contractors are held responsible for uninterrupted services of utilities and are charged for re-routing costs in construction projects. As the presence and precise location of utilities are often unknown, on-time realization of projects may be problematic causing severe financial implications to the contractor. Lack of proper registration of such utilities in the past did generate debates on the fairness of this responsibility (Fung 2002).

Constructing multipurpose utility tunnels may reduce the risk of service disruptions and nuisance to the general public from maintenance and reconstruction. Figure 6.3 shows an example.

Normally, local authorities issue building permits for (underground) construction as a constituent of national legislation. But sometimes regional or even national authorities can play a role in issuing permits as well, depending on the size of the construction. Quite significant is the role of international (e.g. EU) legislation but also recommendations and guidance by organizations like UNESCO and the UN on local permitting. An interesting example is the European Agreement on the protection of archaeological values, the Malta

Fig. 6.3. Cable and pipeline tunnel in Amsterdam (The Netherlands) (photographs: E. De Mulder).

Convention or Valletta Treaty (Council of Europe, 1992, see Chapter 3.3.3). This Treaty aims to preserve archaeological heritage sites. Already in 1956, UNESCO published its Recommendation on International Principles Applicable to Archaeological Excavations, urging for international co-operation for the protection and for establishing general principles for archaeological heritage. Another such Agreement was reached at the American nations' Convention of San Salvador in 1976. Today, many nations have national legislation, procedures and regulations in place for archaeological excavations, for example, the National Historic Preservation Act (NHPA), in the USA. That Act resulted in the creation of the National Register of Historic Places listing all cultural sites in the USA worth preserving.

The Valletta Treaty and similar national Acts require proper surveying of the construction site prior to construction. This often results in a wealth of archaeological findings. On the negative side, construction activities should be postponed until the archaeological excavations are completed. That may create major problems in urban areas with a long history of human occupation. That was the case, for example, in Athens (Greece), where a metro line including underground stations was constructed for the 2004 Olympics (Athens 2007). To avoid problems, tunnels were dug while

Table 6.3. *Countries identified with legislation addressing subsurface construction*

Country	Subsurface construction
Denmark	Danish subsurface law. Right to develop subsurface space is given by the State, usually with time-limited concession.
Hungary	Only state owned firms may have permission for underground development except for garages, basements, cellars, metro, utilization of caves, etc.
The Netherlands	The zoning plan based on the Spatial planning Act should not exclude the construction purpose. Housing Act requires permit for construction of a building (this includes subsurface construction)
Norway	In urban areas not allowed to build surface structures deeper than 6 m.
Sweden	Environmental code. Swedish building code will establish standards especially related to negative impacts on groundwater table. Largest cities implemented registration of existing underground installations. Planning and Building Act 1986, requires building permit for most subsurface installation. Construction of tunnels facilitated by specific rules whereby 'a right of disposition of subsurface space can be awarded through other owners' estates.
Venezuela	Underground apartments will not be approved.
China	Minimization of disturbance of vegetation and area designation for surplus material required based on Soil and Water Conservation Law. Water Law requires remedial measures and/or compensation of losses resulting from mining, subsurface construction due to subsidence and other impacts.
Japan	Special Measures Act for Public Use of Deep Underground. Subsurface space owned by owner of the mining rights.

the abundant ancient remains on top were saved. For archaeological reasons the location of the Monastiraki metro station was even moved, which was an important reason for the two year construction delay (Leoutsakos 2005).

As stated before, specific legislation regarding subsurface construction as such is rare. Table 6.3 contains information for countries where legislation exists to some extent. In most countries, however requirements for subsurface construction are derived from building- and mining acts.

6.1.3. The environment

Environmental legislation covers a wide area from protection to regulating the effects of human activities. It has had a major impact on existing and sometimes ancient rules and customs, in particular concerning natural resources. Worldwide legislation to promote environmental protection has been developed, some very generic and some even at the level of the Constitution, others more specific and focused on particular activities. Initially, policy makers have split the environment artificially into three compartments: air, water and ground (i.e. soil and/or rock), thereby ignoring the integrated, cyclical character of natural processes and their driving geological forces. That phase in legislation reflected political and economic sensitivities of those days and resulted in a distinct but rather conservative approach. In this section legal aspects specifically directed to environmental issues such as soil protection, (subsurface) waste disposal and storage, remediation and reclamation are discussed.

Since it was recognized that human activities may have significant impact on the environment, Environmental Impact Assessment (EIA) was developed as a tool to describe such potential impacts. EIA was first introduced in the USA through the National Environmental Policy Act of 1970. Today, legislation requires EIAs for new projects almost everywhere in the world, however, the exact requirements and legal implementation may differ from place to place. Conclusions and recommendations following from EIAs have mandatory legal status in some countries whereas in others they just form advice that can be followed (or not).

The European Committee drafted its first Environmental Action Plan in 1972. Since adoption, some 250 items of legislation, mainly to limit pollution by setting minimum standards for waste management, and water and air pollution. Initially, this resulted in quite a few unbalanced decisions on clean-up activities. A coherent, overall vision on how to manage environmental issues developed only slowly. One of the main reasons for this was that in the past the European Union legislators favoured a 'level playing field' standard, insisting for example on accepting single maximum values allowed for arsenic, nickel or acidification in water and soils across the entire territory of the European Union. This was done despite of the vast area and the geological variety resulting from the natural chemical and physical geodiversity in the subsurface. The EU rarely touched soil and subsurface contamination issues until 2002 for reason of the subsidiarity (or 'localization') principle, as these were considered to be of national interest only (see e.g. Table 6.4 for a list of EU-countries with legislation on subsurface protection). The subsidiarity principle states that the central government should leave decisions to the local level if this would be better suited for decision making on a particular topic.

Table 6.4. *Regulation on subsurface protection in various EU countries (modified from Vermeulen 2002)*

	Yes	No
Integrated law on protecting the quality of the subsurface	Denmark, Germany, United Kingdom, The Netherlands, Italy	Sweden, Finland, Ireland, Luxemburg, France, Portugal, Spain, Greece

Between 1973 and 1991, a large quantity of EU Directives on environmental issues was adopted. Several of the Directives split the same or related issues into individual air, water and soil compartments, while producing myriads of comparable legislative texts, including for example more than 60 pieces of onshore water legislation. The overall environmental policy in the European Union comprises the following principles (Article 174 EG, 2nd paragraph): high protection level, prevention, source-based approach, and the 'polluter pays' principle.

In 1998, increased public concern about the environment resulted in adoption of the Århus Convention, urging public authorities to make environmental information for decision making available to the general public. This UN/ECE Convention on access to information, public participation in decision-making and access to justice was signed by all EU Member States and several other nations.

By 2012, still no integrated EU policy on soils or ground exists but several relevant regulations are in place, including the:

- European Framework Directive Water and the 'daughter' Directive on groundwater (surface water and groundwater)
- Regulations regarding disposal (waste disposal, sewage sludge)
- IPPC Integrated Pollution Prevention Control guideline
- Environmental Impact Assessment guidelines
- Environmental Liability Directive

Moreover, an EU Soil Framework Directive was proposed (EC 2006a, b) and discussed by the European Parliament. Soil threats to be addressed include erosion, loss of organic matter, compaction, salinization, landslides, contamination and sealing. A first opinion was adopted by the end of 2007, but no political agreement was reached to date (early 2012) yet.

It is not just the existence of environmental legislation that protects the environment and reduces the potentially adverse impacts that development of the subsurface may have. Enforcement is essential to ensure effective implementation. Beside penalty-systems included in existing legislation, an Environmental Liability Directive (2004/35/CE) was adopted to establish a common legal framework in which 'the polluter pays' principle to prevent and remedy environmental damage is implemented. Environmental damage to land (including the subsurface) is covered by this Directive.

The European Commission and Parliament concluded that the existing instruments did not suffice given an increase in environmental offences (EC 2008). Therefore, an Environmental Crime Directive was proposed and adopted in October 2008 to bring to court intentional offences or cases with serious neglect under criminal law. This Directive had to be implemented in the national legislation systems of EU Member States by the end of 2010.

6.1.3.1. (Subsurface) waste disposal and storage

In the European Union, separate legislation addresses different types of waste and the medium into which the waste is disposed or stored. As a rule, underground discharging of (liquid) waste is prohibited since 1979, but under certain circumstances permits are granted in the EU. The European Directive 99/31 recognizes negative environmental impact of storing waste materials in the subsurface. Therefore, the Directive aims to diminish total mass of waste produced in the EU. That may be realized by waste prevention and reuse stimulation through separation in recyclable categories (paper, glass, organic waste) on the one hand, and by taking soil protection measures through barriers and percolating water (leachate) collection systems on the other. The Directive 2000/76/EC on waste incineration also aims to prevent or limit emissions into the environment, including soil and groundwater.

In 2006, a Directive 2006/21/EC on waste resulting from mineral resource production and operation in quarries was adopted (EC 2006c). The earlier EC Directive 99/31 addressing disposal waste is no longer applicable to this category of waste. The new Directive requires suitable measures to be taken and procedures and monitoring to be put in place preventing or reducing adverse environmental effects and risks to human health from waste management of extractive industries by the operator. Competent authorities should also have an emergency plan in place for off-site measures in case of an accident. The Directive covers full-cycle waste management in extractive industries, from prospecting, extraction, treatment and storage of mineral resources to quarry operation, but excluding waste from offshore activities and activities related to injection of water and re-injection of pumped groundwater.

The combination of the Environmental Protection Law, the Soil Protection Law and the Dutch Directive on Soil Protection, together with local development plans, regulate waste disposal in dumpsites in The Netherlands. Recently, underground waste collection systems and temporary storage facilities reducing odours and erratic waste and optimizing treatment logistics, managed by local authorities, became operational in some Dutch cities (Fig. 6.4).

Since the 1986 nuclear power plant accident at Chernobyl (Ukraine), underground disposal of dangerous chemical and nuclear waste (mostly deep geological disposal) was reconsidered resulting in effect in a worldwide moratorium on building new nuclear power plants in Europe and the USA. That debate became more focused on policy matters rather than on legislation. Increased public concern on CO_2 emissions by conventional plants revitalized discussion on installing (almost) CO_2 free nuclear power production since 2006. The devasting effects of the 2011 tsunami in Japan and the Fukushima Daiichi nuclear incident resulting from it again put a halt to this discussion.

In the USA, the Resource Conservation and Recovery Act (RCRA) of 1976 and the Hazardous and Solid Waste Amendments of 1984 regulate treatment, storage and disposal of hazardous waste. Furthermore, the Safe Drinking Water Act (1974) sets the framework for underground injection of wastes into groundwater. The Environmental Protection Agency (EPA) is responsible for implementing this programme, but frequently delegates this responsibility to individual States. As injection normally occurs deep in the subsurface, this issue will not be addressed here. But in Florida, large quantities of wastewater are injected at depths less than 250 m, as no suitable active flowing surface water systems (rivers and the like) are existing (Keith *et al.* 2005).

Energy-related storage capacity of the subsurface gained interest since the Kyoto Treaty (1997). Legislation plays a role in exploring energy-efficient technologies, for example related to heat and cold storage for buildings. To avoid disturbance of these systems by future activities, related spatial claims should be registered and permitted. Carbon dioxide Capture and Sequestration (CCS) is normally done in deep oil and natural gas reservoirs and in salt water aquifers, as briefly discussed in Box 3.1. Worldwide, legislation is being adapted to allow specifically for carbon sequestration. Aspects that need to be addressed include ownership in view of long term storage, risks/liabilities related to movement to neighbouring properties, triggering of (small) earth tremors, and escape to the surface. In addition, mining legislation and that controlling injection of waste/hazardous substances are being reviewed in nations that consider applying CCS in their policies to control 'green house' gas-emissions.

6.1.3.2. Remediation and reclamation

In the USA, ground remediation has been addressed on a national level since the early 1980s; shortly after the discoveries of hazardous waste deposits such as Missouri's Times Beach and the New York Love Canal. Missouri's Times Beach is a former small town in Missouri, USA, where in the 1970s, to combat the dust from roads, they had been sprayed with waste oil. This oil was also mixed with waste containing toxic, high-concentrations of dioxin. After complaints and investigations of dead horses it was found in the early 1980s that the subsurface was highly polluted with dioxin and the town had to be evacuated. The houses were demolished and the ground remediated in the 1990s. Love Canal was a neighbourhood in Niagara Falls, New York, USA, where a municipal and chemical waste dump site was developed into a school and housing area in the 1950s. In the 1970s it was found that highly toxic chemicals oozed from the ground, entering basements and inhabitants became ill with diseases likely related to chemical wastes dumped at the site. Most houses have been demolished and the site partially remediated. The federal US government responded with the development of new statutes such as the Comprehensive Environmental Response, Compensation, and Liability Act (CERCLA = Superfund, in 1980). CERCLA was developed to finance clean-up activities of hazardous waste sites in case the responsible parties could not be identified or when they were unable to pay their share in the clean-up operation. CERCLA enhanced and increased enforcement of regulations already in place under the Resource Conservation and Recovery Act (RCRA), the Toxic Substances Control Act (TSCA), and the Clean Water Act (CWA). Russell *et al.* (1991) estimate that through these Acts, regulations

Fig. 6.4. Underground urban repositories for various types of urban waste in Haarlem, The Netherlands (photographs: E.F.J. De Mulder).

and funds, between 480 billion and 1 trillion US$ will be spent over three decades to remediate hazardous waste sites in the USA. Most of these sites are controlled by the Department of Energy (DOE), Department of Defense (DOD), other federal agencies, the States, or are at the direction of the EPA.

In The Netherlands, awareness of ground and groundwater contamination initially resulted in legislation based on a Reference and Target value system for a list of specific contaminants. That process started with the Interim Act on Soil remediation in 1983, later incorporated in the Soil Protection Act of 1995. Currently, with the Dutch Environmental Management Act in place, functional land use is also relevant to soil remediation decisions using a risk-based approach. Clean-up levels are determined by the type of land use and the potential exposure risks to the contaminants. Land zoned for industrial use is commonly considered less sensitive than urban use and may therefore require less remediation. The approach of setting different action levels for different contaminant concentrations was followed in Canada and Japan. In Japan, the Water Pollution Control Law was introduced in 1997. This law mandated prefecture authorities ordering companies to clean-up groundwater contamination. Clean-up levels are determined by setting environmental quality standards (EQS) for specific compounds in groundwater. For soil contamination EQS have also been set but legislation to enforce these is still lacking in Japan.

Many countries prohibit any infrastructure development which may harm the environment. Legislation aims at maintaining or restoring air quality, surface water or contaminated sites to natural background levels, although these levels are often not supported by proper scientific evidence. Often, such new environmental legislation developed only based on knowledge stemming from life sciences, neglecting the Earth sciences. The understanding life scientists have of the normally long term Earth processes proved to be quite insufficient. The heterogeneity and the dynamic aspects of the subsurface (Sections 4.2 and 4.3) were virtually unconsidered, hence, features such as weathering, soil and crust forming, slow but persistent groundwater circulation, or the causes of natural hazards were not included. Countries as The Netherlands, Denmark, Germany and the United Kingdom based maximum levels for concentrations of chemical compounds on such 'natural background levels'. Exceeding such levels leads to the legal obligation of remediation (or clean-up). Geochemical mapping during the 1990s demonstrated the wide variety of natural background concentrations of chemical components throughout Europe (Salminen et al. 2005). These studies proved that the natural variability of background values is controlled by geological rather than by human factors. In conjunction with the worldwide economic downturn, this approach attributed to stagnation in overall infrastructure development in the late 1980s and early 1990s.

Land reclamation of former mining areas or waste dump sites to restore or rehabilitate the natural environment requires significant efforts, as discussed in Section 3.1. That resulted in extensive legislation increasingly affecting the mining industry, (Pring 2000). In the European Union, the European Directive 85/337/EEC for Environmental Impact Assessment resulted in considerable harmonization in licensing prior to authorization of development projects. However, no serious attempts are being made yet to harmonize laws or to develop a grand natural resources supply strategy for Europe (Van der Moolen et al. 1998). In 2005, the European Union published a communication on a Thematic Strategy on sustainable use of natural resources (EC 2005). It aims at a more effective and sustainable use of natural resources throughout their life cycle. This communication, however, has no legislative power and no target figures were set. Data collection and aiming to introduce life cycle thinking in this field and strengthen policy making by collecting and providing data on impacts and resource use and efficiencies (indicator-development) are sought after by the European Union.

In the USA, the Surface Mining Control and Reclamation Act of 1977 is another example where effects of surface and underground mining of coal are regulated to reduce environmental impacts and protect society from negative effects.

6.1.3.3. Risks and liabilities

Beside the assets in the subsurface there are also risks and liabilities related to underground development. Apart from legal instruments to plan and control aspects of the use of the subsurface, it is important to identify parties liable to and responsible for managing risks related to subsurface activities beforehand. Lack of clarity in legislation may seriously hamper any underground development.

Legislation related to risks and liabilities in the subsurface already exists. In general, the owner of land, a construction, a mine or a groundwater well is responsible for damage to third parties and therefore carries the liability in legislation worldwide. This assumes that the landowner exists as a person or as a legal entity, they can be identified from past or present activities and that the adverse effects can be connected to their activities. The main drivers determining legal options for developing the subsurface are:

- Risks of and resulting from the (intended) use of the subsurface
- Liabilities to potential damages and consequential losses (economic or otherwise).

Different types of underground activities, as mining or subsurface construction, may have different impact on people, goods or the environment and bear varying risks. Human-induced geohazards and their associated risks and liabilities play a crucial role in addressing legal aspects related to the use of the subsurface. Among others, such hazards include (unexpected) surface settlement and groundwater contamination from waste water. Another important issue is the right of access to a site in case of emergency, both at the surface and underground. Normally, full ownership is considered best to secure that right. A Dutch study (COB/RAVI 2000) revealed that in case of doubt, governmental organizations will normally prefer to buy the respective pieces of land rather than entering into other types of contracts for subsurface use.

Groundwater extraction may cause land subsidence and subsequent stability problems in cities. The same holds for roof top collapse in former mining sites (see also Section 4.5.5) a frequent problem in the United Kingdom. There, the Coal Mining (Subsidence) Act of 1957 and the Coal Industry Act (1975) were issued to address this problem. Insurance claims for subsidence damage to private properties may amount to some 100 million GB£ a^{-1}. In the United Kingdom, reliable mining records exist since the late 19th century (Department of Environment (DoE) 1990). Examples of land subsidence in The Netherlands are due to the extraction of natural gas disturbing regional water management systems. The total annual costs of land subsidence in The Netherlands amount to about €1.65 billion (De Mulder *et al.* 2003). Bangkok, Venice, Jakarta and Mexico City are (sinking) cities that suffer from land subsidence and subsequent flooding partially due to (excessive) groundwater extraction. In Japan, the Industrial Water Law and the Law for Groundwater Use in Buildings forbids withdrawing groundwater in subsidence prone areas without a permit (World Bank 2006).

Environmental liability has become a major issue in protection and subsurface management, mainly due to health risks and the major financial consequences of clean-up activities.

An EU Directive (EC 2004c) particularly relevant to the environment and the subsurface, concerns liability for remediation of environmental damage. In February 2000, the European Commission adopted a White Paper on environmental liability. Its objective was to explore how the 'polluter pays' principle may best be applied to serve the aims of EU environmental policy, mainly to avoid environmental damage. So far, the EU Member States assessed national environmental liability regimes that cover damage to persons and goods. Moreover, they introduced laws to deal with liability for clean-up of contaminated sites. However, nations did not yet address liability for (ecological) damage to nature. This may explain why economic actors in the EU have focused on their responsibilities to other people's health or property, but did not consider their responsibilities for damage to the 'wider environment'. The latter was traditionally considered a 'public good' for which 'society as a whole' should be held responsible rather than individuals. 'Traditional damage', such as personal injury and damage to goods and property, even if caused by 'risky and potentially risky' activities covered by the Environmental Liability Directive, will now be dealt with under national civil liability legislation in the EU. The Environmental Liability Directive deals with damage to the 'wider environment' only. Introduction of liability for damage to nature could generate a change of attitude that can result in an increased level of prevention and precaution to environmental issues.

In the USA, remediation cost of contaminated sites may well be the largest unrecorded liability for corporations associated with these sites. This makes contingent environmental liabilities one of the most significant unavailable pieces of financial information to a corporation. Therefore, the US Securities and Exchange Commission (SEC) and the Financial Standards Board demanded responsible recognition and disclosure of environmental liabilities. As USA-based corporations started operating to comply with SEC requirements, the extent of corporate liabilities for environmental issues became more transparent, amounting to as much as 250 Billion USD in 1997. In the late 1980s and early 1990s, the public perception in the USA was that more time and money was spent on litigation and studies than on actual clean-up activities. In 1992, the New York Times reported that out of 470 Million USD processed by insurance companies from claims relating to the Comprehensive Environmental Response, Compensation and Liability Act (Superfund), 410 Million USD went to legal cost in 1989.

6.1.4. Evaluation

The overview in the preceding chapters are far from complete. Within the scope of this book it aims to give a sense of the complexity, the variation and the trends in the legal aspects concerning the subsurface in various parts of the world.

The legal system, whether civil, common or religious law, has an impact on the range of legislation and its set-up. Under Civil law the need to produce legislation may be felt earlier than under Common law. Legislation concerning landownership, natural resources and the environment is normal but legislation in the field of spatial planning related to the subsurface including construction is far from ordinary. Most information concerning legal aspects of the subsurface is from Europe and North America. Except for South Africa, information is lacking for the rest of the African continent. This also holds for parts of the world where the legal system is based on religious laws where no specific legislative information was found on development of the shallow subsurface with the exception of water rights.

At a global level, legislation on environmental issues has probably been one of the most dynamic fields in jurisdiction since the mid 1970s. Evolution of legislation over time, efforts to exclude overlap in jurisdiction, development of new legislation based on new insights in regulatory needs and technical capabilities, all affected existing legislative frameworks. Legal aspects of ownership, mineral rights and mining, cultural heritage and water resources underwent quite significant modifications. Very little development, however, occurred in legislation on the role of the subsurface in spatial planning and urban development. This by no means implies that a coherent structure does exist with clear definitions, objectives and responsibilities of governing bodies in areas where extensive legislation exists (see e.g. Table 6.5 for regulations on subsurface activities in The Netherlands). Given the segmentation in legislation for different subjects (water, air, soil, mining and environment), the involvement of different levels of government (national, regional and local) that need to align their governance, and

Table 6.5. *Overview of regulations on subsurface activities in The Netherlands (modified from RPD 2000, with incorporated related EU-legislation)*

Activities	Regulation	Short summary	EU-legislation & policies
General	Civil law book: article 20	Ownership rights	Charter of fundamental rights
	Civil law book: article 21	User rights	Charter of fundamental rights
Structures and infrastructure	Town and Country Planning Act	In revision, not explicitly focused on the subsurface, three scales are identified: national, regional (structural plans) and local (zoning plans)	Thematic Strategy on Urban Environment
	Structure plan Logistics and Transport	Policy plan relating to underground pipelines for oil, gas and chemicals	NA
	Building materials decree in the law on soil protection	Use of building materials in the subsurface, to prevent emissions	NA
Mining natural resources	Mining Act	Regulates permits for mining activities	Indirectly through environmental legislation
	Law on the prospect for resources	Permit grants for exploration drillings	Thematic Strategy on the sustainable use of natural resources
	Groundwater Act	Groundwater extraction	Groundwater Daughter Directive
Storage	Nuclear Energy Act	Storage of nuclear products	European Atomic Energy Community (Euratom)
	Waste disposal decree in the Soil Protection Act	Regulates the disposal of waste to prevent contamination of soil and groundwater	Directive on hazardous waste
Remediation and protection	Environmental Management Act	Framework law for protection of the environment, amongst other dealing with: environmental impact assessment, chemical waste, establishments	Environmental Impact assessment Directive; Environmental liability Directive
	Soil Protection Act	Goal: to restrict developments in the subsurface that negatively influence subsurface characteristics and that are irreversible. Included are orders on infiltration of groundwater, drainage of waste water, waste disposal, building materials, obligatory site investigation and underground tanks	Soil Thematic Strategy, proposed Soil Directive
	Pollution of Surface Waters Act		Water Framework Directive
	Water Management Act	Protection of the surface and groundwater, mainly concerned with quantity	Water Framework Directive; Groundwater Daughter Directive

the need to balance the resource requirements of for example rural and urban areas, a long term strategy is called for that encompasses the separate regulations into an overall framework.

Although EU-wide Directives may significantly influence national legislation, major differences in regulation concerning the subsurface still exist among EU nations.

Some authors believe that the existing legal framework in most countries may be used to regulate development of the subsurface (COB 2004a). But specific local conditions might complicate implementation in practice. Perceived lack of control and the challenge of balancing risks with liabilities from other issues often makes full ownership of the land the preferred option for subsurface construction. With anticipation of an increased use of the subsurface, expropriation or the option to separate underground use from that at the surface, in conjunction with financial compensation to landowners, may become an increasingly important issue. This is necessary to balance the rights of individual owners with the more general interests of society.

For natural resources sustainability aspects of their exploitation are most relevant. This is especially true for groundwater as this resource lost most of its original asset as private property in most countries. Water acts transferred the rights of groundwater use to governmental bodies who were authorized to grant permits. Water as a resource is seen as critical to health, food production and the quality of life in general for a significant part of the world population in the next decades.

Space is another important asset of the subsurface that may help combating urban sprawl and improving quality of the urban environment. Clear, legal frameworks on permitting, safety, risks and liabilities are necessary to support further underground urban development. An important aspect of subsurface construction deals with funding and the need for banks to protect their assets when lending money to developers. Full ownership will give banks more clarity and avoid uncertainty. However, preference for full ownership will unnecessarily limit the range of options for developing the subsurface. Therefore, the authors would rather stimulate the option to subdivide ownership of the subsurface into three-dimensional units. That may be done through standardized methods and registration in a fully 3D-cadastre. That approach will also encourage development of better surveying and mapping technologies to delineate the boundaries between such units (Sandberg 2003).

Until now, an integrated, system-based approach, that takes principles of geodiversity and different process scales into account is missing. Instead, a wide range of legislation needs to be taken into account when undertaking activities in the underground. An example of a current legal framework for national legislation concerning the subsurface is given in Table 6.5, for The Netherlands. This table is modified from RPD (2000) by adding a column concerning the European legislation and policies linked to the national legislation or is part of an active policy-area of the Member States.

Concerning development of soil- and groundwater protection, considerable overlap and interaction between regulations may exist, often aiming at securing maximum protection of the quality of the subsurface.

It is not implied that all complexities of natural systems can be addressed in legal terms, even when sufficient data and information on these systems are available and accessible, but natural scientists should be consulted to identify gaps and inconsistencies in current legislative frameworks and to provide necessary evaluation criteria to develop proper policy and management strategies. Earth scientists should assess the impacts on natural resources, and Earth systems, and should produce alternative development strategies.

6.2. Policy

Not only the legal framework, but also policies related to spatial planning, exploitation of natural resources and environmental protection may influence development of the subsurface. Moreover, economic factors, population density, and physiographic characteristics play a role in such development. Decisions to go underground depend on many factors, including vision and personal drive by individuals in office. In this section, some policy aspects with a wider bearing on the subsurface are presented, together with some specific examples, mainly from the European Union.

6.2.1. General policies

At a global level, the United Nations developed a policy aiming at environmental sustainability by reversing the loss of environmental resources and improving, for example, access to safe drinking water and basic sanitation. These UN Millennium Development Goals (UN 2000) have a significant dimension with respect to the subsurface. Also the UN Intergovernmental Panel on Climate Change (IPCC) reports include policy aspects that are directly or indirectly related to the subsurface, such as the option to capture and store carbon dioxide (CCS) in the (deep) subsurface.

At a continental level, the European Union produced several policy documents based on legislations and guidelines, directly or indirectly influencing the subsurface (See Section 6.1). Generic EU policy lines include:

- The 'Polluter pays' principle on remediation. A variety of EU subsidy systems make application of this principle sometimes quite problematic and even conflicting. For example: subsidies to support food production through agriculture v. environmental impacts (use of pesticides and manure);
- The Valletta Treaty on cultural heritage; this Treaty resulted in a policy that parties responsible for potential disturbance of the subsurface, including archaeological heritage sites, should bear the cost for exploration and conservation of relevant remains.

Most prominent with respect to the subsurface are policies on environmental issues. In 1995, the European Environment Agency (EEA 1995) published the Dobříš Report. That Report assessed the State of the Environment in Europe and aimed to set a baseline reference level for periodic updating. It confirmed the rather poor quality of the European environment, particularly in parts of Central and Eastern Europe. It listed 56 environmental problems, grouped into 12 prominent European issues, including climate change, stratospheric ozone depletion, loss of biodiversity, major accidents, acidification, troposphere ozone and other photochemical oxidants, management of fresh water, forest degradation, coastal zone threats and management, waste production and management, urban stress, and chemical risk.

In 1998, the Dobříš Report was already revised (Second Assessment). All 12 environmental problems were analysed in terms of a 'DPSIR' cycle of Driving forces (economic sectors, human activities), Pressures (emissions, waste), States (physical, chemical and biological), Impacts (on ecosystems, human health and functions) and Responses (mostly political – prioritization, target setting, indicators) (Fig. 6.5).

As most environmental policies were directed at 'end-of-pipe' measures, little improvement as to the state of the

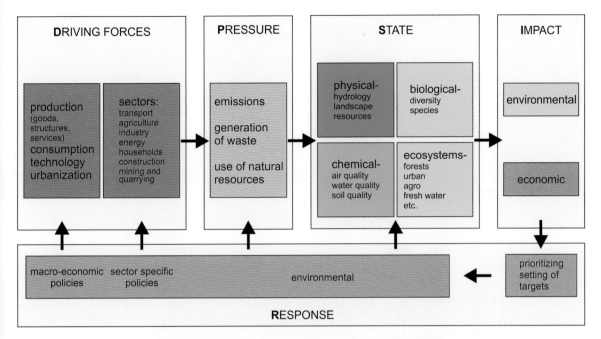

Fig. 6.5. DPSIR-framework (modified from Hartlén & Bendz 2004).

environment was registered in the second assessment. Such measures were considered insufficient to address increased infrastructure development, production and consumption.

Another forward looking EEA document: 'Environment in the European Union at the turn of the century' was produced two years later (EEA 1999). That document had the ambition to analyse if the sectoral and environmental policies anticipated for the next decade would promote or rather hamper social and economical progress in society. This report advocated the need for integrated environmental assessment. In both sectoral (industry, agriculture, energy, transport and tourism) and economic (related to creating a European market) decisions the environmental impact should be evaluated according to the DPSIR-framework.

Again two years later, the European Parliament and Council adopted The Sixth Environment Action Programme of the European Community 2002–2012 for environmental policy making. Seven thematic strategies were to be developed, including a Thematic Strategy on the sustainable use of natural resources (EC 2005) and one on the urban environment (EC 2004b). Both are relevant policy documents with regard to the subsurface. Examples of relevant policies will be discussed in the next paragraphs.

6.2.2. Mineral resources

Nations around the world have different policies to benefit from the natural resources present in their subsurface. In broad terms, two questions related to mineral and oil and natural gas resources dominate national policies: 'How to balance benefit for the State with profit for the extractor?' and: 'Who has the right to explore and exploit what?' Access to sufficient natural resources is of strategic importance to all nations. If no such resources have been identified in their subsurface, countries need to take adequate measures. Japan, for example, has very limited mineral and oil and natural gas resources in its own territory. Therefore, the Japanese Ministry of Economy, Trade and Industry created a government body in 2004 to secure stable supply of oil, natural gas and non-ferrous minerals while supporting companies to promote and acquire overseas exploration and exploitation rights. In China and India, with high demands for natural resources, comparable situations occur.

Although only 0.05–0.1% of the European land surface is used for opencast mining, this industry is considered to severely impact the environment. That holds, for example, for the locally >400 m deep lignite mines in Germany. However, policy is in place to reclaim such areas (Fig. 3.9). To secure the supply of construction materials, the national or regional government has to produce long term policies, including options for the use of alternative or secondary materials. The European Union developed a policy on natural resources for the next 25 years (EC 2005). In this policy, sustainability is the key for long term prosperity, at the same time preventing negative environmental impacts ('decoupling'). Actions derived from this policy include developing a knowledge base, indicators to

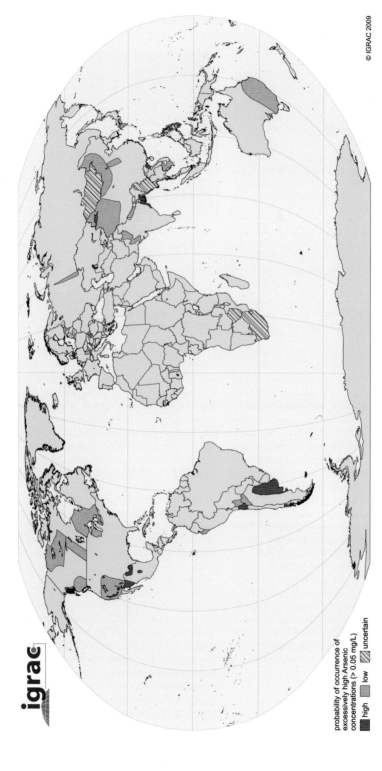

Fig. 6.6. Probability of occurrence of excessively high Arsenic concentrations (>0.05 mg L^{-1}) (IGRAC 2011a). Probability of occurrence of excessive concentrations is indicated by three different classes: High: areas have a relatively high probability to contain arsenic in the groundwater. The locations of these areas are often well defined and many groundwater samples of these areas contain high concentrations of arsenic. Low: areas have a relatively low probability to contain arsenic contaminated groundwater. The precise location of the contaminated groundwater is not known and only few groundwater samples contain high concentrations of arsenic. Uncertain: areas are mentioned in the literature as contaminated areas, but no detailed information is available. For the remaining areas, no information about arsenic contamination has been found so far.

measure progress, and establishment of an international forum to facilitate identification of (national) measures.

6.2.3. (Ground) water policy

There is a need to develop policies to cope with the disparity between demand and availability of and access to groundwater as a resource. This relates to water quantity and quality, but also to availability for rural, urban and industrial use and regional or even national redistribution. As discussed in Section 6.1.2.2, cross-border water and groundwater problems may cause serious international conflicts. Resolving such controversies requires international or supranational policies. Policies on water allocation are in place in Israel, where priorities are set for water use, with domestic use as the highest priority (World Bank 2006).

Groundwater is not a static resource that may just be mined. For groundwater policy making, natural processes (geological and chemical) should be taken into account. Groundwater is not by definition a safe resource, as exploitation may also generate hazards (Keller *et al.* 1998). An example of adverse effects due to falling groundwater tables is the production of arsenic in some regions in the world (see Section 4.3.9). Worldwide, arsenic levels exceed WHO maximum levels (of 0.05 mg l^{-1}) in a total area of 34 000 km^2 (Das *et al.* 1995), particularly in India (West Bengal) and Bangladesh. Figure 6.6 shows the areas worldwide with a possible impact of arsenic levels on groundwater quality (IGRAC 2011*a*).

Land subsidence problems due to falling water tables have been discussed in Chapter 3 and Section 6.1.2. Rising groundwater tables may lead to water logging, overflowing sewage systems with health risks and spreading of pollution. That may happen after closure of mining operations or other water demanding industries when pumping has ceased. In cities, buildings and infrastructure constructed in periods of low groundwater levels, may suffer from leakage or from structural damage when water tables rise. Water policies should be in place to cope with such hazards, in particular when large rivers are involved that may create cross-border problems, such as the Yangtze (China) and the Ganges (India).

6.2.4. Transport, infrastructure and planning

Policies in directing development and land occupancy may differ from country to country but most policies include the following elements (Landahl 1995; Marker *et al.* 2003):

- Land use planning – type of land use which ensures the perceived best balance between social, economic and environmental costs in the use of space. This normally involves strategic or spatial planning at regional levels, development planning at the more local level (including allocation or 'zoning' of land for development and conservation), and detailed planning applications for specific development (permitting). There is usually no policy framework and little published planning guidance for use of the subsurface. It is, therefore, difficult for planners to consider the potential of subsurface development or for permitting authorities to assess applications for planning permission. An important, but largely overlooked aspect of planning for the development of the subsurface is the need to integrate surface and underground proposals. As stated in Section 7.4, it could be argued that policies should be developed to safeguard the subsurface from developments that may inhibit future options or lead to serious risks of suboptimal use (consumption of underground space). Heat and cold storage is an example where more strategic planning is required to prevent systems for interfering with each other or with other functions like water supply.
- Building control – establishes whether the structure is adequate for safe occupancy and use, but focuses on surface structures and foundations. However, standards and specifications for example lighting, heating, ventilation, and noise transmission may also be applied for subsurface development. For transport tunnels design guidance is growing, which also serves as example for other type of subsurface developments.
- Health and safety provisions – safeguard the workforces and visitors on or in construction sites during and after construction. In many countries, a well-developed set of procedures and laws exists, developed for surface and underground mining. These are also applicable to tunnelling and other large-scale underground constructions. Underground conditions create a working environment including features and hazards that do not exist at surface. The most common incidents (e.g. falling from height) may not differ but there are other, less frequent incidents that need to be addressed properly. Beside cave-ins, the enclosed situation may also lead to accumulation of natural gases (e.g. methane) and explosion risks or depletion of oxygen from inadequate air-circulation (Sections 4.5.8 and 4.7.4.2)
- Environmental protection strategies – provide for control of emissions, waste management, and safeguarding of air, water, soils and natural habitats. This requires understanding of potential hazards and their mitigation and the potential vulnerability and sensitivity of different aspects of the environment to change. This often requires baseline monitoring and modelling of effects of proposals in both the short and long term.

All of these aspects have a bearing on whether, where and how underground development is undertaken and applied (Fig. 6.7). Permit applicants are legally required to consult all authorities that administer relevant regulations. Moreover, all authorities should consult one another to ensure that all relevant matters are covered and that no conflicting conditions are applied. The right information (including geo-data) is needed at the appropriate level of detail and at the proper stage in the administrative process (Chapter 5).

Land use planning in three-dimensions and over time is essential for long term sustainable development incorporating

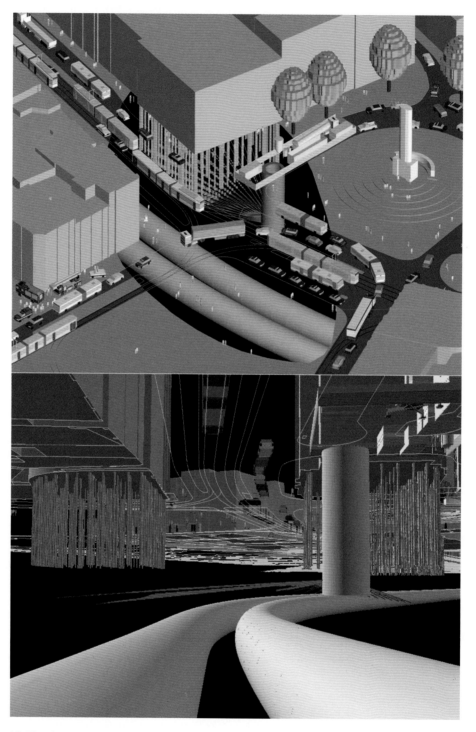

Fig. 6.7. Very busy urban surface and underground space (yellow lines are foundation piles, red lines are tubes a manchettes for compensation grouting, blue spheres are injection points; Kaalberg *et al.* 2012 reproduced with permission CRC Press/Balkema, figures courtesy Witteveen + Bos).

> **Box 6.2. Coordination of subsurface plans in Helsinki, Finland**
>
> Since 1984, all underground plans and activities by municipal authorities and private construction projects in Helsinki are coordinated (ITA 2000). Helsinki has developed over 7 million m^3 of underground space within an Underground Space Allocation Plan devised by the City Planning Department. According to Chow (2002) development and implementation of such planning guidelines by urban planning authorities should be encouraged. In 1956, the Helsinki City Real Estate Department began to collect, collate and present information on ground (soil and rock) and groundwater conditions within the city area. This Helsinki Geotechnical Database currently contains information from over 200 000 boreholes and 4000 groundwater-monitoring locations. Building foundations and tunnels are also included in the GIS database that is accessible to consultants, construction companies and the public. Availability of this data allows development proposals to be assessed, site investigations to be designed, and the re-development of sites to take place based on sound knowledge about the foundations and potential obstructions that may exist under the site (Chow 2002).

the subsurface as a (in many aspects non-renewable) resource. It should be realized that spatial planning is often more recent than the actual constructions at a number of sites. Moreover, constructions may occur that were built without the proper authorization.

A new approach in spatial planning was introduced by distinguishing three (information) layers: the subsurface (soil, water), networks (infrastructure) and occupation (housing, working, recreation, etc.), as was done in The Netherlands in the late 1990s. For the first time, the subsurface was recognized as a relevant feature in Dutch spatial planning policies. Parallel to this layer approach, separate levels of authority may be distinguished. In this approach, municipalities may regulate activities in the shallow subsurface (local infrastructure, remediation), while regional authorities can regulate activities at somewhat deeper levels (groundwater protection). Finally, national governments may then regulate activities in the deep subsurface (e.g. mining) (RPD 2000). The need to integrate information and different types of registration systems on the subsurface has in The Netherlands lead to a program called Base Registration Subsurface, which started in 2009. Its aims are to combine not only geological, geohydrological and soil information, but also infrastructure and user rights (KWR 2010; BRO 2012).

Co-ordinated subsurface planning is known at some local levels. Conflicts between different subsurface interests are most pressing in cities. That makes policy development on subsurface use and management more urgent for larger cities. An example of subsurface planning is given in Box 6.2.

Normally, national policies do not include the subsurface except for soil, groundwater protection and mineral exploitation. Almost everywhere, regulations on subsurface activities are fragmented and follow a sectoral approach. There is little understanding of the interaction between subsurface functions, and insufficient allowance about how sub-surface and surface affect each other. Extensive literature searches revealed that valuable elements may occur in current planning policies that address some aspect of the subsurface but that more co-ordination is required to ensure that the subsurface is managed in a more sustainable manner. In an extensive study, Foster et al. (1998) observed a distinct need for long-term policies and legislation on water resources for rapidly growing cities in countries such as Bolivia, China, Mexico and Thailand.

Subsurface functions such as storage and space for transport are normally organized on a regional/local level. Municipalities handle most specific legislation in the field of construction, infrastructure and environmental protection, while national governments normally only produce general guidelines. An overview of levels of authority in various sectors of subsurface management in The Netherlands is given in Table 6.6.

6.3. Management

Management relates to the sequence of human activities to reach predetermined goals in an organized and controlled way. With respect to the subsurface, no such management concept exists yet. This section discusses several managerial aspects and principles that can become part of a management concept for the subsurface. As stated earlier, laws and regulations provide the legal basis, while policies contain ambitions, plans or actions steering decision-making in a preferred direction and thus influence management strategies. Continuing with sectoral approaches will increasingly lead to conflicts between subsurface functions and hamper more sustainable approaches, particularly with the global urbanization trend. Better-integrated development and management of the subsurface may allow functions that are kept separate today, to concur on the same surface area with no additional burden to the environment. Superposing functions, like underground transport, heat and cold storage, groundwater remediation and use of mineral resources from construction work are promising potential combinations. Opting to bring functions underground may significantly reduce energy consumption as temperatures underground are more stable and require less heating or cooling in cold and warm seasons (see Section 7.4).

Table 6.6. *Levels of authority related to various functions of the subsurface in The Netherlands*

	EU/Federal	State/National	Province	Municipality
Construction and infrastructure		+		++
Mining quarries, oil and gas exploitation	(+)	++		
(Waste) Storage		+	++	
Ownership		++		
Archaeology		+		
Environmental protection		+		+
Spatial planning		+	+	++

++, leading and/or responsible for permitting.
+, to be consulted and/or involved to lesser extent.

Proper management relies on organizational capabilities such as the existence of institutions empowered to decision making. Optimal decision making also builds on the availability of adequate and relevant information, in particular with respect to the contents of the subsurface and the rights to use the land in all its aspects (ownership, exploitation, and functions). Subsurface development may provide opportunities but may also jeopardize life, health and properties. As underground structures would normally have very long life times, long term management needs should be addressed in planning, maintenance and control. Information gaps stemming from the sometimes ancient past also need to be filled. Management strategies are based on policy ambitions and legislation. That also holds for managing the subsurface. The authors adhere to the vision that underground development should be economically and environmentally sustainable and should thus attribute to the well-being of present and future generations.

6.3.1. Sustainable development

Humans have always been interested in controlling their environment and to avoid unexpected surprises. This also holds for the subsurface, but as this invisible realm seems quite complex, priority was given to safety and health, followed by prosperity and quality of life issues at the Earth's surface.

The concept 'Sustainable Development' was first published by the International Union for the Conservation of Nature (IUCN) in 1980. This term came into common use after publication of the 1987 report of the UN World Commission on Environment and Development (the Brundtland Commission). That Commission adopted what was to become the most frequently-quoted definition of sustainable development as development that 'meets the needs of the present without compromising the ability of future generations to meet their own needs' (Brundtland 1987). The concept of 'sustainable development' may be divided into three or four constituent parts: environmental sustainability, economic sustainability, and social sustainability, sometimes supplemented by political sustainability.

Quite a wide variety of definitions on sustainable development exist, often biased by stakeholders. Such definitions individually point to specific elements that may jointly contribute to a management framework for sustainable use of the subsurface:

- To use a descriptive 'systems' approach determining the present and future continuity by taking natural processes into account (determining the performance of the ecosystem).
- To focus on renewable resources, budgeting of non-renewable resources to safeguard adequate stocks for the future.
- To take dynamics and scales (3D, time-dependent) into account.
- To provide society with a safe and healthy environment with adequate options for food supply, (drinking) water and shelter from adverse weather and other natural conditions (social and economical aspects).
- To enact a legal framework addressing integrated development of natural resources and underground space.

Essentially, all definitions of sustainable development include the 'three pillars of sustainability': People, Planet and Profit, and support the view that sustainable development should strike a proper balance between them (Cramer 1999). *People* relates to social aspects and the aim for a just and safe society. *Planet* addresses the environment or ecosystems, providing essential goods and benefits to People (Bergkamp & Cross 2006). *Profit* represents monetary value of the environment and relates to economic prosperity. From this point of view, development of the subsurface may be considered if the environmental, social and economic benefits are higher than the respective costs.

For benefit and risk assessment in subsurface development, time and scale effects are essential elements. Both short term (years to decades) and long term (centuries and longer) effects are relevant in this respect. For example, for underground storage of nuclear waste not only risks concerning current environmental conditions and human activities should be considered but also those relevant on longer (geologic) time scales, as (real) climatic change, erosion and tectonic displacement (Commissie Opberging te Land (OPLA) 1989, 1993).

Under what conditions will development of the subsurface be sustainable and when not? In terms of resources, the subsurface is partly non-renewable (e.g. if excavated) and partly renewable (e.g. groundwater and soil formation) within (extended) human life spans (Bergkamp & Cross 2006). Geological processes are often quite slow on human timescales but may generate dramatic changes in the landscape at geological timescales. A *dynamic systems approach* in describing the subsurface and its relevant processes can bridge this gap. Although these processes and their interaction with other parts of the ecosystem are not fully understood yet, this approach may provide a better understanding of links and feedbacks. Discrepancies between actual and projected system behaviour will uncover essential gaps in knowledge. Therefore, geological knowledge may provide essential information on impact and feedback mechanisms, for example through *natural analogues*, while contributing to addressing questions on sustainability. As stated in Section 3.4, the Earth is and has always been dynamic and the numerous interacting, often slow Earth processes are not and never will be in full balance.

In line with sustainable development principles, potential impact of any new underground structure on the subsurface environment (virgin or earlier developed) should be assessed. For proper and transparent decision making, it is recommended to develop a profile with relevant characteristics of the proposed underground structure and to check that in terms of sustainable development. Such a profile should incorporate the structure's function, its life time, the disorder created as well as future technological developments and other factors (Hudson & Hudson 2003). Such an approach complies with the philosophy expressed in the Plan of Implementation of the World Summit on Sustainable Development (WSSD).

Thus it is recommended to establish an *auditing framework* for use of the underground analagous to that used for the surface (Hudson & Hudson 2003). If use of underground space does not compromise alternative options for the future, either because it does not significantly prejudice the available space or it may be re-used, such use may be considered as an acceptable consumption of resources (Roberts 1996). Underground constructions may very well keep ecosystems at the surface unchanged; if building in the subsurface also does not have adverse socio-economic impacts on society, then it might be more acceptable/sustainable to build underground.

An auditing framework will also provide a common tool for all parties to check and assess sustainable development aspects of any proposal to create underground space. To address the various hierarchical levels of sustainable development, an integrated procedure will be required. That will make a trade-off between such levels possible, for example between improved sustainability at national level possibly at the expense of local sustainability ambitions. Such an auditing framework can make all motivations and interactions of the proposal explicit and thus contribute to more transparent decision making.

6.3.2. Management tools

6.3.2.1. Land registration systems

A cadastre is 'a complete and up-to-date official register or inventory of land parcels in any state or jurisdiction containing information about the parcels regarding ownership, valuation, location, area, land use and usually buildings or structures thereon' (Williamson 1983). Not all nations have a cadastre. In modern Greece, for example, the development of a cadastre did not start until 1995, following a previous attempt by King Othon in Greece to introduce a land register in 1836 (Stamatiou 2002). Despite of their significance to integrate national, state and provincial land policies, 'modern' land administration systems are often out of date. Another challenge is to integrate these with indigenous cultures and tenure systems in developing countries (Williamson 2001).

The Torrens System of title registration, developed in Australia in the mid 19th century, is an example of a 'modern' system (Simpson 1976). It was introduced in many of the British colonies in the late 19th or early 20th century, as well as in Thailand, Brazil and Hawaii. Originally, the Torrens System was a registration system for ship ownership rights, in use by the British Ship Registry. In that system, each ship owner is given a certificate with information about the ship. Upon selling, the seller surrenders the certificate for cancellation and the ship is then registered in the name of the new owner who receives a new certificate. Sir Robert Richard Torrens (1814–1884), an Irish emigrant to Australia and a land law reformer, developed this system for land.

For registering subsurface property rights, a 3D Cadastre system will be needed together with proper surveying practices and techniques. Stoter *et al.* (2004) describes and evaluates the approaches on 3D Cadastres in Norway, Sweden, Queensland (Australia), British Columbia (Canada) and Israel. These studies prove that no complete solution exists for 3D cadastral registration. Moreover, property units are not yet represented in full 3D-geometries from which overlap and dimensions can be checked and viewed at different angles. Sometimes, just 2D-foot prints on the surface parcels are presented (in British Columbia). In Norway and Sweden, registration of 3D property units is only possible for built constructions (Stoter 2004).

6.3.2.2. Permitting

Permitting is an important instrument in the management process. Most common are building permits, groundwater withdrawal permits, exploration permits, mining permits and environmental permits. A permit states that the law was applied to a certain activity or process. Permitting procedures ensure that minimum standards are maintained. These are used to collect relevant information from the permit applicant on his activities. In turn, this procedure supports authorities in setting permit requirements and reaching policy-goals. Permitting may also protect the holder from infringements on his rights and serves to protect third parties from harm by unregulated activities.

Depending on the legislative framework, more than one permit may be required for a single activity, in particular if large-scale and complex. Lack of proper coordination mechanisms among authorities in charge may lead to inefficiencies or even conflicting requirements in permitting. Management and control of the various underground infrastructure systems may be in the hands of a variety of parties. This leads to inefficient use of the subsurface and to risk of damages.

6.3.2.3. Data-collection

The state of the subsurface is normally described in maps and publications. National geological, groundwater and soils surveys are charged to produce such documents. For modifications in the state of the subsurface over time, in particular for groundwater, normally monitoring systems are installed (Sections 4.6.3 & 5.4.3). Data collection from permitting procedures may provide an important additional source of information. All such data are relevant and necessary to initiate, steer and guide activities from an economical, environmental and societal point of view. Although extensive databases and mapping systems exist, common definitions and tools are often lacking. Therefore, it is important to know at least about (systematic) differences when linking databases between regions and countries. That is necessary when it comes to the development of policies and management that exceed local levels. Information systems for water wells exist in almost every province in the USA and Canada, but these are not accessible at a national level and not centrally maintained.

6.3.3. Mineral resources

Worldwide, national geological surveys produce maps and reports with the mineral contents of the subsurface which are vital tools in mineral resource management. Mining companies normally use such information as a basis for more detailed exploration of prospective areas. Exploration permits are required for obtaining additional data. A next step can be application for mining permits. For example, the authority of the United States Geological Survey (USGS) is based on many pieces of legislation and regulations linking it to management of natural resources. This mandate of the USGS may serve as an example to show in which legislative area information about the subsurface in terms of geology and mineral content may play a role. Foremost, this legislation includes the Mining and Mineral Policy Act from which the responsibility to assess the national mineral resources is derived. The USGS Organic Act instructs how to classify public lands, to examine the geological structure, the mineral resources and other products. National or local authorities may also exert their management power on mineral resources by, for example, promoting the use of secondary construction materials, such as recycled aggregate minerals to save on exploiting primary resources (USGS 2006).

6.3.4. Groundwater resources

Water resources management is a major issue worldwide today in view of population and urbanization growth and the increasing number of megacities. Foster *et al.* (1998) describe the interrelations between water supply, waste-water and waste disposal. In expanding cities the challenge is to get long term and sustainable access to clean water supplies (often groundwater) while simultaneously developing waste-water and solid waste disposal systems. In practice, legislation is only part of the solution. Fair policies and setting long term management targets are also required to repair decades-long failures in groundwater management.

Groundwater is of major significance for global access to safe drinking water and is part of the UN Millennium Development Goal on ensuring Environmental Sustainability. As groundwater resources in and near urban centres in developing countries often provide low-cost and high quality municipal and domestic water supplies, proper management and preventing spillage of such resources is of utmost importance (Foster *et al.* 1998). Management strategies should recognize and address the complex linkages between groundwater supplies, urban land use, and effluent disposal. Groundwater management should also include groundwater level control, infrastructure maintenance, and allocation of resources to various groups of stakeholders.

On a global scale, failing groundwater management becomes apparent, for example in depletion, salinization and pollution. Groundwater pollution by fluoride in West India is attributed to inadequate management of groundwater resources through excessive irrigation (IWMI 2000). That may result in upward groundwater flows from deeper, brackish aquifers which may contain natural contaminants, like fluoride (Abrol *et al.* 2004). Fluoride levels may also rise in soils by geochemical processes related to irrigation in agricultural practice. Figure 6.8 shows the areas worldwide in which there is a possibility that high fluoride levels exceed the general WHO guideline of 1.5 mg L^{-1} in drinking water (IGRAC 2011*b*).

Groundwater management failed in parts of Yemen, where groundwater resources were depleted by excessive abstraction while exceeding nationwide recharge (Shah *et al.* 2000). According to these authors, proper water resource management should include:

- information systems, resource planning
- demand-side management
- supply-side management
- groundwater management in a river basin context

For proper decision-making, a solid understanding of the groundwater system, including its delicate interactive processes with the surrounding ground, is compulsory. In turn, groundwater systems are the key drivers of many ecosystem services, as groundwater plays an important role in terrestrial vegetation, river base flow, cave ecosystems and wetlands (Bergkamp & Cross 2006).

Pricing of water may contribute to sustainable management of water resources, including groundwater. In some European countries groundwater extraction is taxed. In The Netherlands, two types of such taxes exist. In Germany, such tax revenues are used to compensate for costs resulting from reduced fertilizer use in water catchment areas. Finland, Sweden, Austria and Norway discourage the use of fertilizers through taxation to protect (ground) water quality (Cao 2006). Water pricing may also be imposed to reduce pressures on water availability for agricultural purposes by urbanization-related industries. However, such measures might jeopardize ecosystems as it may steer water to human consumption rather than to feeding wetlands because of lack of funds for just maintaining these ecosystems (De Savenije 2002).

Food security may also depend on groundwater management. In India, half of the irrigated area is supplied by groundwater, causing groundwater tables to fall $2-3$ m a^{-1}. From a sustainability point of view, reduction of groundwater use is not the solution but groundwater recharge may increase by water recycling. In turn, this can cause increased risk of pollution by fertilizers and pesticides. Forty percent of the world's food production relies on irrigation, mainly supplied by groundwater. Temporary storage of fresh water in aquifers is an important option to cope with strong variations in supply and demand of water over time in arid to semi-arid regions.

6.3.5. Subsurface structure and infrastructure

Management of subsurface structures and infrastructure is largely determined by spatial planning and zoning decisions and by the type of structures involved. Some require long term management strategies including complex stakeholder consultation processes due to their societal significance, scale and expected lifetime. This holds for tunnels, water mains, sewer systems and national grids. Smaller constructions and infrastructure of mainly local significance require less complex management processes.

For optimal results of subsurface development and management, proper communication between civil engineers, geoscientists, developers, planners, regulators and the public is vital. To that end, advantages and disadvantages of subsurface development should be communicated to stakeholders, including the public, in clear, non-technical terms. Underground constructions do impact local land-use once an application for a construction permit is lodged. Therefore, prospective developers are advised to discuss any such proposal with the regulatory authorities in the earliest stages and to involve the public in consultation. Too often, proposals remain confidential until a formal decision process begins. Lack of transparency leads to suspicion and eventually mistrust amongst nearby residents and the public in general which may negatively impact decision making. Likely questions to arise include costs, reliability of underground service systems and safety aspects in case of emergency. These are best addressed by case studies of existing examples and, ultimately, guidance on good practices. However, the most persuasive examples are often local ones.

Identifying options for underground development may arise when consulting spatial development plans. As a next step, such options can be proposed for integration in such plans. However, few planning authorities currently consider subsurface development in policymaking and there is a general lack of specific planning guidance. Therefore, initiatives are needed to promulgate good practices, while taking into account differences in national or local legislation and procedures (Marker 2003).

Subsurface management also applies to preserving remains in archaeological sites. Such remains may disintegrate when groundwater levels fall and the remains are exposed to aerobic conditions in ground above groundwater. That may occur if water management schemes are modified. Excavation is not necessarily the best option to preserve artefacts. Apart from being prone to vandalism and theft, exposure may cause physical deterioration, such as cracking and spalling of monumental stone structures, weathering and crumbling of mud brick features, erosion and slumping of unexcavated cultural layers. In many countries archaeological sites are considered to be at risk, including Bolivia, Bulgaria, Cambodia, the Czech Republic, Guatemala, India, Israel, Italy, Jordan, Kenya, Lebanon, Mexico, Pakistan and Thailand (ICOMOS 2000). Risk of physical deterioration can be mitigated by permitting, making proper backfilling conditional to the excavation after exploration.

An increasingly significant issue in subsurface management concerns separating groundwater flows related to underground storage of warm and cold water for energy-saving purposes. Therefore, activities in the subsurface should be approached in a truly three-dimensional way (TCB 1996, 1997, 2004).

6.3.6. Comparison at regional and national levels

Management of contaminated land may bridge the gap between current (surface) land and subsurface management, in particular as sustainability issues are at stake. In 1994, the US National Research Council recognized that a permanent remedy for many waste sites was unrealistic. Brady *et al.* (1999) consider spending billions of US dollars to remediation every year a waste of money as this will not improve human or environmental health, not even in the distant future. Until recently, contaminants were considered to stay in the underground forever, leaving active removal by digging, pump-and-treat of the contaminants as the only option to clean-up the subsurface. This, however, has limitations (Nyer *et al.* 2001). The fact that natural processes (natural attenuation) can remove toxicity often faster and more definite than by engineering approaches, was first understood in the USA, in the early 1980s. That notion was based on a better understanding of the interacting geo- and biochemical processes in the subsurface. It introduced an entirely new perception on how to cope with

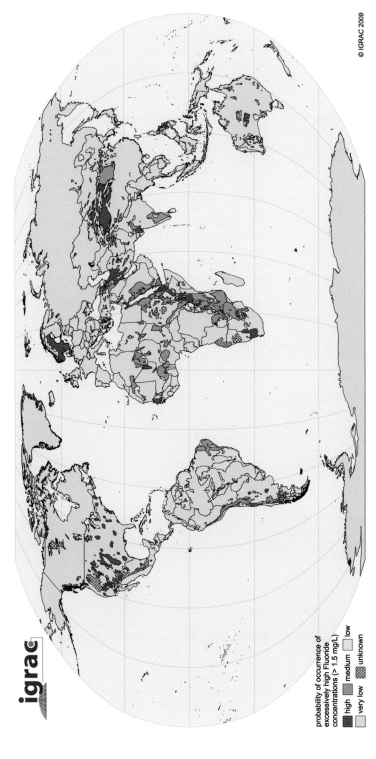

Fig. 6.8. Probability of occurrence of excessively high Fluoride concentrations (> 1.5 mg L^{-1}) (IGRAC 2011b). Five probability classes are distinguished: High-probability: Geological formation with an area of fluoride contaminated groundwater, situated in a hyper-arid or arid zone. Medium-probability: Geological formation with an area of fluoride contaminated groundwater, situated in a semi-arid or dry-subhumid zone. Geological formation or area, which has the characteristics of a potential fluoride rich environment, and is either (a) located in a known fluoride problem country (according to IGRAC and/or Unicef) or (b) adjoined to a fluoride problem country with the same geological formation crossing the border. In the second case no further evidence of fluoride contaminated groundwater is found in the literature so far. The climate is hyper-arid, arid or semi-arid in both cases. Low-probability: Geological formation with an area of fluoride contaminated groundwater, which continues in a moist subhumid or humid climate. Geological formation or area, which has the characteristics of a potential fluoride rich environment, and is either (a) located in a known fluoride problem country (according to IGRAC and/or Unicef) or (b) adjoined to a fluoride problem country with the same geological formation crossing the border. In the second case no further evidence of fluoride contaminated groundwater is found in the literature so far. The climate is dry-subhumid, moist-subhumid or humid in both cases. Unknown, assumed-probability: Geological formation or area, which has the characteristics of a potential fluoride rich environment, but is not located in a known fluoride problem country (according to IGRAC and/or Unicef), and no further evidence of fluoride contaminated groundwater is found in the literature so far. The climate is hyper-arid, arid, semi-arid or dry-subhumid. Very low: Remaining areas.

contaminated lands and on the important role of monitoring as part of management strategies. However, the benefits of the natural attenuation concept remained uncovered for a long time in the world of politics and legislation. There, clean-up methodologies for sites were determined by the Environmental Protection Agency and their methodologies are still heavily technology driven. For example, 'pump-and-treat' is still applied for some 75% of all groundwater contamination sites in the USA.

Moreover, CERCLA, RCRA and USA State Laws are still primarily based on Maximum Contaminant Levels (MCLs). As the rationale for these laws and regulations is not entirely incorrect, it is hard to change legislation even if it is based on sound science. Towards 2000, CERCLA came under attack as it was considered too expensive. Budgetary reasons might well be the driving force for replacing the 'active removal' by the 'natural attenuation' concept in legislation and management in the next years (Wiedemeier *et al.* 1999). The scale and complexity of soil and groundwater contamination cases, the use of natural attenuation and the consequently longer times that may be needed to reach end goals has now been recognized at political levels in The Netherlands where the legislative framework and policy guidelines is being adapted accordingly (VROM 2007*a*, *b*).

With only a few exceptions, management of underground structures and infrastructure is not yet practiced in an integral way anywhere. This is a matter of concern as increasing urbanization exerts more and more pressure on urban lands. There, such pressures may be released by more intense development of the subsurface. In some megacities and in a few densely populated countries as Japan and The Netherlands, the necessity to incorporate the subsurface in spatial plans is now being recognized. However, the dominant sectoral approach in legislation and policy development hampers such development (see Sections 6.1.4 and 6.2.4). Apart from integrating the subsurface in legislation and policy making, input of knowledge and (geo)-expertise in the early phases of urban planning and design will be required for sustainable management and development of the shallow subsurface.

6.4. Concluding remarks

This chapter demonstrates that the legal framework in which development of the subsurface should fit is quite complex. Based on the origins of the legal systems, parallels may be identified between nations on registration of landownership, mining of resources and environmental protection. On top of that, social and cultural aspects may influence the organization of land use, which may, in turn, impact potential development of the subsurface. Today, no country on Earth has a legislative, policy or management scheme in place in which the subsurface is integrated. Many have legislation that incorporates specific aspects or elements (minerals, water, environment, etc.) in separate legislative acts. Figure 6.9 presents an overview of types of legislation identified in countries worldwide.

Today, politicians and decision makers tend to approach the subsurface from a two-dimensional, flat perspective and normally ignore its three-dimensional and multifaceted characteristics, for example in spatial planning activities. But, as the underground inevitably will become a more serious option for development (Chapter 7), legal issues concerning underground construction warrant more attention requiring a more integrated approach. Generally, decision makers focus on water and mineral resources only as these support the (local) economy. Normally they go for short term and informal solutions if problems occur. Over-exploitation of for example groundwater resources may lead to short term economic and even health benefits, but may put society at risk in the long term. This requires proper legislation. As the subsurface is to be considered as an at least partly non-renewable, resource, where processes and impacts rule over long time spans, short term based decision making will inevitably lead to more severe problems in the future. Normally, underground structures last for centuries and thus sterilize options for future functions. For balancing such potentials adequate geoscientific knowledge and information is required (Section 7.4).

With a worldwide growing interest in sustainable development and management of the (urban) subsurface, the authors expect that legal and policy frameworks will significantly evolve in the near future. Current legal frameworks may serve as the basis for such development but more emphasis should be put on an integrated approach. Integration of the subsurface into legal, policy and management frameworks will contribute to a more sustainable development of land, allowing societies to benefit from improved quality of life in urban areas and from reduced risks to the environment. Integrated legal frameworks also facilitate innovative concepts and technologies. Therefore, decision-making should focus more on long term effects and address both current and future sustainability criteria.

One option to integrate subsurface development and management can be realized through the concept of 'multiple use of space' (Section 7.4). This option is meant to generate added value by combining functions of land use in three-dimensions. If analysed at relevant scales, the effects of such multiple use may be measured in direct financial terms, indirect economical effects, environmental and societal benefits. Cost-benefit analyses may generate arguments to support or discourage development of the subsurface and may thus assist in financing a project by various stakeholders (COB 2004*b*).

The potential for using the subsurface depends on relative costs of elements that are related to legal, policy and management issues as well as to sustainability. Marker (2003) summarized these in practical terms:

- *Site assessment and evaluation*: underground constructions require more detailed and more costly site

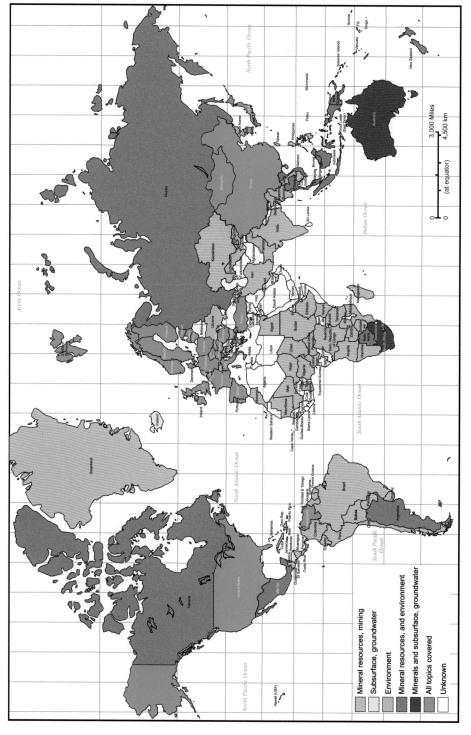

Fig. 6.9. Countries for which legislation relating to the subsurface was identified (basemap: © MapResources).

investigation studies than at the surface (Chapter 5). Major underground constructions often require Environmental Impact Assessment studies;
- *Capital works*: the cost for construction may be up to 50% higher for underground facilities than at the surface (Marker 2003). However, under specific geological conditions building underground facilities may be cheaper (Nordmark 2002). Cost evaluation should include the full life cycle of the construction and other costs, for example environmental needs, multi-use of space, extracted (construction) minerals and reuse of the space which may generate significant benefits;
- *Operational facilities (e.g. ventilation, filtration and other services)*: energy consumption is a significant factor in the operational costs of buildings. Underground facilities require less energy significantly reducing energy costs. Examples show that savings between 35 and 70% may be found (Barker 1986). For other services like waste and emission management, costs are comparable when building at the surface (Marker 2003);
- *Decommissioning*: these costs normally exceed those of removal of constructions at the surface (Marker 2003). However, since underground facilities last (much) longer than buildings at the surface, they have lower maintenance costs and may be re-used over time, the total life cycle cost needs to be considered as well (Sterling & Godard 2001);
- *Earn-back time*: for common subsurface developments, earn-back times are normally between 5 and 10 years mainly because of energy savings depending on local climate conditions (Dames & Moore 1983). The return on investment for subsurface constructions may vary with use, life-span and depth range resulting in different acceptable earn-back times compared to surface developments (Rogers & Horseman 1999);
- *Legal requirements*: Collecting permits for underground construction and use is relatively simple in case of single landownership where roles and responsibilities between stakeholders and authorities are evident. This may become more complicated if privately owned mineral rights have been severed from the land surface ownership. The aim of (sub)surface development and the possible interference with future (mining) activities may pose legal difficulties due to potentially conflicting interests. This may delay or prevent development, particularly of the subsurface.
- *Cost-benefits to society*: development of the subsurface should not be considered in isolation but in connection with other types of land use, including multiple-use. This is particularly true if aimed at relieving environmental pressures at the surface. Moreover, underground development may not be economic on its own, but can be if linked to other activities. For example, open pit mining can physically prepare the way to otherwise uneconomic underground mining and make it economic; or, if mining results in underground cavities that successively could be (re-)used for storage of, for example, strategic goods.

In general, building underground constructions is considerably more complex and expensive than building at the surface. This makes funding underground construction a challenge. However, that may change significantly if the cost of full life cycle, including longer term energy savings and lower depreciation costs over long time spans are taken into account, regardless of non-monetary values such as reduced environmental damages. That may encourage development of public–private partnerships and lead to sharing of investments and profits from underground facilities.

The authors expect that pressures on space, natural resources and quality of life issues will boost development of innovative concepts on sustainable development of the subsurface in the near future. Strategic visions, societal pressures and needs, together with successful examples, will inevitably lead to increased interest in subsurface options. Eventually, the initial higher costs will not hamper such development any longer. Appropriate legal frameworks will then emerge. Unequivocal registration of ownership or user-rights in three-dimensions and strategic planning of subsurface activities and use are priority topics in the legislative framework. Technological developments are expected to lead the way. Evolving new concepts and ideas for future use will be discussed in Chapter 7.

Chapter 7 Future use

This last chapter discusses sustainable management and development of the subsurface for today's and next generations. It builds on current and planned activities but also attempts to look further into the twenty-first century. The authors realize that such long term developments are quite to very, uncertain which makes this chapter not free from some speculation. Some of such uncertainties will be discussed in this chapter.

What developments in sustainable use and management of the shallow subsurface may be expected in the course of this century? In an attempt to answer this question some trends on land-use pressure as briefly described in Chapter 1 are addressed. In addition, public, industrial and governmental perceptions may steer development in a significant way. This is particularly true for living, working, recreation and transporting in the underground realm. But what options for hosting human generated functions may be offered by the underground itself, and will the underground be prepared for that? In other words, do properties of the ground comply with the specific physical requirements for the desired applications, and if so: where?

7.1. Future developments and trends

Five developments and trends concerning the need of physical space are distinguished: population, urbanization, quality of life, environmental awareness and technological development.

7.1.1. Population

The UN Population Division (2011) predicts that 9.3 (middle scenario, high: 10.6 and low: 8.1) billion people will live on this planet by 2050 (Fig. 7.1). From 2050 onwards, this number is expected to increase up to 10.1 billion in 2100 and thereafter the increase is likely to level out and the population will remain approximately constant for the following next centuries until at least 2300 (UN 2004). However, the predictions over longer periods in the future are highly dependent on the fertility rate which is uncertain. Consequently, the spread in the forecasted population is large and ranges between 6.2 and 15.8 billion people in 2100. Whatever the population figures beyond 2050, the planet should prepare for some 30% growth of its population in the next 40 years and provide space to accommodate its new inhabitants.

7.1.2. Urbanization

In its 2011 Review of the World Urbanization Prospects, the UN Population Division (2012) presents estimates and projections of the total number of people living in the world's urban areas until 2050. According to this review, urban people will increase with 2.6 billion between 2011 and 2050 while the entire world population is estimated to grow with only 2.3 billion (cf. middle scenario). This will result in a world urbanization degree of almost 70% in 2050 (around 30% in 1950 and 50% in 2010). Progressing and accelerating urbanization will put a major pressure on urban space as cities cannot always continue expanding horizontally. In combination, both trends will cause urban land prices to rise, particularly in the city centres. It is hard to predict how long this urbanization trend will continue to rise. However, there is no reason to expect this growth to significantly diminish before some stabilization in world population is expected to be reached by 2050 (Fig. 7.1).

7.1.3. Quality of life

Since 1820, GDP (Gross Domestic Product) per capita has risen significantly for all parts of the world. From 1950 to 1995, average incomes in the industrial nations rose 218% and in the developing countries 201% (Lomborg 2001). Simultaneously, life expectancy increased spectacularly alongside with the amount of calories per capita while the proportion of people living in poverty roughly halved. Today, individuals eat more and better quality food and use more natural resources than ever before. Wealthier people normally demand for better and larger housing, while economic growth demands for expanding industrial areas and economic zones. In the past half century, increased prosperity resulted in production of very large volumes of domestic and industrial waste. Cities are surrounded by waste disposal sites, often successively built-over by later urban expansion. Industrial expansion and waste disposal, together with increased food and natural resource production have also increased the demand for physical space. Although the sharply increased commodity prices since 2002 may have had an impact on economic growth patterns, these trends are likely to continue in the future. This assumption is supported by the drop in commodity prices since 2007 and the gradual increase since 1970 in the world's GDP per capita from an average of 2.2% to slightly above 3% in 2008 (IMF). Moreover, global fertility rates (children per woman) is rapidly declining from 4.92 in 1950/1955,

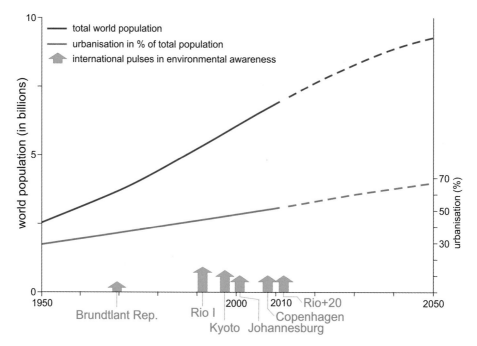

Fig. 7.1. Past development and future perspectives of three trends described above: world population (UN Population Division 2011), urbanization in % of world population (UN Population Division 2012) and pulses of environmental awareness.

via 2.52 in 2000/2005 to (medium scenario) 2.17 in 2045/2050 (UN Population Division 2011). However, the current (2009–2012) global economic crisis will impact economies and GDPs but the authors see no reason yet to believe that the long term economic development towards higher GDPs be reverted for a longer period.

7.1.4. Environmental awareness

Translating public environmental awareness into political action has put additional pressure on available space since the 1970s. Increasingly large areas were converted into natural parks or nature reserves and are closed for (other) development. Simultaneously, significantly large areas of arable land have degraded and became unsuitable for food production and for many other functions. Both land degradation and environmental protection have increased the pressure on land that can be developed and availability of physical space for growing urban populations. As the authors believe that environmental awareness among the public and politicians will continue to be a relevant factor in policy making the pressure is expected to last for at least a few more decades (Fig. 7.1).

7.1.5. Technological development

Finally, technological developments contributed strongly to mitigate some of the impacts of space consumption trends mentioned above. The green revolution dramatically increased food production per hectare and re-using natural resources began to continue to reduce volumes of mine waste materials. Technological development also supported major land reclamations providing additional space in and around urban areas, and led to improved and more concentrated waste disposal sites and to more effective incineration techniques. New technological developments significantly contributed to the tunnelling boom since the 1970s. New methodologies and techniques such as a new generation of Tunnel Boring Machines made underground construction safer, cheaper and faster since the early 1990s (Section 4.4.6). It is expected that these trends will show a fluctuating, but overall increase over the course of the 21st century if no major breaks in the trend of GDP development occur.

4D and 5D analysis tools and database structures will assist in more accurate modelling of the subsurface leading to further improvements in reliability of predicting behaviour of the underground environment (Section 5.2.6).

7.1.6. Future developments

The first four trends described above (either on their own or in combination) will reduce the amount of physically-available space and land will become a more precious commodity, particularly in urban areas. Many of the present and future (mega-)cities border the sea or major rivers limiting their potentials for lateral expansion. The same holds for cities adjacent to mountain ranges. To a certain extent, cities along water can expand on reclaimed land and in mountain areas slopes can be developed for constructions. However, if the pressure is high enough and if sufficient capital is available, underground construction will develop in the downtown areas and in mountainous urban fringe zones.

High-rise buildings may continue to be the dominant construction type in the city centres for the next few decades. But as they have their own intrinsic infrastructural and economic limitations, the authors believe that the underground will become an increasingly interesting domain for solving space problems at the surface in the near future. The subsurface offers an almost unlimited realm of potentials for development as only a very minor part of it is in use yet. Continued improvement of tunnel boring and other excavation techniques in combination with improved geo-prediction tools will further expand underground constructions, as takes place in China today (Qian & Chen 2007).

7.1.7. Uncertainties concerning underground development

Despite the positive indications to further develop underground space as described above, the underground is not the preferred domain for most people. Spending longer times underground will require breaking psychological barriers, cultures and mind sets. From research on public perception of using the subsurface it appears that Indonesians have a more positive perception of using underground space than Koreans and Japanese (Nishida *et al.* 2007).

Another uncertainty deals with the limited knowledge of the subsurface as compared to other compartments in the environment (Chapter 5). How knowledge may be matched with functions for the underground will be discussed in Section 7.3.

Other uncertainties concern public perception vis-à-vis that of local governments. In their research, Ellen *et al.* (2002) found that the best options for subsurface development emerge if strong, local governments take the lead, backed by widespread public and political interest for a better quality of (urban) space. That creates a climate in which the private sector is inclined to support such development. However, underground development may be severely hampered when local governments are reluctant to invest in underground structures in times of because the public interest in a better quality of (urban) space is low.

Options for using underground space may be checked against national, international and intergovernmental policy plans (Chapter 6). Below, potential future development of the subsurface will be discussed for mining and storage (Section 7.2) and for infrastructure and public space (Section 7.3).

7.2. Extraction and storage

Traditionally, the extraction industry applies straightforward, mechanical technology to extract ore, coal, water or oil from the ground. Today, the industry more and more collaborates with technological research centres to radically improve extraction methods. That also holds for more efficient management of subsurface resources, including smart drilling and underground processing of crude products (ISAPP 2011). These new approaches will most probably result in significantly better production schemes and (much) higher yields based on the full life cycle of the operation, from exploration to reclamation. Moreover, such technical and logistic improvements will reduce impact on the visible landscape and thus contribute to landscape preservation and other sustainable development issues.

7.2.1. Mining

In terms of sustainable use of the subsurface, open pit mines may continue to be a major challenge. Reclamation and redevelopment of abandoned mines will probably remain high on political agendas in the near future. Such reclamation may include (re)development of recreational areas and turning these into, for example, lakes. This has been done often in the past when pumping stopped in abandoned mines with high water tables.

Reclamation and landscaping require medium- to long-term investments. This may reduce or even rule out private capital for such investments and instead require tax money. In France, such taxes are currently due for exploitation of construction materials (Gallego & Vadillo 1992). Producers pay a percentage from their earnings on construction materials sold on the market. That money is set aside for financing reclamation plans to be approved by local governments prior to exploitation. Abandoned and depleted open pit mines may also have a future as industrial plants or storage sites. If situated in densely populated or recreational areas, such sites can be covered. Additional gains to the environment may include proper regulation of emissions and using emitted gasses as fuel for heating of buildings.

If society continues to require more construction materials, industrial minerals, metals, and energy, and if their extraction should be economically viable, open pit mining is expected to continue for many decades. That would halt only if many more commodities can be replaced by non-Earth materials and/or by recycling of construction materials. On a limited scale, recycling is applied in mineral resource-poor countries (de Jong & De Mulder 1998), but significant replacement of such materials is not foreseen before the second half of the 21st century. Until then, political pressure may push such extraction sites either underground or into remote, less densely populated areas. Moreover, this development might locally contribute to scarcity of such commodities and, consequently, to higher market prices.

7.2.2. Energy

For the remainder of the 21st century, Lomborg (2001) does not foresee serious global shortages in energy resources. Other authors disagree, stating that Earth resources are and will be exploited faster than new resources can be found. However, availability of non-energy resources appears to be strongly economically and politically controlled. Energy prices are also (and sometimes more seriously) controlled by availability of processing facilities. Traditionally, more resources are found during high price intervals provided that sufficient geo-expertise is available to identify such new resources. Over time, the need for some of the current commodities will decline while that of others, yet unknown, will increase.

Very large volumes of tar sands in Venezuela and Canada are waiting for exploitation as alternatives to oil and gas as energy sources. In Alberta (Canada) alone, these reserves amount to 280–300 billion barrels (Sexton 2011). As they occur at very shallow depths, open pit mining is likely the first option for their extraction. However, sustainable exploitation of these resources comes at quite substantial environmental costs. The tar sands are only excavated if the market price allows the expensive operations and environmental remediation, as was the case when oil prices exceeded 100 US$ barrel^{-1} in 2008.

Other potential energy sources occur in the form of methane hydrates of which very large volumes are found in the Polar Regions and in open water and seabed sediments at depths of over 300 m. As exploitation of methane hydrates may potentially destabilize continental slopes and generate submarine slides, their extraction has been and still is subject to extensive studies. Although Lowrie et al. (2004) believed that exploitation could start by now (2010) in the Gulf of Mexico, most authors believe that regular, safe exploitation may not be possible for at least another decade.

In addition to the methane hydrates, very large natural gas resources occur largely untapped in porous, relatively shallow deposits and dissolved in deep, saline lakes. Worldwide, more than 120 such potentially productive lakes are identified (Sinding-Larsen et al. 2006). Another potential energy resource is formed by shale gas, natural gas trapped in shale formations often at large depth. In the USA it forms already a significant source for natural gas, but in other areas it is also expected to be exploited. Exploitation is by methods similar to production of natural gas from surface by bore holes. To be able to exploit shale gas the naturally very low permeability of the shale has to be increased by hydraulic fracturing, so-called 'fracking' (creating discontinuities in the shale to allow the gas to flow). Production of shale gas is currently under discussion because the methods for fracking (requiring large quantities of (surface) water which is pumped into the shale mixed with chemicals) are thought by some to have a major impact on the overlying (hydro-)geology and surface environment (Kargbo et al. 2010).

Finally, enormous volumes of deep-seated black coal resources are widespread worldwide. However, these are increasingly difficult to exploit in an economically sustainable way with traditional methods. Application of remote controlled Coal Bed Methane (CBM) methods by underground degassing of buried coal beds may become a viable option for future energy supply.

These new potential energy resources must all be extracted from the (sometimes deep) subsurface. Their geological occurrences have to be identified and their production needs proper geological expertise. Market and environmental demands require more sophisticated and smarter production schemes and management plans than normally available today. The extraction industry has joined forces to cope with this challenge (ICMM 2011; IPIECA 2011).

7.2.3. (Ground) water

To comply with one of the UN Millennium Goals: public access to drinking water, major attention should be given to sustainable groundwater exploitation and water management (Struckmeijer et al. 2005). Integrated urban water models may provide a good basis for such development (Buma et al. 2003).

7.2.4. Storage

In the coming decades probably more materials will be stored underground as less storage space at the surface will become available and thus more expensive. Moreover, underground storage can be facilitated in a safer and more controlled environment than at the surface where materials and structures are subject to weather influences, accidents and political disturbances. This is supported by the significantly reduced insurance costs for underground stored food in comparison to storage at the surface in Norway (Broch 2007).

Underground storage may comprise strategic materials, such as energy resources, drinking water, and food. Also waste materials, as radioactive, chemical and other toxic wastes may be stored in the subsurface. Some of these materials will need permanent underground disposal while others should be temporarily stored in accessible repositories or in (confined) aquifers. All types of underground storage, however, do require proper subsurface management.

7.3. Infrastructure and public space

The authors expect that public facilities are the principal asset for developing the subsurface. For reasons given above, it is anticipated that a significant share of public services, economic zones, transport facilities and industrial activities in urban areas will eventually move into the subsurface. First, a brief overview is given of infrastructural works (mainly tunnels) that are anticipated to be commissioned or are built presently or in the near future.

With many long tunnels currently under construction, it seems that the rising trend in tunnelling in Europe (see Section 3.2.1) will continue for at least one more decade. The world's longest railway tunnel, the Gotthard Base tunnel, with a length of 57 km, is scheduled for completion in 2017 (Section 4.5.7; Leybold-Johnson 2010). In the USA, reconstruction of the New York World Trade Center includes substantial underground compartments. The 22.2 km Îyama railway tunnel in Japan is scheduled for completion in 2013.

China is world-leading in tunnelling today. Given that there are significantly more longer traffic tunnels under construction here than in all of Europe, it is likely that China will keep this position for at least another decade. The subway network in Beijing, for example, is expected to grow to 1000 km by 2020 (Qian & Chen 2007). It will then have surpassed New York's subways in length and be the longest underground network in the world. In the next 20–30 years, 5 cross-sea tunnels are planned: the Bohai Bay cross-harbour tunnel from Dalian to Yantai, the Hanzhou Gulf project from Shanghai to Ningbo, the Lingdingyang cross-sea project connecting Hong Kong, Macao, Guangzhou, Shenzhen and Zhuhai, the Qiongzhou Strait cross-sea project joining Guangdong and Hainan and the Taiwan Strait cross-harbour project linking Fujian with Taiwan.

Increase in tunnel construction projects is partly due to the availability of much better tunnelling techniques, in particular for soft underground conditions. But it also seems to reflect a more positive public and political perception of underground construction in general.

7.4. Future subsurface management

7.4.1. Geological quality

Even if all indicators and trends would point to the need for underground construction, local subsurface conditions might prevent doing so. Further use of the subsurface largely depends on technical opportunities and constraints determined by geological factors, for example the quality of subsurface rocks and its constituents. Occurrences (or the lack) of resources, bearing capacity, and thermo-insulation capacity of rocks for underground storage are all determined by geological factors. Often, they also control processes affecting the environment (e.g. groundwater flow). Although the geological quality of the subsurface is not the only decisive factor for underground activities, it does play a key role. For example, fresh groundwater cannot be extracted from dry or impermeable rocks. Proper knowledge of spatial distribution of geological characteristics is required for any subsurface activity. The intrinsic quality of subsurface rocks (Chapter 4) should be determined by geoscientists.

Potentials and constraints of ground units for hosting a wide series of functions may be checked as a first step. Results of the exercise depend on the type of subsurface geological structure and its solid, liquid or gaseous components. The following functions or applications were checked against rock properties for The Netherlands (De Mulder *et al.* 2003):

- Potential extraction of natural gas, oil, (black) coal (in two forms), lignite, peat, hot water, fresh groundwater, rock salt, gravel, sand, clay, and limestone;
- Potential storage of carbon dioxide, temporary storage of natural gas and oil, geothermal (water), fresh water, waste material;
- Infrastructure;
- Buffer for energy storage.

Figure 7.2 illustrates an outcome of such an exercise for The Netherlands. The most relevant host rock properties for extraction include: occurrence and quality of resource material, hydrological isolation capacity of caprock, accessibility and technical extractability. For storage, these requirements include: hydrological isolation and thermal insulation capacity, stability, and accessibility; and for underground infrastructure and public space: accessibility, bearing capacity, potential for tunnel boring machines, stability and hydrological isolation capacity.

When host rock potentials are known, their spatial distribution in the subsurface in terms of occurrence and thickness should be determined. This is normally done through geological mapping programmes by geological survey organizations. Although availability and public accessibility of (preferably digital) maps may vary from country to country (Chapter 5), most geological surveys do have general maps on the distribution of a variety of subsurface strata. 3D systems, including digital maps for all subsurface strata for entire nations are still exceptional (Chapter 5). Today, geological survey organizations invest much in digitization, and by 2025, the subsurface of a quarter of the world's megacities may well be available in digital format (De Mulder & Annells 2000). However, global digital geological map information may be available significantly earlier (De Mulder & Jackson 2007).

7.4.2. Combined use and planning

As a next step, options for future use of the subsurface may be checked against possible geological constraints, such as natural hazards, unfavourable groundwater flows; and against assets such as resources. Often, underground construction projects are less vulnerable to natural disasters if compared to their equivalents at the surface (Section 4.7.4). Such evaluations may result in suitability maps for specific uses, for example Maring *et al.* (2003).

In densely populated areas with high land prices, the subsurface of a single parcel of land might be claimed by more than one party, opening the way for ventures for multiple usage of the subsurface. Not all functions may be adequately

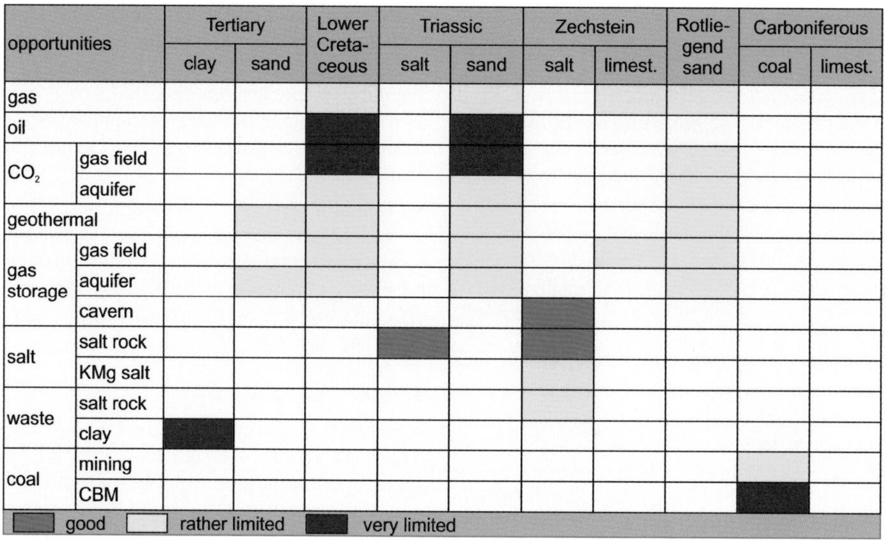

opportunities		Tertiary		Lower Creta-ceous	Triassic		Zechstein		Rotlie-gend sand	Carboniferous	
		clay	sand		salt	sand	salt	limest.		coal	limest.
gas											
oil											
CO$_2$	gas field										
	aquifer										
geothermal											
gas storage	gas field										
	aquifer										
	cavern										
salt	salt rock										
	KMg salt										
waste	salt rock										
	clay										
coal	mining										
	CBM										

Legend: good / rather limited / very limited

Fig. 7.2. Matrix of potential functions (horizontal) checked against host rock properties (vertical) in the subsurface of The Netherlands. From this figure it appears that most rock strata may host extraction, storage or infrastructure functions (CBM: Coal Bed Methane; K/Mg salt: Potassium (K)/Magnesium (Mg) salt).

combined in the subsurface. Functions requiring strongly different ground properties might mutually exclude each other. This applies, for example, for production of fresh groundwater and storage of waste, even if in different aquifers. However, functions exploiting the same rock properties, such as production and storage of natural gas may very well be combined but at successive moments in time. Decision-making on options for multiple use of the subsurface requires expertise on both ground properties and on the boundary conditions for the planned functions.

Again, this was tested for various options in the subsurface of The Netherlands (Fig. 7.3).

Contaminated sites may offer excellent opportunities for underground construction once assessed as new development sites. Decontamination might require quite intensive excavation resulting in creating underground space. Decontamination combined with underground development may significantly cut costs for both types of investment, making relatively expensive underground development economically feasible. This requires substantial planning and

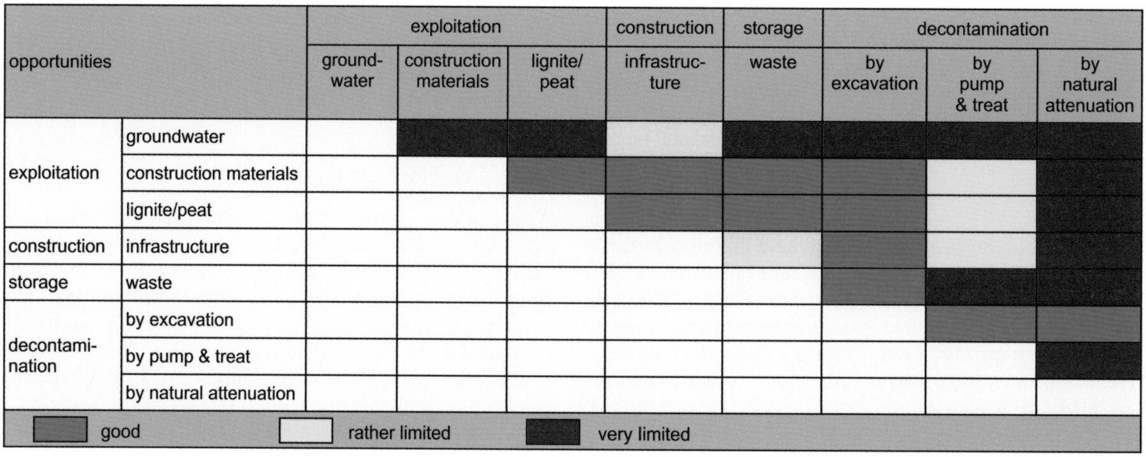

opportunities		exploitation			construction	storage	decontamination		
		ground-water	construction materials	lignite/peat	infrastruc-ture	waste	by excavation	by pump & treat	by natural attenuation
exploitation	groundwater								
	construction materials								
	lignite/peat								
construction	infrastructure								
storage	waste								
decontami-nation	by excavation								
	by pump & treat								
	by natural attenuation								

Legend: good / rather limited / very limited

Fig. 7.3. Options and exclusions of multiple, combined usages of the subsurface for various functions in The Netherlands (updated after Remmelts 1997 in De Mulder *et al.* 2003).

management activities and often a leading role for the (local) government.

7.4.3. Urban planning

The same principles apply for 3D planning of urban development. Based on a proper understanding of the potential assets and challenges of the subsurface, areas containing geological units with particularly favourable properties for certain applications, for example thick layers of impermeable rocks under waste disposal sites, should be given priority over those with unfavourable geological conditions. The same holds for urban expansion where quite significant savings may be generated if construction is located on solid ground rather than in former swamps (De Mulder 1986). High quality construction materials should preferably be mined before they become 'sterilized' by urban sprawl (Lumsden 1992). Such operations may become common practice rather soon as more and more cities start storing and accessing their subsurface data in dedicated databases. If coupled to Geographical Information Systems and handled by urban geoscientists, these will provide urban planners with all necessary information for transparent decision making (Section 5.3.4).

7.4.4. Future underground constructions

Geoscientific information will be indispensable for future decision making for subsurface development. Such information might be inserted into Decision Support Systems for conventional types of underground construction as parking garages, pipelines and shallow tunnels, but also for innovative ways to use the subsurface such as deep dome-type structures, 'Geo-domes' (Taselaar 2002; Box 7.1 and Fig. 7.4). However, such large structures will have a major impact on the subsurface. Initially, the construction of such major underground objects will disturb the groundwater flows, even in a wider realm. Dependent on groundwater flow velocities new balances in flow directions will be established around such structures within years to decades. This also holds for the prevailing biogeochemical processes in the subsurface. If designed properly, such underground structures may even be applied for hydrological engineering and (bio)remediation practices (see Section 3.3).

7.4.5. Non-construction future use of the subsurface

As discussed in Chapter 3, the subsurface may be used for many more purposes than for construction only. Hydrological engineering is a way of managing the subsurface, particularly when dealing with contaminated soils. By manipulating groundwater flow rates and directions, contaminated groundwater plumes may be steered into the direction of pumping stations. This methodology has the potential of becoming an important tool for water managers. It may not only be applied for cleaning-up contaminated soils, but also for combating natural or man-induced groundwater salinization, for drought prevention, or for enhancing fresh water supply.

Temporary cold and hot water storage is already common practice in some places, mainly for energy saving purposes. This application requires careful groundwater management as interference or mixture of water flows will reduce yields. In Beijing, shallow warm and cold water storage is being applied on a large scale aiming at 35 million m^2 by the closure of the 11th Five-Years Plan (2006–2010). This will save 36 000 tonnes of black coal for heating and cooling per year while producing 840 000 tonnes less CO_2 (Shi 2009). If subsurface conditions are known for an entire hydrogeological system, this methodology can be applied on a large scale. This will become reality in the European Union once all geohydrological systems have been investigated and mapped, providing opportunities for large-scale international groundwater management in the future.

As most of the biodiversity on-shore lives in the topmost soil layers (Dent et al. 2005) and in accordance with current trends in subsurface management, subsurface biota are anticipated to be given a much more significant role in development than today. Bioremediation is already applied for decontaminating polluted soils and groundwater. Along with a better understanding of the intrinsic properties of the subsurface biota and buried biomasses, a wealth of new applications in the subsurface may be developed. This would require quite significant additional research than underway today. Taking into account the very considerable progress made in this field during the past two decades, it does not seem unrealistic predicting revolutionary changes in day-to-day management practice for the next two to three decades on the basis of smart applications of specific segments of the living and non-living biomasses in the subsurface.

For these and other reasons, some caution should be expressed on dedicating major portions of the subsurface to underground construction purposes only. By doing so, the subsurface might become sterilized for other, future and yet unforeseen or unknown options. Wise management of the subsurface should also include keeping particular realms free from any development and preserve them for future, perhaps yet unknown, options. Today, this is practiced occasionally to avoid disturbance of natural attenuation processes for decontaminating parts of the subsurface (Brady et al. 1999).

7.5. Closing remarks

From this chapter it might appear that opportunities for future use of the subsurface are almost endless. However, legislative (Chapter 6), knowledge (Chapters 4 & 5), psychological and technological constraints (Chapter 7.1) hamper the full exploitation of this dimension. It is

Box 7.1. Geodomes

A new, prize-winning concept of building in the subsurface is the 'geodome', a rather large, high dome-like structure, excavated in the subsurface (Fig. 7.4). It is built relatively deep (roof top between 500 and 50 m below the surface) where ground pressures maintain the dome-like structures with only minimum support structures if situated in an appropriate ground unit. The roofs are made of concrete or synthetic materials and prevent groundwater leakages. Geodomes are mutually connected and linked to the surface through relatively narrow tubes.

Many space consuming or environmental disturbing functions might be accommodated in geodomes. With constant temperatures of about 18 °C and a relative humidity of 55% (at 300 m depth) geodomes are ideal places to store goods, almost cancelling costs for heating. Geodomes might host industrial plants providing excellent options for closing energy and waste loops. Geodomes may also host social functions such as cinemas, libraries, pools and recreation centres.

Geodomes might have potential for developing the subsurface, in particular under cities and industrial sites where strategic components, such as control rooms, might well be constructed. Development may be preceded by constructing test sites quite soon because no major technical obstacles are envisaged.

Fig. 7.4. Geodomes, a new concept for underground construction (courtesy of Hompe & Taselaar 2007).

reasonable to assume that at least some of these constraints will be overcome in the next decade, at places where pressures on subsurface development are highest, as in cities. By creating comfortable and light environments in underground rooms with sufficiently reassuring safety and security measures, also psychological hesitation by the public may be overcome as demonstrated in several underground sites (Duffaut 2008).

Penetration depth is one of the technical limitations for deeper underground development. As a reference, the deepest borehole on Earth reached a depth of 12 261 m below the surface in 1984 at Kola Peninsula (Russia, Fig. 7.5), when high temperatures (235 °C) and pressures prevented any deeper mechanical penetration. More advanced technologies should allow up to 20 km deep holes in the next two decades (pers. comm., Kümpel KTB). Initially, the need for such deep boreholes would be driven by geological and technological curiosity. Economic prospects, in terms of exploration of specific resources, will drive wider application. New penetration techniques will emerge and may

Fig. 7.5. The Kola super-deep bore hole drilling site in NW Russia. Maximum depth penetrated: 12 261 m (photograph: E. De Mulder).

include non-mechanical methods. Tunnel Boring Machines (TBMs) will become more versatile and cope better with variable and unexpected ground types. Freezing and *in-situ* grouting support with road header type of excavations are also likely to be further developed due to their flexibility and low investment costs. Use of chemical and biological agents to modify ground characteristics, for example *in-situ* sand cementation by specific bacteria, have good perspectives to be further developed and applied at larger scales (Van Paassen *et al.* 2009).

Current trends in technological development, spatial planning and psychological barriers point to development of the topmost 100 m first. Deeper (100–500 m) parts of the subsurface may offer comfortable working and energy saving conditions and will probably be targeted next. However, simultaneous development of both realms, but for different reasons, seems most likely.

Underground construction activities will significantly and (almost) permanently impact the delicate balances in and between geological, hydrological, biological, physical and chemical processes in the subsurface. With growing knowledge of these processes and their complex interactions, together with significantly increased modelling capabilities, it is anticipated that such disturbances and establishment of new balances in the subsurface may be predicted more and more adequately. As knowledge and practical expertise in hydrological engineering increases rapidly as well, we expect that new and (environmentally) more favourable balances may locally replace existing ones.

Knowledge should play a vital role in the current debates on the impact of climate change to the environment. Climate change is nothing new to our planet and the ever more detailed geological records demonstrate that climate continuously changed throughout the 4.5 billion years of the Earth history (Section 3.3.3). How would new or accelerated climate change impact the underground? In general, such change has (much) less impact on the underground than at the surface. Climate change may primarily influence processes that are active at the surface (exogenic processes), such as weathering, sedimentation and erosion. Normally, their impact in the subsurface (>10 m) would be marginal. In some extreme cases, however, as during glaciations, subglacial scouring might locally affect the subsurface as deep as 500 m (Wildenborg *et al.* 1990). Changes in exogenic processes may alter groundwater conditions. As the subsurface is normally saturated with groundwater, changes in groundwater condition will influence the delicate balances in the subsurface as well. Such changes will be more gradual, less pronounced and retarded as compared to the surface. The full impact of climate change might become manifest in the subsurface thousands of years after the event took place in the atmosphere. As a rule, impact of any change at the surface, including climate change, decreases with depth.

As prices of land in downtown areas are (and probably remain) highest, significant underground development may start in (mega)cities first, as demonstrated in Canada and China. Unfavourable climatic conditions are powerful drivers of underground development too (Box 3.3). Interest by investors and visionary municipality leaders are also important factors in this respect. In combination, these factors point to big cities in continental climatic and economic powerful zones with strong and visionary municipal or national leadership as the forerunners of substantial global subsurface development. Such conditions might be found along the East coasts of the USA and Canada, Central USA, in Northern, Central and Eastern Europe, and in China. Qian & Chen (2007) expect that underground Beijing will have tripled in size from 30 km^2 (in 2007) to 90 km^2 in 2020. By then, 20–30% of the total floor area in Beijing's central urban districts will be underground.

Fig. 7.6. Artist impression of a megacity by the end of the 21st century (illustration Beeldleveranciers).

Taking into account the long term benefits of underground development for urban dwellers and their municipalities, it is probable that other nations will follow the Chinese example. The challenge is to build underground cities in harmony with the geological, chemical, biological, and physical subsurface conditions; in harmony with social and physical development at and above the surface; and in harmony with future options. This provides interesting perspectives for cooperation between geo-experts, engineers and subsurface managers.

7.5.1. City of the future

Based on the trends described in Section 7.1, on historic data for urban development and on our still limited but expanding understanding of the Earth processes, it is tempting to speculate how a city may look like by the end of the 21st century. As discussed above, such cities may have a significant underground realm where people work, recreate and eventually even live. Underground urban development may expand in periods of relatively strong, visionary municipal leadership driven by perspectives for sustainable cities, supported by public environmental awareness and commercial opportunities. If inevitable accidents in underground construction would be limited in scale and positive experiences prevail, psychological barriers against staying underground will eventually evaporate. Moreover, if the knowledge base expands and becomes accessible to the public domain, if legislation adapts to this new reality, if technology further improves making underground construction safer and cheaper, the city of the future will be significantly indoors and underground (Fig. 7.6).

References

ABDULAI, R. T. 2006. *Is land title registration the answer to insecure and uncertain property rights in sub-Saharan Africa?* RICS Research paper series, **6/6**. University of Wolverhampton, UK.

ABDALLATIF, T. A., ABDEL RAHMAN, A. A. ET AL. 2009. Groundwater geophysics: a tool for hydrogeology. *In*: KIRSCH, R. (ed.) 2nd edn. Springer-Verlag, Berlin, Heidelberg.

ABROL, I. P., SHARMA, B. R. & SEKHON, G. S. 2004. *Groundwater use in North-West India. Workshop Papers, 202*. Centre for Advancement of Sustainable Agriculture, New Delhi.

AD. 2010. Medewerkers onwel door damp Noord-Zuidlijn. Algemeen Dagblad. http://www.ad.nl/ad/nl/1041/Amsterdam/article/detail/1910042/2010/10/08/Medewerkers-onwel-door-damp-Noord-Zuidlijn.dhtml [in Dutch] [accessed 12 February 2012].

AGS. 2012. *AGS data format*. Association of Geotechnical and Geoenvironmental Specialists (AGS), UK. http://www.ags.org.uk/site/datatransfer/intro.cfm [accessed 7 March 2012].

AKCIN, H., KUTOGLU, H. S., KEMALDERE, H., DEGUCHI, T. & KOKSAL, E. 2010. Monitoring subsidence effects in the urban area of Zonguldak Hardcoal Basin of Turkey by InSAR-GIS integration. *Natural Hazards and Earth System Sciences;* An Open Access Journal of the European Geosciences Union, **10**, 1807–1814.

AKSOY, C. O. 2008. Review of rock mass rating classification: historical developments, applications and restrictions. *Journal of Mining Science*, **44**, 51–63.

ALDRIDGE, D. 2007. *The world's largest foamed concrete project*. AllBusiness.com. http://www.allbusincss.com/mining-extraction/mines-mining-nonmetallic-mineral/8907810-1.html [accessed 10 October 2010].

ALEJANO, L. R., TABOADA, J., GARCÍA-BASTANTE, F. & RODRIGUEZ, P. 2008. Multi-approach back-analysis of a roof bed collapse in a mining room excavated in stratified rock. *International Journal of Rock Mechanics and Mining Sciences*, **45**, 899–913.

ALONSO-ZARZA, A. M. & WRIGHT, V. P. 2010. Chapter 5: Calcretes. *In*: ALONSO-ZARZA, V. P. & TANNER, L. H. (eds) *Carbonates in Continental Settings: Geochemistry, Diagenesis and Applications*. Developments in Sedimentology, **61**, Elsevier, Amsterdam, 225–267.

ANDERSSON, J. C. 2010. *Rock Mass Response to Coupled Mechanical Thermal Loading. Äspö Pillar Stability Experiment, Rocha Medal*. Publ. ISRM, Sweden. http://www.isrm.net/fotos/gca/1244221304abstract_andersson.pdf [accessed 2 November 2010].

ANON. 1990. Rock joints. *In*: BARTON, N., STEPHANSSON, O. (eds) *Proceedings of the International Symposium on Rock Joints*. Loen, Norway, Balkema, Rotterdam.

ANON. 2004. *Tunnel Lining Design Guide*. British Tunnelling Society, Institution of Civil Engineers. Thomas Telford Publishing, London.

ANON. 2010. *Optionenvergleich Asse. Fachliche Bewertung der Stilllegungsoptionen für die Schachtanlage Asse II*. Bundesamt für Strahlenschutz, Salzgitter, Germany. January. BfS-19/10. URN: urn:nbn:de:0221-201004141430. http://www.bfs.de [in German].

ANTIQUITIES. 2011. http://www.antiquities.org.il/article_Item_eng.asp?sec_id=14&subj_id=139

ARENSON, L. U., ALMASI, N. & SPRINGMAN, S. M. 2003. Shearing response of ice-rich rock glacier material. *In*: PHILLIPS, M., SPRINGMAN, S. M. & ARENSON, L. U. (eds) *Permafrost: Proceedings of the 8th International Conference on Permafrost*. Swets & Zeitlinger, Lisse, Zurich, 39–44.

ARNOLD, A. B., BISIO, R. P., HEYES, D. G. & WILSON, A. O. 1972. Case histories of three tunnel-support failures, California aqueduct. *Bulletin of Association Engineering Geologists*, **9**, 265–299.

ARSENAULT, J.-L. & CHOUTEAU, M. 2002. Application of the TISAR technique to the investigations of transportation facilities and detection of utilities. *In: Proceedings of the Geophysics 2002. The 2nd Annual Conference on the Application of Geophysical and NDT Methodologies to Transportation Facilities and Infrastructure*. 15–19 April, Los Angeles. Federal Highway Administration (FHWA-WRC-02-001), Transportation Research Board, California Department of Transportation Federal Highway Administration. Federal Highway Administration, Washington [CD-ROM].

ARTS, R., EIKEN, O., CHADWICK, A., ZWEIGEL, P., VAN DER MEER, L. & KIRBY, G. 2004. Seismic monitoring at the Sleipner underground CO_2 storage site (North Sea). *In*: BAINES, S. J. & WORDEN, R. H. (eds) *Geological Storage of CO_2 for Emissions Reduction*. Geological Society, London, Special Publications, **233**, 181–191.

ASTM. 2011. *ASTM International* (formerly American Society for Testing and Materials). http://www.astm.org/ [accessed 6 February 2011].

ASTRIUM. 2011. ASTRIUM (EADS). http://www.astrium.eads.net/ [accessed 10 January 2011].

ATHENS. 2007. *The Athens Metro and Archaeology*. http://www.athensinfoguide.com/gettingaround.htm#1 [accessed 27 March 2012].

AYAYDIN, N. & LEITNER, A. 2009. Tauern tunnel first and second tubes from the consultant's viewpoint. *Geomechanics and Tunnelling*, **2**, 24–32.

AYDAN, Ö., OHTA, Y., GENIŞ, M., TOKASHIKI, N. & OHKUBO, K. 2010. Response and stability of underground structures in rock mass during earthquakes. *Rock Mechanics and Rock Engineering*, **43** (6), 857–875.

AYKAR, E., ARIOĞLU, B., ERDIRIK, N., YÜKSEL, A., ÖZBAYIR, T., ARIOĞLU, E. & YOLDAS, R. 2005. Tunnel excavation works in Taksim–Kabataş funicular system project. *In*: ERDEM, Y. & SOLAK, T. (eds) *Underground Space Use: Analysis of the Past and Lessons for the Future*. Taylor & Francis Group, London, 973–979.

B&A GROEP. 1997a. Deel 1: Gebruiksmogelijkheden van de ondergrond, interacties en conflicterende belangen. *In: Haalbaarheidsstudie Ruimtelijke Planning van het Gebruik van de Ondergrond* [in Dutch].

B&A GROEP. 1997b. Deel 2: Belangenafweging binnen de huidige regelgeving. *In: Haalbaarheidsstudie Ruimtelijke Planning van het Gebruik van de Ondergrond* [in Dutch].

REFERENCES

BABENDERERDE, S., HOEK, E., MARINOS, P. & CARDOSO, A. S. 2004. Characterization of granite and the underground construction in Metro do Porto, Portugal. *In*: VIANA DA FONSECA, A. & MAYNE, P. (eds) *Proceedings of the 2nd International Conference on Site Characterization, ISC'2*, Porto, Millpress, Rotterdam, 39–47.

BAI, Y., YAN, J. L., GE, K. J. & LI, X. Y. 2006. RBJ for sizable underpass construction in saturated soft soils. *In*: LEE, I.-M., YOO, C. & YOU, K.-H. (eds) *Safety in the Underground Space. Proceedings of the ITA-AITES 2006 World Tunnel Congress & 32nd ITA General Assembly*. 22–27 April, Seoul. Tunnelling and Underground Space Technology. **21/3-4**, 353 (abstract with full article as supplementary content on-line with Science Direct).

BALOBAYEV, V. T., SKACHKOVA, YU. B. & SHENDERA, N. I. 2009. Forecasting climate changes and the permafrost thickness for Central Yakutia into the year 2200. *Geography and Natural Resources*, **30**, 141–145.

BAOJUN, W., BIN, S. & ZHEN, S. 2009. A simple approach to 3D geological modelling and visualization. *Bulletin of Engineering Geology and the Environment*, **68**, 559–565.

BARBER, M. B. & HUBBARD, D. A. 1997. Overview of the human use of caves in Virginia: a 10 500 year History. National Speleological Society. *Journal of Cave and Karst Studies*, **5**, 132–136.

BARKER, M. B. 1986. Using the earth to save energy: four underground buildings. *Tunnelling and Underground Space Technology*, **1**, 59–65.

BARLA, G., BONINI, M. & DEBERNARDI, D. 2008. Time dependent deformations in squeezing tunnels. *In*: JADHAV, M. N., BHENDIGERI, O. B., RAO, B. H., PATEL, A., SHANTHAKUMAR, S., GUMASTE, S. D. & SHINDE, S. B. (eds) *Geomechanics in the Emerging Social & Technological Age, Proceedings 12th International Conference of the International Association for Computer Methods and Advances in Geomechanics (IACMAG)*. 1–6 October, Goa, India. Indian Institute of Technology (IIT), Bombay, India. 4265–4275. Available on CD & http://www.iacmag.org/ [accessed 12 March 2012].

BARTON, N. R. 2000. *TBM Tunnelling in Jointed and Faulted Rock*. Taylor & Francis.

BARTON, N. R. 2002. Some new Q-value correlations to assist in site characterisation and tunnel design. *International Journal of Rock Mechanics and Mining Sciences*, **39** (2), 185–216.

BARTON, N. R., LIEN, R. & LUNDE, J. 1974. Engineering classification of rock masses for the design of tunnel support. *Rock Mechanics and Rock Engineering*, **6** (4), 189–236.

BASTIDA, E. 2002. *Integrating Sustainability into Legal Frameworks for Mining in Some Selected Latin American Countries*. Centre for Energy, Petroleum and Mineral Law & Policy, University of Dundee, UK. Mining, Minerals and Sustainable Development project, **Report 120**, International Institute for Environment and Development (IIED), World Business Council for Sustainable Development.

BAUER. 2010. *Bauer Specialtiefbau GmbH*. http://www.bauer.de/export/sites/http://www.bauer.de/[accessed 30 January 2011].

BBC. 2007. *Mine-shaft hit Metro line reopens*. British Broadcasting Corporation. http://www.news.bbc.co.uk/2/hi/uk_news/england/ tyne/7060683.stm [accessed 12 February 2012].

BEAVER, P. 1973. *A History of Tunnels*. The Citadel Press, Secaucus, New Jersey, USA.

BEER, G. 2009. Introduction (illustration page 6). *In*: BEER, G. (ed.) *Technology Innovation in Underground Construction*. CRC Press/Balkema, Taylor & Francis Group, London.

BEER, G. 2010. The European project 'Technology innovation in underground construction' – Overview of IT results. *In*: TOLL, D. G., ZHU, H. & LI, X. (eds) *Information Technology in Geo-Engineering. Proceedings of the 1st International Conference (ICITG)*. 16–17 September, Shanghai. Joint Technical Committee 2 (JTC2) & Tongji University, Shanghai. IOS Press, Amsterdam.

BEGONHA, A. & SEQUEIRA BRAGA, M. A. 2002. Weathering of the Oporto granite: geotechnical and physical properties. *Catena*, **49**, 57–76.

BEKENDAM, R. F. 2004. Stability and subsidence assessment over shallow abandoned room and pillar limestone mines. *In*: HACK, R., AZZAM, R. & CHARLIER, R. (eds) *Engineering Geology for Infrastructure Planning in Europe: A European Perspective*. Lecture Notes in Earth Sciences, **104**, Springer-Verlag, Berlin, Heidelberg, 657–670.

BELKAYA, H., OZMEN, I. H. & KARAMUT, I. 2008. The Marmaray Project: managing a large scale project with various stake holders. *In: Proceedings of the World Congress on Engineering. WCE 2008*. 2–4 July, London. International Association of Engineers (IAENG). Newswood Ltd, Hong Kong, Vol. 2, 1268–1272.

BELL, F. G. 2007. *Engineering Geology*. Butterworth-Heinemann, Oxford, 2nd edn.

BERGADO, D. T. & PATAWARAN, M. A. B. 2000. Recent developments of ground improvement with pvd on soft Bangkok clay. *In: Proceedings of the International Seminar on Geotechnics in Kochi (ISGK2000)*. 24–25 September, Kochi, Japan. The Japanese Geotechnical Society/Kochi University of Technology. Available from: http://www.cofra.sk/files/documents/technicke_spravy/bergadoosmose.pdf [accessed 17 October 2010].

BERGKAMP, G. & CROSS, K. 2006. Groundwater and Ecosystem Services: towards their sustainable use. *In: Int. Symposium on Groundwater Sustainability*. ISGWAS, Instituto Geológico y Minero de España (IGME), Madrid (Spain), Alicante, Spain, 177–193.

BERGMAN, S. M. 1986. The development and utilization of subsurface space. U.N. Progress Report. *Tunneling and Underground Space Technology*, **1**, 115–144.

BERKELAAR, R. 2009. Coping with underground risks during the development of a new underground metro station in Rotterdam. *In*: CULSHAW, M. G., REEVES, H. J., JEFFERSON, I. & SPINK, T. W. (eds) *Engineering Geology for Tomorrow's Cities. Proceedings of the 10th Congress International Association for Engineering Geology and the Environment*. 6–10 September 2006, Nottingham. Engineering Geology Special Publication, 22, Geological Society, London, paper 460 [on CD-Rom].

BÉTOURNAY, M. & LEFCHIK, T. 2008. Highway stability considerations and site work associated with abandoned North American mines. *In: Proceedings of the Symposium Post-mining 2008*. 6–8 February, Nancy, France. Groupement d'Intérêt Scientifique de Recherche sur l'Impact et la Sécurité des Ouvrages Souterrains (GISOS)/International Mine Water Association (IMWA). Available on CD ROM and from: http://gisos.ensg.inpl-nancy.fr [accessed 11 February 2012].

BEZUIJEN, A. & KORFF, M. 2009. Building damage and corrective grouting research at North South line project in Amsterdam. *GEOtechniek*, Special Issue, 8–9.

BGR/UNESCO. 2008. *Groundwater Resources of the World 1:25 000 000*. World-Wide Hydrogeological Mapping and Assessment Programme (WHYMAP), Hannover.

BGS. 2012. *British Geological Survey*. http://www.bgs.ac.uk/ [accessed 8 March 2012].

REFERENCES

BGS-NGDC. 2012. *British Geological Survey–National Geoscience Data Centre.* http://bgs.ac.uk/services/ngdc/home.html [accessed 10 March 2012].

BIENIAWSKI, Z. T. 1989. *Engineering Rock Mass Classifications: A Complete Manual for Engineers and Geologists in Mining, Civil, and Petroleum Engineering.* Wiley, New York.

BIENIAWSKI, Z. T. 1990. Tunnel design by rock mass classifications. DACW39-78-M-3114; January 1990. Update of Technical Report GL-79-19 U.S for U.S. Army Corps of Engineers. Pennsylvania State University, Department of Mineral Engineering, University Park, PA, USA. Open Library: OL19998962M. http://openlibrary.org/books/OL19998962M/Tunnel_design_by_rock_mass_classifications [accessed 27 September 2011].

BITTERFELD. 2011. *Groundwater Resources of the World 1:25 000 000.* Extracted from: World-wide Hydrogeological Mapping and Assessment Programme (WHYMAP). Hannover. http://www.rtdf.org/PUBLIC/permbarr/prbsumms/profile.cfm?mid=76 BGR/UNESCO 2008.

BLM. 2001. U.S. Department of the Interior, Bureau of Land Management and Office of the Solicitor (eds) *The Federal Land Policy and Management Act, as Amended.* U.S. Dept. of the Interior, Bureau of Land Management Office of Public Affairs, Washington.

BLUE BOOK. 2007. *The Complete ISRM Suggested Methods for Rock Characterization, Testing and Monitoring: 1974–2006.* ULUSAY, R., HUDSON, J. A. (eds) Commission on Testing Methods ISRM. International Society for Rock Mechanics (ISRM), Turkish National Group, Ankara, Turkey.

BOCK, H. 2001. European practice in geotechnical instrumentation for tunnel construction control. *Tunnels and Tunnelling International*, **33** (4), 51–54.

BOSCH, J. W. & BROERE, W. 2009. Small incidents, big consequences. Leakage of a building pit causes major settlement of adjacent historical houses. Amsterdam North-South metro line project. *In*: KOCSONYA, P. (ed.) *Proceedings of the Safe Tunnelling for the City and Environment.* ITA-AITES World Tunnel Congress, Hungarian Tunnelling Association, Budapest, 1–13.

BP. 2006. *Quantifying energy.* BP Statistical Review of World Energy June 2006. BP p.l.c.

BP. 2010. *BP Statistical Review of World Energy* June 2010. BP p.l.c.

BRADY, P. V., SPALDING, B. P., KRUPKA, K. M., BORNS, D. J., WATERS, R. W. & BRADY, W. D. 1999. *Site-screening and technical guidelines for implementation of monitored natural attenuation at DOE sites.* Sandia National Laboratories. **SAND99-0464**.

BRANCATO, A., GRESTA, S. ET AL. 2012. Quantifying probabilities of eruption at a well-monitored active volcano: an application to Mt. Etna (Sicily, Italy). *Bollettino di Geofisica Teorica ed Applicata*, **53** (1), in press, DOI 10.4430/bgta 0040.

BRATTLI, B. & BROCH, E. 1995. Stability problems in water tunnels caused by expandable minerals. Swelling pressure measurements and mineralogical analysis. *Engineering Geology*, **39**, 151–169.

BRO. 2012. *Basisregistratie Ondergrond.* Rijksoverheid NL (in Dutch) http://www.rijksoverheid.nl/onderwerpen/bodem-en-ondergrond/basisregistratie-ondergrond [accessed 9 March 2012].

BROCH, E. 2007. Use of the underground in the city of Trondheim, Norway. *In*: KALIAMPAKOS, D. & BENARDOS, A. (eds) *Underground Space: Expanding the Frontiers.* 11th ACUUS International Conference, Athens, 175–180.

BROCH, E. & GRØV, E. 2008. Construction of long traffic tunnels in Norway. *In*: KANJLIA, V. K., RAMAMURTHY, T., WAHI, P. P. & GUPTA, A. C. (eds) *World Tunnel Congress 2008: Underground Facilities for Better Environment and Safety – India.* Central Board of Irrigation and Power, New Delhi, 892–901.

BROWN, E. T. 2006. Forensic engineering for underground construction. *In*: LEUNG, C. F. & ZHOU, Y. X. (eds) *Rock Mechanics in Underground Construction, Proceedings of the ISRM International Symposium & 4th Asian Rock Mechanics Symposium.* 8–10 November, Singapore. World Scientific Publishing Co. Pte. Ltd., Singapore, 3–18.

BROWN, P., SUTIKNA, T., MORWOOD, M. J., SOEJONO, R. P., JATMIKO, WAYHU SAPTOMO, E. & ROKUS AWE DUE,. 2004. A new small-bodied hominin from the Late Pleistocene of Flores, Indonesia. *Nature*, **431**, 1055–1061.

BRUGMAN, M. H. A., HACK, H. R. G. K. & DIRKS, W. G. 1999. Een drie-dimensionaal systeem voor de raming van baggerprodukties in rots. *Geotechniek*, **2**, 22–26 [in Dutch].

BRUNDTLAND, G. (ed.) 1987. *Our Common Future: The World Commission on Environment and Development.* Oxford University Press, Oxford.

BRUNET, M., GUY, F. ET AL. 2002. A new homonid from the Upper Miocene of Chad, Central Africa. *Nature*, **418**, 145–151.

BS 5930. 1999. *The Code of Practice for Site Investigations.* British Standards Institution, London.

BSI. 2011. British Standards Institution. http://www.bsigroup.com/ [accessed 20 January 2011].

BT Geoconsult. 2010. *GT Geoconsult BV.* The Hague, The Netherlands. http://www.btgeoconsult.nl/ [accessed 8 August 2010].

BTS. 2010. *Specification for Tunnelling.* 3rd edn. The British Tunnelling Society and The Institution of Civil Engineers. Thomas Telford Limited, London.

BUMA, J., GEHRELS, H. & STUURMAN, R. 2003. *Designing tomorrow's urban hydrological system.* Presentation SDMS Symposium, Utrecht.

BUNNELL, D. 2008. *Caves of Fire: Inside America's Lava Tubes*, 1st edn. National Speleological Society, Inc., Huntsville, AL. **124**.

BUNSCHOTEN, P. 2002. Creating a fairytale in Hong Kong. *In*: COX, R. J. (ed.) *Proceedings of the 30th International Navigation Congress (PIANC 2002).* 22–26 September, Sydney. PIANC General Secretariat, Institution of Engineers, Barton, Australia. Paper S7B P101, 1230–1240.

BURCHI, S. 1999. Chapter 3. National Regulations for Groundwater: Options, Issues and Best Practices. *FAO Legal Papers Online #5*, August 1999, http://www.fao.org/Legal/prs-ol/years/1999/list99.htm. [accessed 11 February 2011].

BUTLER, D. K. 2008. Detection and characterization of subsurface cavities, tunnels and abandoned mines. *In*: XU, Y. & XIA, J. (eds) *Near-Surface Geophysics and Human Activity. Proceedings of the 3rd International Conference on Environmental and Engineering Geophysics (ICEEG).* 15–20 June, Wuhan, China. Science Press USA Inc., Monmouth Junction, New Jersey, 578–584.

CAERS, J. 2011. *Modeling Uncertainty in the Earth Sciences.* Wiley-Blackwell, John Wiley & Sons Ltd., Chichester.

CAO, Y. S. 2006. *Evolution of Integrated Approaches to Water Resource Management in Europe and the United States. Some Lessons from Experience.* World Bank Analytical and Advisory (AAA) Program, China: Addressing Water Scarcity. The World Bank Background Paper 2.

CARSON, R. 1962. *Silent Spring.* Mariner Books, Houghton Mifflin. Boston, New York.

CASAGRANDE, G., CUCCHI, F. & ZINI, L. 2005. Hazard connected to railway tunnel construction in karstic area: applied geomorphological and hydrogeological surveys. *Natural Hazards and Earth System Sciences*, **5**, 243–250.

CAUMON, G. 2009. *Vers une intégration des incertitudes et des processus en géologie numérique*. Mémoire présenté pour obtenir l'habilitation à diriger les recherches. Centre de Recherches Pétrographiques et Géochimiques, École Nationale Supérieure de Géologie de Nancy, France. 19 October. [in French]. http://www.gocad.org/~caumon/HDRCaumon.pdf [accessed 22 December 2010].

CAUMON, G. 2010. Towards stochastic time-varying geological modeling. *Mathematical Geosciences* **42** (5), 555–569.

CAVES HAN-SUR-LESSE. 2010. *Caves of Han-Sur-Lesse*. Wikipedia. http://en.wikipedia.org/wiki/Caves_of_Han-sur-Lesse [accessed 12 September 2010].

CEKEREVAC, C., WOHNLICH, A. & BELLWALD, Ph. 2009. Deriner Hydropower Scheme Geotechnical issues and the particular case of the spillway tunnels design and construction. In: DIEDERICHS, M. & GRASSELLI, G. (eds) *Rock Engineering in Difficult Conditions; ROCKENG09; Proceedings of the 3rd Canada-US Rock Mechanics Symposium & 20th Canadian Rock Mechanics Symposium*. 9–12 May, Toronto. Canadian Rock Mechanics Association/American Rock Mechanics Association, Alexandria, VA, Paper 3997. http://www.geogroup.utoronto.ca/rockeng09/proceedings/ [accessed 3 November 2010].

CEN. 2011. Comité européen de normalisation, European Committee for Standardisation. http://www.cen.eu/cen/ [accessed 1 March 2012].

CGI. 2012. Commission for the Management and Application of Geoscience Information (CGI), Commission of the International Union of Geological Sciences (IUGS). http://www.cgi-iugs.org/ [accessed 10 March 2012].

CHAN, K. F. & STONE, P. C. 2005. Back-analysis of monitoring results at Macquarie Park Station, Epping to Chatswood Rail Line. In: GOURLAY, T. & BUYS, H. (eds) *AGS AUCTA Mini symposium: Geotechnical Aspects of Tunnelling for Infrastructure Projects*. The Sydney Chapter of the Australian Geomechanics Society, Australian Tunnelling Society (ATS), Sydney.

CHAN, K. F., KOTZE, G. P. & STONE, P. C. 2005. Geotechnical modelling of station caverns for the Epping to Chatswood Rail Line project. In: GOURLAY, T. & BUYS, H. (eds) *AGS AUCTA Mini Symposium: Geotechnical Aspects of Tunnelling for Infrastructure Projects*. The Sydney Chapter of the Australian Geomechanics Society. Australian Tunnelling Society (ATS), Sydney.

CHANG, Y. S. & PARK, H. D. 2004. Development of a web-based Geographic Information System for the management of borehole and geological data. *Computers and Geosciences*, **308**, 887–897.

CHEN, A. 2009. KPF crowns an ever-expanding skyline with the Shangai World Financial Center. *Architectural Record, McGraw-Hill Construction*, 184–190.

CHEN, L. T. 2010. Geotechnical aspects of a deep excavation in Doha. In: *Proceedings of the Geotechnical Aspects of Deep Excavation – 30th Annual Seminar*. 6 May. Geotechnical Division, The Hong Kong Institution of Engineers, Hong Kong Geotechnical Society, Hong Kong, 67–73.

CHESWORTH, W. (ed.) 2008. *Encyclopedia of Soil Science*. Springer, Dordrecht, The Netherlands.

CHINA COAL FIRES. 2011. *Mining Accident*. http://en.wikipedia.org/wiki/Mining_accident [accessed 27 March 2012].

CHMELINA, K. 2010. A new information system for underground construction projects. In: TOLL, D. G., ZHU, H. & LI, X. (eds) *Information Technology in Geo-Engineering. Proceedings of the 1st International Conference (ICITG)*. 16–17 September, Shanghai. Joint Technical Committee 2 (JTC2) & Tongji University, Shanghai. IOS Press, Amsterdam.

CHOE, E., VAN DER MEER, F. D., VAN RUITENBEEK, F. J. A., VAN DER WERFF, H. M. A., DE SMETH, J. B. & KYOUNG-WOONG, K. 2008. Mapping of heavy metal pollution in stream sediments using combined geochemistry, field spectroscopy, and hyperspectral remote sensing: a case study of the Rodalquilar mining area, SE Spain. *Remote Sensing of Environment*, **112/7**, 3222–3233.

CHOI, Y., YOON, S. Y. & PARK, H. D. 2009. Tunnelling analyst: a 3D GIS extension for rock mass classification and fault zone analysis in tunnelling. *Computers and Geosciences*, **35**, 1322–1333.

CHOU, H. S., YANG, C. Y., HSIEH, B. J. & CHANG, S. S. 2001. A study of liquefaction related damages on shield tunnels. *Tunnelling and Underground Space Technology*, **16**, 185–193.

CHOW, F. C. 2002. Underground space: the final frontier? *Ingenia, Infrastructure*, **14**, 15–20.

CIAMEI, A. & MOCCICHINO, M. 2009. EPB TBM under the city centre of Vancouver: risk management and settlement control. In: KOCSONYA, P. (ed.) *Proceedings of the ITA-AITES World Tunnel Congress 2009 and the 35th ITA-AITES General Assembly (WTC 2009) – Safe Tunneling for the City and for the Environment*. Hungarian Tunnelling Association/ITA-AITES, Budapest. http://www.selitunnel.com/pdf_articoli/Canada_line_per_sito.pdf

CIANCIA, M. & HEIKEN, G. 2006. Geotechnical properties of tuffs at Yucca Mountain, Nevada. In: HEIKEN, G. (ed.) *Tuffs – Their Properties, Uses, Hydrology, and Resources*. Geological Society of America, Boulder, CO, Special Papers 408, 33–89.

CIRCEO, L. J. & MARTIN, R. C., JR. 2001. In situ Plasma Vitrification of buried wastes. In: *Proceedings of the International Containment & Remediation Technology Conference and Exhibition*, 10–13 June, Orlando. Florida State University, Tallahassee, Paper no. 132. Available from: http://www.containment.fsu.edu/cd/content/srch_f_a.htm [accessed 13 February 2012].

CityGML. 2012. CityGML. http://kugel.bv.tu-berlin.de/typo3-igg/index.php?id=1523 [accessed 9 June 2012].

CLAESSEN, F. A. M., VAN BRUCHEM, A. J., HANNINK, E., HULSBERGEN, J. G. & DE MULDER, E. F. J. 1987. Secondary effect of the reclamation of the Markerwaard Polder. *Geologie & Mijnbouw*, **67**, 283–291.

CLARK, E. F. 1965. *Camp Century: Evolution of Concept and History of Design, Construction and Performance*. Technical Report 174. DA Project IV025001A130. U.S. Army Materiel Command, Cold Regions Research & Engineering Laboratory, Hannover, New Hampshire, USA. http://www.dtic.mil/cgi-bin/GetTRDoc?AD=AD477706&Location=U2&doc=GetTRDoc.pdf [accessed 12 October 2010].

CLARKE, S. M. 2004. *Confidence in geological interpretation. A methodology for evaluating uncertainty in common two and three-dimensional representations of subsurface geology*. British Geological Survey Internal Report, **IR/04/164, 29**.

CLEMENCE, S. P. & FINBARR, A. O. 1981. Design considerations for collapsible soils. *Journal of the Geotechnical Engineering Division*, **ASCE 107**, (GT3, Proc. Paper, 16106), 305–317.

COB. 2004a. *Ondergrondse ordening. Naar een meerdimensionale benadering van bestaande praktijken*. **COB B212-W-04-129** [in Dutch].

COB. 2004b. *Waardering van de ondergrond*. **COB E110-W-04-128** [in Dutch].

COB/RAVI. 2000. *COB-Ravi-onderzoeksrapport. Privaatrechtelijke Aspecten van Ondergronds Ruimtegebruik*. CUR Centrum Ondergronds Bouwen, Gouda, The Netherlands [in Dutch].

COMMISSIE OPBERGING TE LAND (OPLA). 1989. *Onderzoek naar geologische opberging van radioactief afval in Nederland*. Eindrapport Fase 1, Bijlage 2, Samenvatting van de deelstudies Geologie, Geohydrologie, Gesteentemechanica, Stralingseffecten, Mijnbouwkunde, Uitgave Ministerie van Economische Zaken, Den Haag [in Dutch].

COMMISSIE OPBERGING TE LAND (OPLA). 1993. *Bijlage Eindrapport aanvullend onderzoek van fase 1*. Samenvattingen van de deelstudies', Uitgave ministerie van Economische Zaken, Den Haag [in Dutch].

COMPENDIUM. 2011. *Compendium voor de leefomgeving*. [in Dutch] http://www.compendiumvoordeleefomgeving.nl/indicatoren/nl0067-Winning-en-verbruik-van-oppervlaktedelfstoffen.html?i=20-78 [accessed 9 September 2011].

COOBER PEDY. 2011. District Council of Coober Pedy. http://en.wikipedia.org/wiki/District_Council_of_Coober_Pedy. [accessed 27 March 2012].

COULTER, S. & MARTIN, C. D. 2004. Ground deformations above a large shallow tunnel excavated using jet grouting. *In*: SCHUBERT, W. (ed.) *Rock Engineering. Theory and Practice. Proceedings of the ISRM Regional Symposium EUROCK 2004 and 53rd Geomechanics Colloquy.* 7–9 October, Salzburg. Austrian Society for Geomechanics, VGE Verlag Glückauf, Essen, Germany, 155–160.

COUNCIL OF EUROPE. 1992. *European Convention on the Protection of the Archaeological Heritage (Revised)*. European Treaty Series 143.

CRAIG, R. F. 1978. *Soil Mechanics*. Van Nostrand Reinhold Company, New York.

CRAMER, J. 1999. *Towards Sustainable Business; Connecting Environment and Market*, SMO-1999, Stichting Maatschappij en Onderneming, Den Haag.

CROSSRAIL BILL. 2010. *Railways: Crossrail; Question Blackheath/Adonis*. House of Lords, UK. 18 March 2010, Column 657. http://www.publications.parliament.uk/pa/ld200910/ldhansrd/text/100318-0001.htm [accessed 20 January 2011].

CROWSON, P. 1998. *Minerals Handbook 1998–99*. Mining Journal Books Ltd, London.

CRREL. 2010. Permafrost Tunnel Research Centre. U.S. Army's Cold Regions Research and Engineering Laboratory, U.S. Army Corps of Engineers Engineer Research and Development Center. http://permafrosttunnel.crrel.usace.army.mil/ [accessed 14 October 2010].

CUI, Y.-J., LE, T. T., TANG, A. M., DELAGE, P. & LI, X. L. 2009. Investigating the time-dependent behaviour of Boom clay under thermomechanical loading. *Geotechnique*, **59**, 319–329.

CULSHAW, M. 2003. Bridging the gap between geoscience providers and the user community. *In*: ROSENBAUM, M. S. & TURNER, A. K. (eds) *New Paradigms in Subsurface Prediction. Characterization of the Shallow Subsurface; Implications for Urban Infrastructure and Environmental Assessment.* Lecture Notes in Earth Sciences, **99**, Springer, Berlin, Heidelberg, 7–26.

CULSHAW, M. 2005. From concept towards reality: developing the attributed 3D geological model of the shallow subsurface. *Quarterly Journal of Engineering Geology and Hydrogeology*, **38**, 231–284.

CULSHAW, M., DONNELLY, L. & MCCANN, D. 2004. Location of buried mineshafts and adits using reconnaissance geophysical methods. *In*: HACK, R., AZZAM, R. & CHARLIER, R. (eds) *Engineering Geology for Infrastructure Planning in Europe: A European Perspective*. Lecture Notes in Earth Sciences, 104, Springer-Verlag, Berlin, Heidelberg, 565–573.

CUMMINGS, R. A., KENDORSKI, F. S. & BIENIAWSKI, Z. T. 1982. *Caving Rock Mass Classification and Support Estimation*. U.S. Bureau of Mines Contract Report #J0100103. Engineers International Inc., Chicago.

D'AGNESE, F. & O'BRIEN, G. 2003. Impact of geo-informatics on the emerging geoscience knowledge integration paradigm. *In*: ROSENBAUM, M. S. & TURNER, A. K. (eds) *New Paradigms in Subsurface Prediction. Characterization of the Shallow Subsurface; Implications for Urban Infrastructure and Environmental Assessment.* Lecture Notes in Earth Sciences, **99**, Springer, Berlin, Heidelberg, 303–312.

DAMES & MOORE. 1983. *Use of Underground Space for Storage and Disposal*. Report for the U.K. Department of the Environment, London.

DAS, D., CHATTERJEE, A., MANDAL, B. K., CHOWDHURY, T. R., SAMANTA, G. & CHAKRABORTI, D. 1995. Arsenic in Ground Water in Six Districts of West Bengal, India: the Biggest Arsenic Calamity in the World. Part I: Arsenic Species in Drinking Water and Urine of the Affected People. *The Analyst*, **120**, 643–650.

DAVISON, L., FOOKES, P. & BAYNES, F. 2003. Total geological history: a web-based modelling approach to the anticipation, observation and understanding of site conditions. *In*: ROSENBAUM, M. S. & TURNER, A. K. (eds) *New Paradigms in Subsurface Prediction. Characterization of the Shallow Subsurface; Implications for Urban Infrastructure and Environmental Assessment.* Lecture Notes in Earth Sciences, **99**, Springer, Berlin, Heidelberg, 237–252.

DAY, M. J. 2004. Karstic problems in the construction of Milwaukee's Deep Tunnels. *Environmental Geology*, **45**, 859–863.

DEARMAN, W. R. 1991. *Engineering Geology Mapping*. Butterworth & Heinemann, Oxford.

DE GRAAF, L. W. & RUPKE, J. 1999. *Analyse der Felssturzbedrohung am Breiten Berg, Austria*. Alpine Geomorphology Research Group, University Amsterdam, The Netherlands [in German].

DE JONG, B. & DE MULDER, E. F. J. 1998. Construction materials in the Netherlands: resources and policy. *In*: BOBROWSKY, P. (ed.) *Aggregate Resources, a Global Perspective*. Uitgave, Balkema, Rotterdam, 203–214.

DE KOK, M., DIRKS, W. & HESSELS, R. 1997. The Øresund fixed link: dredging reclamation. *Terra et Aqua*, **66**, 16–24. http://www.terra-et-aqua.com/dmdocuments/Terra-et-Aqua_nr66_03.pdf.

DELATTE, N., CHEN, S. *ET AL.* 2003. Application of non-destructive evaluation to subway tunnel systems. *In: Transportation research record*. National Research Council, Washington, 127–135.

DE MULDER, E. F. J. 1986. *Subsoil Uncovered*. Geological Survey of the Netherlands, Institute for Applied Geosciences & National Institute for Soil Mapping, Haarlem, Delft, Wageningen.

DE MULDER, E. F. J. 1988. Thematic geological maps for urban management and planning. *In: Urban Geology in Asia and the Pacific; Proceedings of the International Symposium on Urban Geology*. Shanghai, Atlas of Urban Geology Series, **2**. ST/ESCAP/586, UNESCAP, Bangkok, Thailand, 46–59.

DE MULDER, E. F. J. 1999. *Ruimte voor de Bodem*. Technische Universiteit Delft, Delft, The Netherlands [in Dutch].

DE MULDER, E. F. J. 2005. IUGS and the State of the Geosciences. *EPISODES*, **28**, 224–225.

DE MULDER, E. F. J. & ANNELLS, R. N. 2000. Geosciences for Europe's environment in the 21st century. *In*: CORDANI, U. G., MILANI, E. J., THOMAZ-FILHO, A. & CAMPOS, D. A. (eds) *Proceedings of the 31st International Geological Congress*, Rio de Janeiro.

REFERENCES

DE MULDER, E. F. J. & JACKSON, I. 2007. *Data and information in the International Year of Planet Earth (2007/2009). EOS Trans*, 88, Jt. Assembly. Suppl., Abstract **IN31A-02**.

DE MULDER, E. F. J., BAARDMAN, B. A. M. & TEN KATE, A. M. 1997. The Underground Municipal Information System (UMIS). *In*: MARINOS, P. G., KOUKIS, G. C., TSIAMBAROS, G. C. & STOUMARAS, G. C. (eds) *Engineering Geology and the Environment*. Balkema, Rotterdam, The Netherlands, 1387–1394.

DE MULDER, E. F. J., GELUK, M. C., RITSEMA, I., WESTERHOFF, W. E. & WONG, T. E. 2003. *De ondergrond van Nederland*. Geologie van Nederland 7. Nederlands Instituut voor Toegepaste Geowetenschappen TNO [in Dutch].

DE MULDER, E. F. J., MCCALL, G. J. H. & MARKER, B. R. 2001. Geosciences for Urban Planning and Management. *In*: MARINOS, P. G., KOUKIS, G. C., TSIAMBAOS, G. C. & STOUMARAS, G. C. (eds) *Engineering Geology and the Environment*. Swets & Zeitlinger, Lisse, 3417–3438.

DENT, D., HARTEMINK, A. & KIMBLE, J. 2005. *Soil–Earth' living skin*. International Year of Planet Earth Corp. **10**.

DEPARTMENT OF THE ENVIRONMENT (DOE) 1990. *Planning Policy Guidance Note 14: Development on Unstable Land*. Department of the Environment, HMSO, London.

DE SAVENIJE, H. 2002. Why water is not an ordinary economic good, or why the girl is special. *Physics and Chemistry of the Earth*, Parts A/B/C, **27**, 741–744.

DIGGS. 2012. Data Interchange for Geotechnical and GeoEnvironmental Specialists (DIGGS). http://www.diggsml.com/ [accessed 10 March 2012].

DIN. 2011. *Deutsches Institut für Normung*. http://www.din.de/ [accessed 6 February 2011].

DINIS DA GAMA, C. & NAVARRO TORRES, V. 2002. Prediction of EDZ (excavation damaged zone) from explosive detonation in underground openings. *In*: DINIS DA GAMA, C. & RIBEIRO E SOUSA, L. (eds) *ISRM International Symposium on Rock Engineering for Mountainous Regions, Eurock*. 25–28 November, Funchal. Sociedade Portuguesa de Geotecnia, Lisbon, 401–408.

DINO. 2012. DINO*Loket*. Data en Informatie van de Nederlandse Ondergrond (DINO). http://www.dinoloket.nl/en/DINOLoket.html [accessed 10 March 2012].

DÖLL, P. & FIEDLER, K. 2008. Global-scale modeling of groundwater recharge. *Hydrology and Earth System Sciences*, **12**, 863–885.

DONNELLY, L. J. 2009. Investigation of geological hazards and mining risks, Gallowgate, Newcastle-upon-Tyne. *In*: CULSHAW, M. G., REEVES, H. J., JEFFERSON, I. & SPINK, T. W. (eds) *Engineering Geology for Tomorrow's Cities. Proceedings 10th Congress International Association for Engineering Geology and the Environment*. 6–10 September 2006, Nottingham, UK. Engineering Geology Special Publication, **22**. Geological Society, London, paper no. 113 [on CD-Rom].

DONNELLY, L. J., CULSHAW, M. G. & BELL, F. G. 2008. Longwall mining-induced fault reactivation and delayed subsidence ground movement in British coalfields. *Quarterly Journal of Engineering Geology and Hydrogeology*, **41**, 301–314.

DOYLE, B. R. 2001. *Hazardous Gases Underground: Applications to Tunnel Engineering (Civil & Environmental Engineering)*. CRC Press, Marcel Dekker, Inc., New York.

DUBEY, R. K. 2006. Mechanical response of Vindhyan sandstones under drained and confined conditions. *In*: LEUNG, C. F. & ZHOU, Y. X. (eds) *Rock Mechanics in Underground Construction, Proceedings of the ISRM International Symposium & 4th Asian Rock Mechanics Symposium*. 8–10 November, Singapore. World Scientific Publishing Co. Pte. Ltd., Singapore, 343–343 (extended abstract).

DUFFAUT, P. 2008. *L'espace souterrain, un patremoine à valoriser*. Géosciences, Special Publications, BRGM, Orléans, France, **7 & 8**, 224–235.

DUNCAN, S. D. & WILSON, W. 1988. Summit tunnel – post fire remedial works. *In*: *5th International Symposium (Tunnelling '88)*. 18–21 April. Institution of Mining and Metallurgy/British Tunnelling Society/Institution of Mining Engineers/Transport and Road Research Laboratory, Department of Transport. Institution of Mining & Metallurgy, London, 87–95.

DUNCAN, I. J., ANDERSON, S. & NICOT, J.-P. 2009. Pore space ownership issues for CO_2 sequestration in the U.S. *Elsevier Energy Procedia*, **1**, 4427–4431.

DUSSEAULT, M. B. & FORDHAM, C. J. 1993. Time-dependent behaviour of rocks. *In*: HUDSON, J. A. (ed.) *Comprehensive Rock Engineering: Principles, Practice & Projects*, **2**, Pergamon Press, Oxford, 119–149.

DW. 2010. *Deutsche Welle. Workers stole steel parts before Cologne archive collapse, prosecutors say*. (date: 09.02.2010) http://www.dw.de/dw/article/0,5232539,00.html [accessed 15 February 2012].

EC. 2002. *Proposal for a Directive of the European Parliament and the Council on minimum safety requirements for tunnels in the Trans-European Road Network*. **COM 769**.

EC. 2004a. *Minimum safety requirements for tunnels in the Trans-European Road Network*. **Directive 2004/54/EC**.

EC. 2004b. *Towards a thematic strategy on the urban environment*. **COM 60**.

EC. 2004c. *Environmental liability with regard to the prevention and remedying of environmental damage*. **Directive 2004/35/EC**.

EC. 2005. *Thematic Strategy on the sustainable use of natural resources*. **COM 670**.

EC. 2006a. *Thematic Strategy for Soil Protection*. **COM 231**.

EC. 2006b. *Establishing a framework for the protection of soil and amending Directive 2004/35/EC*. **COM 232**.

EC. 2006c. *Management of waste from extractive industries and amending Directive 2004/35/EC*. **Directive 2006/21/EC**.

EC. 2008. Protection of the Environment through criminal law. Position of the European Parliament adopted at first reading on 21 May 2008 with a view to the adoption of Directive 2008/.../EC of the European Parliament and of the council on the protection of the environment through criminal law. EP-PE_TC1-COD(2007)0022, 21.5.2008.

EC. 2009. *The world in 2025. Rising Asia and socio-ecological transition*. European Commission, Directorate-General for Research Socio-economic Sciences and Humanities. **EUR 23921 EN**.

EC. 2010. DECISION No 661/2010/EU OF THE EUROPEAN PARLIAMENT AND OF THE COUNCIL of 7 July 2010 on Union guidelines for the development of the trans-European transport network (recast). *Official Journal of the European Union*, L 204/1, 5.8.2010.

EDWARDS, J. S., WHITTAKER, B. N. & DURUCAN, S. 1988. Methane hazards in tunnelling operations. *In*: *Tunnelling '88, Proceedings 5th International Symposium*. 18–21 April, London. Institution of Mining and Metallurgy/British Tunnelling Society/Institution of Mining Engineers/Transport and Road Research Laboratory, Department of Transport. Institution of Mining & Metallurgy, London, 97–110.

EEA. 1995. *Europe's Environment: The Dobris Assessment*. *In*: STANNNERS, D. & BOURDEAU, PH. (ed.) Office for Official Publications of the European Communities, Luxembourg.

REFERENCES

EEA. 1998. *The Second Assessment*. Elsevier Science Ltd. European Environment Agency, Copenhagen, Denmark.

EEA. 1999. *Environment in the European Union at the turn of the century*. European Environment Agency, Copenhagen, Denmark.

EEA. 2005a. *Sustainable use and management of natural resources*. **EEA-report 9**. European Environment Agency, Copenhagen, Denmark.

EEA. 2005b. *The European Environment. State and Outlook 2005*. European Environment Agency, Copenhagen, Denmark.

EHLEN, J. 1999. Fracture characteristics in weathered granites. *Geomorphology*, **31**, Elsevier, 29–45.

EHLEN, J. 2002. Some effects of weathering on joints in granitic rocks. *Catena*, **49**, 91–109.

EHLKE, Th. 2003. *Die unterirdische Stadt Oppenheim. Von der Schattenwelt zum Erlebnisraum*. Emons Verlag, Köln, Germany.

EHM. 2011. *Earthquake Hazard Maps*. United States Geological Survey (USGS). http://earthquake.usgs.gov/hazards/products/conterminous/2008/maps/ [accessed 10 February 2011].

EHRBAR, H. 2008. Gotthard Base Tunnel, Switzerland: Experiences with different tunnelling methods. *In*: *Proceedings 2nd Congresso Brasileiro de Túneis e Estruturas Subterrâneas Seminário Internacional (CBT'2008). International Seminar South American Tunneling (SAT'2008)*. 23–25 June, Sao Paulo. Brazilian Association of Soil Mechanics and Geotechnical Engineering (ABMS); Brazilian Tunnelling Committee (CBT), Sao Paulo, Brasil. Available from: http://www.acquacon.com.br/ [accessed 12 February 2012].

EIKEN, O., BREVIK, I., ARTS, R., LINDEBERG, E. & FAGERVIK, K. 2000. Seismic monitoring of CO_2 injected into a marine aquifer. *In*: *SEG Calgary 2000. Int. Conf. & 70th Annual Meeting*. 6–11 August, Stampede Park, Canada, The Society of Exploration Geophysicists, Tulsa, USA, Annual Meeting, **19**, paper RC-8.2, 1623–1626.

EINSTEIN, H. 2007. Changing uncertainties through updating from experiments to the observational method. *In*: E SOUSA, L. R. & GROSSMANN, C. O. N. (eds) *The Second Half Century of Rock Mechanics*. **3**. Taylor & Francis/Balkema, Rotterdam, The Netherlands, 1427–1445.

EISPALAST. 2010. *Ice Palace in the Jungfrau Glacier*. http://www.jungfraujoch.ch/pages/de/Tourismus_Aktivitaeten.html [accessed 12 October 2010].

ELLEN, G. J., NIJHUIS, E. W. J. T., WESTERHOF, R., REMMELTS, G. & WERKSMA, H. 2002. *De rol van de ondergrond in de ruimtelijke ordening, een toekomstverkenning*. Report **STB-02-25** [in Dutch].

ENGINEERING. 2006. *Lærdal Tunnel*. http://www.engineering.com/Library/ArticlesPage/tabid/85/articleType/ArticleView/articleId/60/Laerdal-Tunnel.aspx [accessed 14 April 2011].

ENVIRONMENT AGENCY. 2005. High-resolution in situ monitoring of flow between aquifers and surface waters. Science report SC030155/SR4. Environment Agency, Bristol, UK. http://publications.environment-agency.gov.uk/PDF/SCHO0605BJCK-E-E.pdf

ERDEM, E. & ERDEM, Y. 2005. Underground space in Ancient Anatolia: the Cappadocian example. *In*: ERDEM, E. & SOLAK, T. (eds) *Underground Space Use: Analysis of the Past and Lessons for the Future*. Taylor & Francis Group, London, 35–39.

ESA. 2011. *European Space Agency*. ESA Satellite Images – Observing the Earth. http://earth.eo.esa.int/satelliteimages/ [accessed 6 February 2011].

EVANS, R. 2003. Current themes, issues and challenges concerning the prediction of subsurface conditions. *In*: ROSENBAUM, M. S. & TURNER, A. K. (eds) *New Paradigms in Subsurface Prediction. Characterization of the Shallow Subsurface; Implications for Urban Infrastructure and Environmental Assessment*. Lecture Notes in Earth Sciences, **99**, Springer-Verlag, Berlin, Heidelberg, 359–378.

EZIGBALIKE, I. C., RAKAI, M. E. T. & WILLIAMSON, I. P. 1995. Cultural issues in land information systems. Position paper commissioned by the UN Food and Agriculture Organisation, Rome, Italy.

FANAGALO. 2010. *Pidgin mining language in Southern Africa*. http://en.wikipedia.org/wiki/Fanagalo [accessed 15 November 2010].

FEKETE, S., DIEDERICHS, M. & LATO, M. 2010. Geotechnical and operational applications for 3-dimensional laser scanning in drill and blast tunnels. *Tunnelling and Underground Space Technology*, **25**, 614–628.

FELDER, W. M. & BOSCH, P. W. 2000. *Geologie van Nederland, deel 5: Krijt van Zuid-Limburg* NITG-TNO, The Netherlands. [in Dutch].

FELLIN, W. & LACKINGER, B. 2007. Foundations of cable car towers upon alpine glaciers. *Acta Geotechnica*, **2**, 291–300.

FENTON, G. A. & GRIFFITHS, D. V. 2008. *Risk Assessment in Geotechnical Engineering*. John Wiley and Sons, Inc., Hoboken, NJ.

FIELD, D. P., HAWLEY, J. & PHELPS, D. 2005. The North American tunneling method: lessons learned. *In*: HUTTON, J. D. & ROGSTAD, W. D. (eds) *Proceedings of the Rapid Excavation and Tunneling Conference (RETC)*, SME, Littleton, 871–881.

FIELDS, S. 2003. The earth's open wounds: abandoned and orphaned mines. *Environmental Health Perspectives*, **111**, a154–a161. http://dx.doi.org/10.1289/ehp.111-a154.

FIGUEIREDO, B., LAMAS, L. & MURALHA, J. 2010. Determination of in situ stresses using large flat jack tests. *In*: SHARMA, K. G. (ed.) *Advances in Rock Engineering. ISRM International Symposium – 6th Asian Rock Mechanics Symposium (ARMS)*. 23–27 October, New Delhi, India. Central Board of Irrigation and Power (CBIP) & Indian National Group of the ISRM, New Delhi.

FOOKES, P. G. 1997. Geology for engineers: the geological model, prediction and performance. The first Glossop lecture. *Quarterly Journal of Engineering Geology and Hydrogeology*, **30**, 293–424.

FOOTPRINT. 2011. Footprint. http://www.footprintnetwork.org/gfn_sub.php?content=global_footprint) [accessed 22 April 2011].

FORD, J., BURKE, H., ROYSE, K. & MATHERS, S. 2008. The 3D geology of London and the Thames Gateway: a modern approach to geological surveying and its relevance in the urban environment. *In*: *Cities and their Underground Environment (Euroengeo 2008). Proceedings 2nd European Conference of the International Association of Engineering Geology (IAEG)*. 15–20 September, Madrid. Asociación Española de Geología Aplicada a la Ingeniería & La Escuela de Ingeniería Técnica de Obras Públicas, Madrid, Spain (available on CD and from http://www.euroengeo.com/english and http://nora.nerc.ac.uk/3717/).

FORD, J., KESSLER, H., COOPER, A. H., PRICE, S. J. & HUMPAGE, A. J. 2010. *An enhanced classification for artificial ground*. British Geological Survey, Open Report, OR/10/036. http://nora.nerc.ac.uk/10931/1/OR_10_036_updated.pdf [accessed 21 November 2010].

FOSTER, S., LAWRENCE, A. & MORRIS, B. 1998. *Groundwater in Urban Development. Assessing management needs and formulating policy strategies*. World Bank Technical Paper 390.

FRANCKE, J. & UTSI, V. 2009. Advances in long-range GPR systems and their applications to mineral exploration, geotechnical and static correction problems. *First Break*, **27**, 85–93.

FRANDINA, F. P. & WITT, P. 2002. Design of the reconstruction of the Anton Anderson Memorial tunnel. In: MERRILL, K. S. (ed.) *Cold Regions Engineering: Cold Regions Impacts on Transportation and Infrastructure. Proceedings of the 11th International Conference*. 20–22 May, Anchorage, AK. American Society of Civil Engineers (ASCE), Reston, VA, 207–218.

FUGRO. 2010. *Bathymetric Lidar* http://www.fugro-pelagos.com/lidar/tech/index.html [accessed 15 December 2010].

FUGRO AIRBORNE SURVEYS. 2011. Fugro Airborne Surveys: DIGHEM. http://www.fugroairborne.com/services/geophysical services/bysurvey/electromagnetics/helicopter-electromagnetic/dighem [accessed 23 April 2011].

FUNG, C. Y. 2002. Risk Allocation of Unforeseen Ground Conditions and Underground Utilities in Construction Contracts-Time for a Rethink. *Hong Kong Surveyor*, **14**. http://www.jrk.com.hk/pdf/CYF Article 1.pdf [accessed 2 July 2008].

GAEA TECHNOLOGIES LTD. 2010. Gaea Technologies Ltd. http://www.gaeatech.com/index.html [accessed 21 April 2011].

GAGNON, R. E. & GAMMON, P. H. 1995. Triaxial experiments on iceberg and glacier ice. *Journal of Glaciology*, **41**, 528–540.

GALIPEAU, G. & BESNER, J. 2003. The Underground City of Montreal: a win-win approach in the development of a city. In: *Presentation at the 1st International Conference 'Sustainable Development & Management of the Subsurface*, Utrecht, The Netherlands.

GALLEGO, E. & VADILLO, L. 1992. Reclaiming areas degraded by mining operations. In: CENDRERO, A., LÜTTIG, G. & WOLFF, F. C. (eds) *Planning the Use of the Earth Surface*. Springer Verlag, Berlin, Lecture Notes in Earth Sciences, **42**, 393–408.

GAZIEV, E. G. & ERLIKHMAN, S. A. 1971. Stresses and strains in anisotropic foundation (model studies). In: *Rock Fracture: Proceedings of the International Symposium on Rock Mechanics, ISRM*. 4–6 October, Nancy. Ecole Nationale Supérieure de Géologie Appliqué et de Prospection Minière: Ecole Nationale Supérieure de la Métallurgie et de l'Industrie des Mines, Rubrecht, Nancy, France, paper II-1.

GCO. 1990. *Foundation Properties of Marble and Other Rocks in the Yuen Long-Tuen Mun Area*. Geotechnical Control Office, Civil Engineering Services Department Hong Kong. **2/90**, GCO Publication.

GEF. 2012. *Geotechnical Exchange Format*. http://www.geffiles.nl/ [accessed 7 March 2012].

GENDLER, S. G. 2008. Ventilation of the Northern Mujsky Railway Tunnel. In: WALLACE, K. G. (ed.) *Proceedings 12th U.S./North American Mine Ventilation Symposium*. 9–11 June, Reno, NV. University of Nevada, Reno & Mine Ventilation Services, Inc., Clovis, CA, 407–413.

GENE CONSERVATION. 2011. Svalbard Global Seed Vault. www.wikipedia.org/wiki/Svalbard_Global_Seed_Vault. [accessed 27 March 2012].

GENS, A., DI MARIANO, A., GESTO, J. M. & SCHWARZ, H. 2006. Ground movement control in the construction of a new metro line in Barcelona. In: BAKKER, K. J., BEZUIJEN, A., BROERE, W. & KWAST, E. A. (eds) *Geotechnical Aspects of Underground Construction in Soft Ground*. Taylor & Francis Group, London, 389–396.

GEO. 2011. *Technical guidance documents*. Geotechnical Engineering Office (GEO). Civil Engineering and Development Department. The Government of the Hong Kong Special Administrative Region. http://www.cedd.gov.hk/eng/publications/guidance_notes/ [accessed 22 January 2011].

GEODATABANK. 2012. Geodatabank. http://www.geodatabank.nl/intro.html [accessed 8 March 2012].

GeoSciML. 2012. GeoSciML. http://www.cgi-iugs.org/tech_collaboration/geosciml.html [accessed 7 March 2012].

GeotechLinks. 2010. *Free internet link directory for geotechnical engineers and researchers*. http://www.geotechlinks.com/index.php [accessed 15 November 2010].

GEOTECH-XML. 2012. *The Geotech-XML Project*. W3G – The World Wide Web of Geotechnial Engineers. http://www.ejge.com/GML/ [accessed 7 March 2012].

GHOSH, N. C. & SINGH, R. D. 2009. Groundwater arsenic contamination in India: vulnerability and scope for remedy. In: *5th Asian Regional Conference of INCID, Special Session on Groundwater. Int. Commission on Irrigation and Drainage (ICID)*. 9–11 December, Vigyan Bhawan, New Delhi. Available from: http://cgwb.gov.in/documents/papers/INCID.html [accessed 6 February 2012].

GIMCIW. 2012. Geo-Informatie Management voor Civieltechnische Infrastructurele Werken (GIMCIW) (partially in Dutch). Available from: http://kennis.rgi.nl/?page=projects&sub=details&id=102 [accessed 10 March 2012].

GLUCKMAN, R. 1995. Home under the range. *Wall Street Journal*. http://gluckman.com/CooberPedy.Australia.html [accessed 2 April 2012].

GONZÁLEZ-DRIGO, R., PÉREZ-GRACIA, V., DI CAPUA, D. & PUJADES, L. G. 2007. GPR survey applied to Modernista buildings in Barcelona: the cultural heritage of the College of Industrial Engineering. *Journal of Cultural Heritage*, **9**, 196–202.

GOODMAN, R. E. 1989. *Introduction to Rock Mechanics*. Wiley, New York.

GOOGLE EARTH. 2011. Google Earth. http://www.google.com/intl/nl/earth/index.html [accessed 6 February 2011].

GOOGLE EARTH/AERODATA INTERNATIONAL SURVEYS. 2012. Google earth/Aerodata Internation Surveys. http://www.google.com/earth/index.html [accessed 1 March 2012].

GOOGLE EARTH/TERRA METRICS. 2011. Google Earth/Terra Metrics. http://www.google.com/http://www.google.com/ [accessed 6 February 2011].

GOULDEN, J. 2006. *Worldwide Exploration Trends – 2005*. Metals Economics Group, Halifax, Canada.

GRESCHIK, C. E. Gy. 1975. Engineering-geological problems in constructing the Budapest metro. *Bulletin of the International Association of Engineering Geology (IAEG)*, **12**, 75–78.

GRM. 2011. *Groundwater Resources Maps of Europe*. European Commission- JRC, Institute of Environment and Sustainability, Italy. http://eusoils.jrc.ec.europa.eu/esdb_archive/eusoils_docs/other/GroundwaterCD/Start.html [accessed 12 February 2011].

GSI. 2012. *National Geotechnical Borehole Database*. Geological Survey of Ireland (GSI) http://www.gsi.ie/Programmes/Quaternary+Geotechnical/Databases/ [accessed 8 March 2012].

GUDEHUS, G. 2001. Answer on review of Kovari in TUNNEL 4/2001 on a book by Kolymbas: *Geotechnik – Tunnelbau und Tunnelmechanik*. http://www.uibk.ac.at/geotechnik/publ/gu2kovari_english.pdf [accessed 7 November 2010].

HA, H. S., KIM, D. S. & PARK, I. J. 2010. Application of electrical resistivity techniques to detect weak and fracture zones during underground construction. *Environmental Earth Sciences*, **60**, 723–731.

HACK, H. R. G. K. 1998. *Slope stability probability classification: SSPC (2nd version)*. International Institute for Aerospace Survey and Earth Sciences Enschede XVIII.

HACK, H. R. G. K. 2000. Geophysics for slope stability. *Surveys in Geophysics*, **21**, 423–448.

HACK, H. R. G. K. 2002. An evaluation of slope stability classification. *In*: DINIS DA GAMA, C. & RIBEIRA E SOUSA, L. (eds) *Proceedings of the ISRM EUROCK' 2002*. Publ. Sociedade Portuguesa de Geotecnia, Lisbon, 3–32.

HACK, H. R. G. K. 2009. Advances in the use of geodata for the urban environment. *In*: CULSHAW, M. G., REEVES, H. J., JEFFERSON, I. & SPINK, T. W. (eds) *Engineering Geology for Tomorrow's Cities*. Engineering Geology Special Publication, **22**, Geological Society, London, 201–208.

HACK, H. R. G. K. 2010. Integration of surface and subsurface data for civil engineering. *In*: TOLL, D. G., ZHU, H. & LI, X. (eds) *Information Technology in Geo-Engineering. Proceedings of the 1st International Conference (ICITG)*. 16–17 September, Shanghai. Joint Technical Committee 2 (JTC2) & Tongji University, Shanghai. IOS Press, Amsterdam, 37–49.

HACK, H. R. G. K. 2012. *Discontinuous Rock Mechanics: Lecture Notes*. Univ. Twente-ITC, Enschede, The Netherlands.

HACK, H. R. G. K. & HUISMAN, M. 2002. Estimating the intact rock strength of a rock mass by simple means. *In*: VAN ROOY, J. L. & JERMY, C. A. (eds) *Engineering Geology for Developing Countries – Proceedings of the 9th Congress of the International Association for Engineering Geology and the Environment (IAEG)*. 16–20 September, Durban, South Africa. IAEG & South African Institute for Engineering and Environmental Geologists (SAIEG), Houghton, South Africa, 1971–1977.

HACK, H. R. G. K. & PRICE, D. G. 1997. Quantification of weathering. *In*: MARINOS, P. G., KOUKIS, G. C., STOURNA, G. C. & TSIAMBAOS, G. C. (eds) *Engineering Geology and the Environment*. Balkema, Rotterdam, 145–150.

HACK, H. R. G. K. & SIDES, E. 1994. Three-dimensional GIS: recent developments. *Proceedings of the Symposium, ITC-Journal*, **1994-1**, 64–72.

HACK, H. R. G. K. & SLOB, S. 2008. Feasibility study on Ukrainian railway tunnel: valuable consulting experience for ITC advisory services. *GeoInformatics*, **11**, 44–47.

HACK, H. R. G. K., ORLIC, B., OZMUTLU, S., ZHU, S. & RENGERS, N. 2006. Three and More Dimensional Modelling in Geo-engineering. *Bulletin of. Engineering Geology and the Environment*. 1435–9537 (Online), **65**, 143–153.

HACK, H. R. G. K., PRICE, D. G. & RENGERS, N. 2003. A new approach to rock slope stability: a probability classification SSPC. *Bulletin of Engineering Geology and the Environment (IAEG)*, **62**, 167–184.

HANNUM, W. H., MARSH, G. E. & STANFORD, G. S. 2005. Smarter Use of Nuclear Waste. *Scientific American*, **293**, 84–91.

HARRIS, C. S., HART, M. B., VARLEY, P. M. & WARREN, C. D. 1996. *Engineering Geology of the Channel Tunnel*. Thomas Telford, London.

HARRISON, P. & PEARCE, F. 2000. *AAAS Atlas of Population and Environment*. *In*: MARKHAM, V. D. (ed.) AAAS & University of California Press, London.

HART, R. & FAIRHURST, C. 2000. *In*: FAIRHURST, C. & CARRANZA-TORRES, C. 2002. Closing the circle – Some comments on design procedures for tunnel supports in rock. *In*: LABUZ, J. F. & BENTLER, J. G. (eds) *Proceedings of the 50th Annual Geotechnical Engineering Conference*. 22 February, St-Paul, MN. The Minnesota Geotechnical Society, University of Minnesota, Minneapolis, MN, 21–84.

HARTLÉN, J. & BENDZ, D. 2004. Geotechnet. Workpackage 5: Geotechnical Working Methods v. Environmental Impact with specific reference to the Water Framework Directive. Report 2: Evaluation of pressures and impact. Geotechnet; European Geotechnical Thematic Network. http://www.geotechnet.org/ [accessed 2 September 2011].

HASSAN, T. M., ATTIA, F. A. & EL-ATTFY, H. A. 2004. Groundwater Potentiality Map of the Nile Basin Countries – A step towards integrated water Management. *In*: *Proceedings of the International Conference and Exhibition on Groundwater in Ethiopia; Providing Water for Millions*. Addis Abbeba, United Nations Economic Commission for Africa. http://www.uneca.org/groundwater/Docs/Egypt-%20NO-%2080.pdf [accessed 11 April 2012].

HAY, R. L. 1960. Rate of clay formation and mineral alteration in a 4000-year-old volcanic ash soil on Saint Vincent. *Bulletin of the W.I. American Journal Science*, **258**, 354–368.

HEALING, T. D., HOFFMAN, P. N. & YOUNG, S. E. J. 1995. The infection hazards of human cadavers. Communicable Disease Report. CDR Review, **5**, R62–R68. http://www.hpa.org.uk/web/HPAwebFile/HPAweb_C/1200660055286 [accessed 18 November 2010].

HEARN, E. H. & FIALKO, Y. 2009. Can compliant fault zones be used to measure absolute stresses in the upper crust? *Journal of Geophysical Research*, **114**, B04403, doi:10.1029/2008JB005901.

HEDEDAL, O. & STRANDGAARD, T. 2008. *3D elasto-plastic spring element for pipe-soil interaction analysis. Offshore Pipeline Technology Conference, OPT2008*, http://www.dtu.dk/English/Service/Phonebook.aspx?lg=showcommon&id=222382

HENCHER, S. R. & MCNICHOLL, D. P. 1995. Engineering in weathered rock. *Quarterly Journal of Engineering Geology and Hydrogeology*, **28**, 253–266, doi: 10.1144/GSL.QJEGH.1995.028.P3.04.

HENDRIKS, M. R. 2010. *Introduction to Physical Hydrology*. Oxford University Press, Oxford.

HERBICH, J. B. 2000. *Handbook of Dredging Engineering*. 2nd edn., McGraw-Hill Professional, New York.

HERRENKNECHT AG. 2011. http://www.herrenknecht.com/ [accessed 15 April 2011].

HEWETT, B. H. M. & JOHANNESSON, S. 1922. *Shield and Compressed Air Tunneling*. 1st edn. Mcgraw-Hill Book Company, Inc, New York. http://www.archive.org.

HOEK, E. & BROWN, E. T. 1990. *Underground Excavations in Rock*. Institute of Mining and Metallurgy, London; Taylor and Francis, Abingdon.

HOEK, E. & MARINOS, P. 2000. Predicting squeeze. *Tunnels and Tunnelling International, Part I and 2*, **32** (11), 45–46, 48–51, and **32** (12), 33–36.

HOEK, E., CARRANZA-TORRES, C. & CORKUM, B. 2002. Hoek–Brown failure criterion, 2002 edition. *In*: BAWDEN, H. R. W., CURRAN, J. & TELSENICKI, M. (eds) *Mining Innovation and Technology, Proceedings 5th North American Rock Mechanics Society and the 17th Tunnelling Association of Canada Conference: NARMS-TAC 2002*. 7–10 July, Toronto. University of Toronto Press, Toronto, Canada, 267–273.

HOEK, E., WOOD, D. & SHAB, S. 1992. A modified Hoek-Brown criterion for jointed rock masses. *In*: HUDSON, J. A. (ed.) *Rock Characterization, Proceedings ISRM Symposium EUROCK'92*. 14–17 September, Chester, UK. British Geotechnical Society, Thomas Telford Ltd, London, 209–214.

HOLLOWAY, S., CHADWICK, A., LINDEBERG, E., CZERNICHOWSKI-LAURIOL, I. & ARTS, R. (eds) 2004. *Best Practice Manual from Saline Aquifer CO_2 Storage Project (SACS)*. http://www.co2store.org/TEK/FOT/SVG03178.nsf/Attachments/SACSBestPractiseManual.pdf/$FILE/SACSBestPractiseManual.pdf [accessed 3 April 2012].

REFERENCES

HOMPE & TASELAAR. 2007. *Hompe en Taselaar.* http://www.hompetaselaar.nl/ [accessed 6 May 2011].

HOULDING, S. W. 2001. XML – an opportunity for <meaningful> data standards in the geosciences. *Computers and Geosciences*, **277**, 839–849.

HOUSES ON WATER. 2011. Floating Houses. http://www.ecoboot.nl/artikelen/floating_houses.php [accessed 27 March 2012].

HOUSTON, S. L., HOUSTON, W. N. & LAWRENCE, C. A. 2002. Collapsible Soil Engineering in Highway Infrastructure Development. *Journal of Transportation Engineering*, **128**, 295–300.

HSE. 2000. *Collapse of NATM tunnels at Heathrow Airport. A Report on the investigation by the Health and Safety Executive into the collapse of New Austrian Tunnelling Method (NATM) tunnels at the Central Terminal Area of Heathrow Airport on 20/21 October 1994.* Health and Safety Executive, HSE Books, Sudbury, UK.

HSU, S.-C. & NELSON, P. P. 2002. Characterization of Eagle Ford Shale. *Engineering Geology*, **67**, 169–183.

HUANG, R. Q., XU, Q. & HUO, J. J. 2009. Mechanism and geomechanics models of landslides triggered by '5.12' Wenchuan Earthquake. *In*: HUANG, R. Q., RENGERS, N., LI, Z. & TANG, C. (eds) *Geological Engineering Problems in Major Construction Projects. Proceedings of the International Symposium and the 7th Asian Regional Conf. of International Association for Engineering Geology and the Environment (IAEG).* 9–11 September, Chengdu. The China National Group of IAEG, Chengdu University of Technology (CDUT), State Key Laboratory of Geohazard Prevention and Geoenvironment Protection (SKLGP), Chengdu, **2**, 845–855.

HUDSON, J. A., BÄCKSTRÖM, A. ET AL. 2009. Characterising and modelling the excavation damaged zone in crystalline rock in the context of radioactive waste disposal. *Environmental Geology*, **57**, 1275–1297.

HUDSON, J. T. & HUDSON, J. A. 2003. Is the exploitation of underground space compatible with the concept of sustainable development? *Paper presented at the First Int. Conf. on Sustainable Development & Management of the Subsurface (SDMS) Conference*, Utrecht, Delft Cluster (unpublished).

HUGHES, M., BONAPACE, P., RIGBEY, S. & CHARALAMBU, H. 2007. An innovative approach to tunnelling in the swelling Queenston formation of Southern Ontario. *In*: TRAYLOR, M. T. & TOWNSEND, J. W. (eds) *Proceedings ot the Rapid Excavation and Tunneling Conference Society for Mining*, Metallurgy & Exploration, Littleton, 901–912.

HUISMAN, M., HACK, H. R. G. K. & NIEUWENHUIS, J. D. 2006. Predicting rock mass decay in engineering lifetimes: the influence of slope aspect and climate. *Environmental and Engineering Geoscience*, **7**, 49–61.

HUNT, R. E. 2005. *Geotechnical Engineering Investigation Handbook.* CRC Press, Taylor & Francis, Boca, FL.

IAEA. 2003. *The Long-term Storage of Radioactive Waste: Safety and Sustainability.* A Position Paper of International Experts, IAEA, Vienna.

IAEG. 2012. *International Association for Engineering Geology and the Environment (IAEG).* http://www.iaeg.info/ [accessed 1 March 2012].

ICC. 2010. Institut Cartografic de Catalonya (ICC). http://www.icc.cat/ [accessed 24 December 2010].

ICMC. 2008. *Mine closure and post-mining management – International state-of-the-art.* International Commission on Mine Closure. Report to the International Society for Rock Mechanics (ISRM). http://www.ineris.fr/centredoc/CDi__mineclosure_29_11_08-ang.pdf [accessed 18 November 2010].

ICMM. 2011. *International Council on Mining and Metals.* http://www.icmm.co.

ICOMOS COMMITTEE ON ARCHAEOLOGICAL HERITAGE MANAGEMENT. 2000. *Heritage at Risk.* ICOMOS World Report 2000 on monuments and sites in danger. KG. Sauer Verlag GmbH & Co.

IGRAC. 2011*a*. Arsenic in groundwater worldwide. http://www.igrac.net/publications/142 [accessed 27 March 2012].

IGRAC. 2011*b*. Fluoride in groundwater worldwide. http://www.igrac.net/publications/151 [accessed 27 March 2012].

IHN. 2011. *Namibian Economic Research. Namibia Macroeconomic Outlook 2011–2012.* Investment House Namibia (Pty) Limited (IHN). Windhoek, Namibia. http://www.investmenthousenamibia.com/IHN_Economic_outlook_2011.pdf [accessed 12 April 2012].

IMF. 2012. http://www.imf.org.external/np/res/commod/index.aspx [accessed 6 February 2012].

INAUDI, D. 2003. State of the Art in Fiber Optic Sensing Technology and EU Structural Health Monitoring Projects. *In*: ABE, M. & WU, Z. S. (eds) *Proceedings of the First International Conference on Structural Health Monitoring and Intelligent Infrastructure.* 13–15 November, Tokyo. Taylor & Francis Group, Tokyo, 191–198.

INAUDI, D. 2007. Fiber optic sensors for structural monitoring. *In*: ANTOINE, P. (ed.) *Proceedings of the 3rd Workshop on Optical Measurement Techniques for Structures and Systems (OPTIMESS 2007).* 28–29 May, Leuven, Belgium. OPTIMESS Scientific Research Network supported by the Fund for Scientific Research, Flanders. European Commission, Joint Research Centre, Brussels. http://www.roctest-group.com/rtgroup/sites/default/files/bibliography/pdf/c153.pdf [accessed 9 November 2010].

INSPIRE. 2012. *Infrastructure for Spatial Information in the European Community (INSPIRE).* http://inspire.jrc.ec.europa.eu/index.cfm [accessed 10 March 2012].

INTI. 2011. *Instituto Nacional de Tecnología Industrial.* http://www.inti.gov.ar/index.html [accessed 26 January 2011].

IPA. 2010. *International permafrost association.* http://ipa.arctic-portal.org/ [accessed 12 October 2010].

IPCC. 2005. *Carbon Dioxide Capture and Storage.* IPCC Special Report, Cambridge University Press.

IPIECA. 2011. *International Petroleum Industry Environmental Conservation Association.* http://www.ipieca.org

ISAPP. 2011. *Integrated System Approach Petroleum Production (ISAPP).* http://www.isapp.nl

ISO. 2011. *International Organization for Standardisation.* http://www.iso.org/iso/home.htm [accessed 6 February 2011].

ISO 14688-1/2:2002/2004. 2002/2004. *Geotechnical investigation and testing.* Identification and classification of soil. Part 1: Identification and description; Part 2: Principles for a classification. International Organization for Standardization.

ISO 14689-1:2003. 2003. *Geotechnical investigation and testing.* Identification and classification of rock. Part 1: Identification and description. International Organization for Standardization.

ISRM. 2012. The International Society for Rock Mechanics (ISRM). http://www.isrm.net/ [accessed 1 March 2012].

ISSMGE. 2012. *International Society for Soil Mechanics and Geotechnical Engineering.* http://www.issmge.org/en/.

ITA. 1991. *Legal and Administrative Issues in Underground Space Use: a Preliminary Survey of ITA Member Nations.* ITA-report. Tunnelling and Underground Space Technology **6/2**, 191–209.

REFERENCES

ITA. 2000. Planning and Mapping of Underground Space – an Overview. Working Group No. 4, International Tunnelling Association. *Tunnelling and Underground Space Technology*, **15**, 271–296.

ITA-AITES. 2012. *International Tunnelling And Underground Space Association (ITA-AITES)*. http://www.ita-aites.org/ [accessed 1 March 2012].

ITASCA. 2011. Itasca International Inc.; Itasca Consulting Group, Inc. http://www.itascacg.com/ [accessed 20 December 2011].

ITIG. 2006. *A code of practice for risk management of tunnel works*, The International Tunnelling Insurance Group. http://www.imia.com/downloads/external_papers/EP24_2006.pdf [accessed 18 November 2010].

IUGS. 2012. *International Union of Geological Sciences*. Tectask. http://tectonique.net/tectask/ [accessed 7 March 2012].

IWMI. 2000. *World Water Supply and Demand: 1995 to 2025*. International Water Management Institute, Colombo, Sri Lanka.

IYPE. 2010. *International Year of Planet Earth*. http://www.yearofplanetearth.org [accessed 24 December 2010].

JAMES, P., CHESTER, D. & DUNCAN, A. 2000. *Volcanic soils: their nature and significance for archaeology*. In: MCGUIRE, W. J., GRIFFITHS, D. R., HANCOCK, P. L. & STEWART, I. S. (eds) The Archaeology of Geological Catastrophes. Geological Society, London, Special Publications, **171**, 317–338.

JESSBERGER, H.-L., JAGOW-KLAFF, R. & BRAUN, B. 2003. Ground freezing. In: SMOLTCZYK, U. (ed.) *Geotechnical Engineering Handbook*, **2**, Ernst & Sohn Verlag, Wiley, Berlin, 117–164.

JETHWA, J. L. 2009. Blast induced rock mass damage in tunnels. *The Indian Mining and Engineering Journal*, **48**, 29–33.

JMA. 2011. *Japan Meteorological Agency*. http://www.jma.go.jp/jma/indexe.html [accessed 6 February 2011].

JOHN, M., SPÖNDLIN, D., AYADIN, N., HUBER, G., WESTERMAYR, H. & MATTLE, B. 2005. Means and methods for tunneling through highly squeezing ground; a case history of the Strenger Tunnel, Austria. In: HUTTON, J. D. & ROGSTAD, W. D. (eds) *Proceedings of the Rapid Excavation and Tunneling Conference, Seattle*. Society for Mining, Metallurgy, and Exploration Ltd (SME), Littleton, CO (CD-Rom).

JOHNSON, E. L. 2010. *Mineral rights. legal systems governing exploration and exploitation*. Doctoral Thesis in Real Estate Planning, Royal Institute of Technology (KTH), Stockholm, Sweden.

JTC2. 2012. *Joint Technical Committee number 2* representing the International Geo-Engineering Societies (International Association for Engineering Geology and the Environment (IAEG), International Society for Rock Mechanics (ISRM) and the International Society for Soil Mechanics and Geotechnical Engineering (ISSMGE)). http://www.geotechml.com [accessed 8 March 2012].

JUUTI, S. P. & TAPIO, S. K. (eds) 2005. *Water, Time and European Cities. History Matters for the Futures*. Tampere University Press, Tampere, Finland.

KÄÄB, A. 2008. Remote Sensing of Permafrost-related Problems and Hazards. *Permafrost and Periglacial Processes*. **192**, 107–136.

KAALBERG, F. J., ESSLER, R. D. & KLEINLUGTENBELT, R. 2012. Compensation grouting of piled foundations to mitigate tunnelling settlements. In: VIGGIANI, G. (ed.) *Geotechnical Aspects of Underground Construction in Soft Ground. Proceedings 7th International Symposium TC28 (ISSMGE-TC 28)*. 16–18 May 2011, Rome. International Society for Soil Mechanics and Geotechnical Engineering (ISSMGE); Technical Committee 204 & Associazione Geotecnica Italiana (AGI). CRC Press/Balkema, Taylor & Francis Group, Rotterdam (in press).

KALIAMPAKOS, D. C. & MAVRIKOS, A. A. 2004. *Underground development in Greece: history, current situations and trends*. Paper presented at the First International Conference on Sustainable Development and Management of the Subsurface. Delft Cluster, Utrecht (unpublished).

KALTENBORN, B. P., NELLEMANN, C. & VISTNES, I. I. (eds) 2010. *High mountain glaciers and climate change – Challenges to human livelihoods and adaptation*. United Nations Environment Programme, UNEP, GRID-Arendal, http://www.grida.no.

KAPTAN, E. 1980. New findings on the mining history of Turkey around Tokat region. *Bulletin of Mineral Research and Exploration Institute of Turkey*, **93–94**, 65–76.

KARAKUŞ, M. & FOWELL, R. J. 2004. An insight into the New Austrian Tunnelling Method (NATM). KAYAMEK'2004-VII. Bölgesel Kaya Mekaniği Sempozyumu/ROCKMEC'2004-VIIth Regional Rock Mechanics Symposium, 21–24 October, Sivas, Turkey. http://www.1insaat.com/uploads/TrbBlogs/pdfs_3/33802_1219844134_535.pdf [accessed 6 November 2010].

KARAKUŞ, M. & FOWELL, R. J. 2005. Back analysis for tunnelling induced ground movements and stress redistribution. *Tunnelling and Underground Space Technology*, **20**, 514–524.

KARGBO, D. M., WILHELM, R. G. & CAMPBELL, D. J. 2010. Natural gas plays in the Marcellus shale: challenges and potential opportunities. *Environmental Science & Technology*, **44**, 5679–5684.

KARSTENS, S. A. M., VAN REE, C. C. D. F., DE MULDER, E. F. J., DE CLEEN, M. & JELLEMA, J. J. 2003. *Sustainable Subsurface Management, Inventory study, International comparison on the use and the management of the urban subsurface*. DC Report 750401/9.

KARTHIKEYAN, M. 2005. *Application of Radioisotope Cone Penetrometer to Characterize a Lumpy Fill*. PhD thesis. Dept. of Civil Engineering, National University of Singapore. https://scholarbank.nus.edu.sg/handle/10635/14705 [accessed 20 December 2010].

KATONGO, C. 2005. Ground conditions and support systems at 1 shaft, Konkola mine, Chililabombwe, Zambia. In: *Third Southern African Base Metals Conference*. 26–29 June, Kitwe, Zambia. South African Institute of Mining and Metallurgy, Johannesburg. Symp. Series S39, 253–279.

KAVVADAS, M. J. 2005. Monitoring ground deformation in tunnelling: current practice in transportation tunnels. *Engineering Geology*, **79**, 93–113.

KAZA, N. 2004. *Spatial representation of property rights. Considerations of Doxastic & Epistemic Systems*. Term paper for GEOG495 B&M: GIS and Society, University of Illinois.

KEITH, D. W., GIARDINA, J. A. & WILSON, E. J. 2005. Regulating the underground injection of CO_2. *Environmental Science and Technology*, **1**, 499A–505A.

KELK, B. 1992. 3D-Modelling with geoscientific information systems: the problem. In: TURNER, A. K. (ed.) *Three- Dimensional Modelling with Geo-scientific Information Systems*. Kluwer Academic Publishers, Dordrecht, 29–37.

KELLER, A., SAKTHIVADEL, R. & SECKLER, D. 1998. *Water scarcity and the role of storage in development*. International Water Management Institute. World Water Supply and Demand. IWMI, Colombo, Sri Lanka.

KESSLER, H., TURNER, A. K., CULSHAW, M. G. & ROYSE, K. R. 2008. Unlocking the potential of digital 3D geological subsurface models for geotechnical engineers. In: *Cities and their Underground Environment (Euroengeo 2008). Proceedings 2nd European Conference of the International Association for*

Engineering Geology and the Environment (IAEG). 15–20 September, Madrid. Asociación Española de Geología Aplicada a la Ingeniería & La Escuela de Ingeniería Técnica de Obras Públicas, Madrid, Spain (available on CD and from http://www.euroengeo.com/english and http://nora.nerc.ac.uk/3717/).

KHALIFA, M. A., KUMON, F. & YOSHIDA, K. 2009. Calcareous duricrust, Al Qasim Province, Saudi Arabia: Occurrence and origin. *Quaternary International*, **209**, 163–174.

KIMAP. 2002. *Mapping in Engineering Geology*. Compiled by GRIFFITHS, J. S. GSL Key Issues in Earth Sciences. The Geological Society, London.

KIRSCH, F. & RICHTER, Th. 2009. Ground freezing for tunnelling under historical structures. *Proc eedings of the 17th International Conference on Soil Mechanics and Geotechnical Engineering ICSMGE*. Alexandria, http://www.gudconsult.de/bilder/00184_d.pdf [accessed 25 January 2011].

KLAASSEN, R. K. W. M. 2009. *Factors that Influence the Speed of Bacterial Wood Degradation*. SHR Timber Research, Wageningen, The Netherlands.

KLEINEBECKEL, A. 1986. *Unternehmen Braunkohle*. Geschichte eines Rohstoffs, eines Reviers, einer Industrie im Rheinland. Rheinische Braunkohlenwerke Aktiengesellschaft. Greven Verlag Köln.

KLIC. 2012. Kadaster Dienst KLIC. http://www.kadaster.nl/window.html?inhoud=/klic/ [accessed 13 March 2012].

KOLYMBAS, D. 1998. *Geotechnik – Tunnelbau und Tunnelmechanik. Eine systematische Einführung mit besonderer Berücksichtigung mechanischer Probleme*. Springer-Verlag, Berlin-Heidelberg.

KOLYMBAS, D. 2008. *Tunnelling and Tunnel Mechanics – A Rational Approach to Tunnelling*. Springer, Berlin, Heidelberg.

KONTOGIANNI, V. A. & STIROS, S. C. 2005. Induced deformation during tunnel excavation: Evidence from geodetic monitoring. *Engineering Geology*, **79**, 115–126.

KOUOKAM, E. 1993. *Compilation of all Falset Data to Final Report on Engineering Geological Map*. ITC, Delft.

KOVÁRI, K. 1993. *Gibt es eine NÖT?* Fehlkonzepte der Neuen Österreichischen Tunnelbauweise. 42th Geomechanik-Kolloquium, Salzburg. Tunnel 1. 16–25 [in German]. Also: Erroneous Concepts behind NATM. [in English]. http://www.igt.ethz.ch/resources/publications/26/papers/6/93-10.pdf [accessed 16 February 2012].

KOVÁRI, K. & FECHTIG, R. 2000. *Historical Tunnels in the Swiss Alps. Gotthard, Simplon, Lötschberg*. Stäubli AG, Zürich & Gesellschaft für Ingenieurbaukunst.

KOVIN, O. N. & ANDERSON, N. L. 2007. Use of 3-d groundpenetrating radar data for fractures imaging. *In*: ANDERSON, N. L. (ed.) *Geophysics 2006, Conference on Applied Geophysics. 2006. FHWA-sponsored Highway Geophysics-NDE Conference*. 4–7 December 2007, St. Louis, MI, USA. U.S. Dept. of Transportation, Washington & University of Missouri, Rolla, MO, Paper 053, 566–573.

KRULC, M. A., MURRAY, J. J., MCRAE, M. T. & SCHULER, K. L. 2007. Construction of a mixed face reach through granitic rocks and conglomerate. *In*: TRAYLOR, M. T. & TOWNSEND, J. W. (eds) *Proceedings of the Rapid Excavation & Tunneling Conference*. 10–13 June, Toronto. Society for Mining, Metallurgy and Exploration, Inc. (SME), Littleton, CO, 928–942. http://www.jacobssf.com/images/uploads/2007_McRae_Murray_San-Vicente_RETC.pdf.

KWR. 2010. *Ordening van de ondergrond. Een fysiek en juridisch afwegingskader*. Hoofdrapport en Bijlagenrapport [in Dutch]. Reportnr. KWR 2010.010, Nieuwegein, The Netherlands.

LAM, S. Y. W. 2010. Recent advances of engineering survey operations for tunnel construction in Hong Kong. *In*: FRIIS-HANSEN, L., MARKUS, B., POTSIOU, C., STAIGER, R. & VIITANEN, K. (eds) *Facing the Challenges – Building the Capacity*. FIG Congress 2010, 11–16 April, Sydney, paper 3995. Available from: http://www.fig.net/pub/fig2010/ [accessed 11 November 2010].

LANDAHL, G. 1995. Planning and building permission underground. *In*: BARLES, S. (ed.) *Underground Space and Urban Planning. Proceedings of the 6th ACUUS International Conference*. 26–29 September, Paris. University of Paris & Associated Research Centers for Urban Underground Space (ACUUS). Dollard-des-Ormeaux (Quebec), Canada.

LAndXML. 2012. LandXML. http://www.landxml.org/ [accessed 7 March 2012].

LAUBSCHER, D. H. 1990. A geomechanics classification system for rating of rock mass in mine design. *Journal South African Inst. of Mining and Metallurgy*, **90**, 257–273.

LAUBSCHER, D. H. & JAKUBEC, J. 2001. The MRMR rock mass classification for jointed rock masses. *In*: HUSTRULID, W. A. & BULLOCK, R. L. (eds) *Underground Mining Methods: Engineering Fundamentals and International Case Studies*. Society for Mining, Metallurgy & Exploration, Inc. (SME), Littleton, CO, 475–481.

LAWRENCE, A. R. & CHENEY, C. 1996. Urban groundwater. *In*: MCCALL, G. J. H., DE MULDER, E. F. J. & MARKER, B. R. (eds) *Urban Geoscience*. Balkema, Rotterdam, 61–80.

LCHC. 2010. *Lebanon County Historical Society*. Lebanon, Pennsylvania, USA. http://lebanoncountyhistoricalsociety.org/canal-tunnel [accessed 2 September 2010].

LEE, S. G. & DE FREITAS, M. H. 1990. Seismic Refraction Surveys for Predicting the Intensity and Depth of Weathering and Fracturing in Granitic Masses. *In*: BELL, F. G., CRIPPS, J. C., CULSHAW, M. G. & COFFEY, J. R. (eds) *Field Testing in Engineering Geology*. Geological Society, London, Engineering Geology Special Publications, **6**, 241–256.

LEE, M. Y., FOSSUM, A., COSTIN, L. S. & BRONOWSKI, D. 2002. *Frozen Soil Material Testing and Constitutive Modeling*. Sandia National Laboratories, Albuquerque. http://www.prod.sandia.gov/techlib/access-control.cgi/2002/020524.pdf [accessed 25 January 2011].

LEHMANN, B., ORLOWSKY, D. & MISIEK, R. 2010. Exploration of tunnel alignment using geophysical methods to increase safety for planning and minimizing risk. *Rock Mechanics and Rock Engineering*, **43**, 105–116.

LEITH, W. 2001. *Geological and Engineering Constraints on the Feasibility of Clandestine Nuclear Testing by Decoupling in Large Underground Cavities*. U.S. Geological Survey, Open file Report **01-28**.

LEOUTSAKOS, G. 2005. Athens Metro base project and extensions – Project structuring and management characteristics. *In*: AHMED, S. M., AHMAD, I., PANTOUVAKIS, J., AZHAR, S. & ZHENG, J. (eds) *Third International Conference on Construction in the 21st Century (CITC-III) 'Advancing Engineering, Management and Technology'*. 15–17 September 2005, Athens. CITCIII, Greece, 31–37. http://www2.fiu.edu/~citc/citc3/CITC-III Proceedings.pdf [accessed 3 April 2012].

LEYBOLD-JOHNSON, I. 2010. *Switzerland has its record-breaking tunnel*. http://www.swissinfo.ch/eng/Specials/Gotthard_base_tunnel/The_tunnel/Switzerland_has_its_record-breaking_tunnel.html?cid=28532002 [accessed 7 November 2010].

LI, T. 2011. Damage to mountain tunnels related to the Wenchuan earthquake and some suggestions for aseismic tunnel construction. *Bulletin of Engineering Geology and the Environment*, doi: 10.1007/s10064-011-0367-6.

LI, K. Y. K. & CHAN, A. T. 2004. Management of Radon in Tunnel Drilling. *Journal Construction Engineering and Management*, **1305**, 699–707.

LINDLEY, A. & SHARP, J. M., JR. 2003. *Urban epi karst*. GEO383C – urban hydrology. University of Texas, Austin.

LIU, D. P., WANG, R. & LIU, G. B. 2009. Research on the effect of buried channels to the differential settlement of building. *In*: NG, C. W. W., HUANG, H. W. & LIU, G. B. (eds) *Geotechnical Aspects of Underground Construction in Soft Ground*. Taylor & Francis Group, IS-Shanghai, 413–418.

LIU, F., BARDET, J.-P. & MOKARRAM, N. 2010. XML-based approach for reporting and exchanging experimental data sets using metadata model. *In*: TOLL, D. G., ZHU, H. & LI, X. (eds) *Information Technology in Geo-Engineering. Proceedings of the 1st International Conference (ICITG)*. 16–17 September, Shanghai. Joint Technical Committee 2 (JTC2) & Tongji University, Shanghai, IOS Press, Amsterdam.

LOMBORG, B. 2001. *The Skeptical Environmentalist: Measuring the Real State of the World*. Cambridge University Press, Cambridge.

LONGWALL MINING. 1995. *Longwall Mining*. Energy Information Administration, Office of Coal, Nuclear, Electric and Alternate Fuels, U.S. Department of Energy. Washington. http://www.tonto.eia.doe.gov/ftproot/coal/tr0588.pdf [accessed 31 January 2011].

LOVELOCK, J. 2003. GAIA: the living earth. *Nature*, **426**, 769–770.

LOVELOCK, J. 2006. Living Planet. *Geoscientist*, **1610**, 4–15.

LOWRIE, A., DEAN, P. A. & LUTKEN, C. B. 2004. *Within Five Years, Hydrate Exploitation can be a Reality in the Northern Gulf of Mexico*. Gulf Coast Association of Geological Societies Transactions, **54**, 371–381.

LUMB, P. 1983. Engineering properties of fresh and decomposed igneous rock from Hong Kong. *Engineering Geology*, **19**, 81–94.

LUMSDEN, G. I. 1992. *Geology and the Environment in Western Europe, a Coordinated Statement by the Western European Geological Surveys*. Clarendon Press, Oxford.

LUNARDI, P. 2000. The design and construction of tunnels using the approach based on the analysis of controlled deformation in rocks and soils. T & T International ADECO-RS approach. A T & T International special supplement in conjunction with Rocksoil spa. *Tunnels and Tunnelling International*, May.

LUNARDI, P. 2008. *Design and Construction of Tunnels. Analysis of Controlled Deformations in Rock and Soils (Adeco-RS)*. Springer, Berlin, Heidelberg.

LUNARDI, P., CASSANI, G. & GATTI, M. C. 2008. Design aspects of the construction of the new Apennines crossing on the A1 Milan-Naples motorway: the base tunnel. *In: Proceedings of the AFTES International Congress "Le souterrain, espace d'avenir'; Journées d'études*. 6–8 October, Monaco. l'Association Française des Tunnels et de l'Espace Souterrain (AFTES), Paris, 147–156.

LUPO, J. F. 2009. Guidelines for stabilizing historic mine workings. *In: Tailings and Mine Waste '08, Proceedings of the 12th International Conference on Tailings and Mine Waste*. 20–22 October 2008, Vail, CO, USA. CRC Press/Balkema, Taylor & Francis Group, Leiden, The Netherlands, 153–164.

MACHADO, S. L., VILAR, O. M. & CARVALHO, M. F. 2008. Constitutive model for long term municipal solid waste mechanical behavior. *Computers and Geotechnics*, **35**, 775–790.

MAIDL, B., SCHMID, L., RITZ, W. & HERRENKNECHT, M. 2008. *Hardrock Tunnel Boring Machines*. STURGE, D. (translator). Ernst & Son, Wiley-VCH, Berlin.

MAIR, R. J. 2008. Tunnelling and geotechnics: new horizons. *Geotechnique*, **58**, 695–736.

MAIR, R. 2009. *What's Going on Underground? Tunnelling into the Future*. Public Lecture. Royal Society, London. http://royalsociety.tv/rsPlayer.aspx?presentationid=355 [accessed 10 June 2012].

MALAYSIA. 2011. Kaiserdom – The Crypt. members.virtualtourist.com/m/p/m/17afde [accessed 27 March 2012].

MANOCHA, J. 2001. *Operational and Regulatory Opportunities in Brine Mining and Underground Storage Facilities*. Spring Meeting, Orlando, USA.

MARING, L., WASSING, B. & KIERKAARD, M. 2003. Chances for the subsoil. *Paper presented at the First International Conference on Sustainable Development and Management of the subsurface*. Delft Cluster, Utrecht (unpublished).

MARINOS, P. & HOEK, E. 2000. GSI: a geologically friendly tool for rock mass strength estimation. *In: Proceedings International Conference on Geotechnical & Geological Engineering, GeoEng2000*. 19–24 November, Melbourne. Technomic Publishing Co., Lancaster, PA, 1422–1446.

MARINOS, P., SAROGLOU, H., NOVACK, M., BENISSI, M. & MARINOS, V. 2004. Site investigation for abandoned lignite mines in urban environment. *In*: HACK, R., AZZAM, R. & CHARLIER, R. (eds) *Engineering Geology for Infrastructure Planning in Europe: A European Perspective*. Lecture Notes in Earth Sciences, **104**, Springer-Verlag, Berlin, Heidelberg, 393–404.

MARINOS, V., MARINOS, P. & HOEK, E. 2005. The geological strength index: applications and limitations. *Bulletin of the Engineering Geology and the Environment*, **64** (1), 55–65.

MARINOS, V., FORTSAKIS, P. & PROUNTZOPOULOS, G. 2009. Estimation of rock mass properties of heavily sheared flysch using data from tunnelling construction. *In*: CULSHAW, M. G., REEVES, H. J., JEFFERSON, I. & SPINK, T. W. (eds) *Engineering Geology for Tomorrow's Cities. Proceedings 10th Congress International Association for Engineering Geology and the Environment*. 6–10 September 2006, Nottingham, UK. Engineering Geology Special Publication, **22**, Geological Society, London, paper no. 314 (on CD-Rom).

MARKER, B. R. 2003. Out of sight, out of mind? Land use planning and development of the subsurface. *Presentation in First International Conference on Sustainable Development & Management of the Subsurface (SDMS) Conference*. Delft Cluster, Utrecht (unpublished).

MARKER, B. R., PEREIRA, J. J. & DE MULDER, E. F. J. 2003. Integrating geological information into urban planning and management: approaches for the 21st century. *In*: HEIKEN, G., FAKUNDINY, R. & SUTTER, J. (eds) *Earth Science in the City: A Reader*. American Geophysical Union, Washington, 379–41.

MARQUES, E. A. G., BARROSO, E. V., MENEZES FILHO, A. P. & VARGAS, E. do A., JR. 2010. Weathering zones on metamorphic rocks from Rio de Janeiro – Physical, mineralogical and geomechanical characterization. *Engineering Geology*, **111**, 1–18.

MARTINO, J. B. & CHANDLER, N. A. 2004. Excavation-induced damage studies at the Underground Research Laboratory. *International Journal of Rock Mechanics and Mining Sciences*, **41**, 1413–1426.

massDOT. 2011. *The project map (completion_lg.jpeg)*. The Central Artery/Tunnel Project – The Big Dig. The Massachusetts Department of Transportation – Highway Division. http://www.massdot.state.ma.us [accessed 13 September 2011].

MATHER, J. D., SPENCE, I. M., LAWRENCE, A. R. & BROWN, M. J. 1996. Man-made hazards. *In*: MCCALL, J., DE MULDER, E. F. J. & MARKER, B. (eds) *Urban Geoscience*. Balkema, Rotterdam, 127–161.

MATSUOKA, N. & MURTON, J. 2008. Frost weathering: recent advances and future directions. *Permafrost and Periglacial Processes*, **19**, 195–210.

MATTEI, U. 1997. Three patterns of law: taxonomy and change in the world's legal systems. *American Journal of Comparative Law*, **45**, 5–44.

MAYNE, P. W. & ELHAKIM, A. F. 2001. In-situ plasma vitrification of geomaterials. *In*: *Proceedings of the 15th International Conference on Soil Mechanics & Geotechnical Engineering*. **3**, Istanbul, Balkema, Rotterdam, 1807–1810.

MCDOWELL, P. W., BARKER, R. D. ET AL. 2002. *Geophysics in Engineering Investigations*. Construction Industry Research and Information Association (CIRIA), London.

MCPHERSON, M. J. 2012. *Subsurface Ventilation and Environmental Engineering*. Chapman & Hall, London, updated printed/on-line version: http://www.mvsengineering.com/ [accessed 17 February 2012].

MEADOWS, D. H., RANDERS, J. & MEADOWS, D. L. 1972. *The Limits to Growth: A Report for the Club of Rome's Project on the Predicament of Mankind*. Universe Books, New York.

MEIER, G. 2007. Historische Tiefkelleranlagen unter urbaner Bebauung an Beispielen Mitteldeutschlands – ingenieurgeologische Probleme und deren Lösungen. Tagungsband 16. Tagung für Ingenieurgeologie. TFH Georg Agricola Bochum. Eigenverlag Bochum, 323–330 [in German with English abstract] http://www.dr-gmeier.de/onlineartikel/oa0031.pdf [accessed 10 October 2010].

MENARD. 2010. MENARD Ground Improvement Specialist. http://www.menardusa.com/index.html [accessed 16 October 2010].

MineralsUK. 2010. *MineralsUK*. British Geological Survey. http://www.bgs.ac.uk/mineralsuk/ [accessed 21 November 2010].

MITHEN, S. 2003. *After the Ice, a Global Human History 20 000–5000 BC*. Phoenix, UK.

MORAN, E. F. 2000. *Human Adaptability. An Introduction to Ecological Anthropology*. 2nd edn., Westview Press Inc., Boulder, CO.

MÜHLL, D. V., HAUCK, C. & GUBLER, H. 2002. Mapping of mountain permafrost using geophysical methods. *Progress in Physical Geography*, **26**, 623–642.

MÜLLER, L. 1978. Removing misconceptions on the New Austrian Tunnelling Method. *Tunnels and Tunneling*, **10**, 29–32.

MUNZ, K. & HARIDAS, G. R. 2000. Marine outfalls project: issues and challenges. *In*: KAUSHISH, S. P. & RAMAMURTHY, T. (eds) *Tunneling Asia 2000*. Central Board of Irrigation and Power, New Delhi, 115–120.

NAITO, K., MYOI, H., OTTO, J., SMITH, D. & KAMITANI, M. 1998. Mineral projects in Asian countries. Geology, regulation, fiscal regimes and the environment. *Resources Policy*, **24**, 87–93.

NAM. 2010. *Bodemdaling Door Aardgaswinning*. Report EP201006302236 September 2010. NAM UIE/T/DPE (Bodembeweging) and NAM ITUI/AW (Geodesie). Nederlandse Aardolie Maatschappij B.V., Assen, The Netherlands.

NAMIBIAN DIAMONDS. 2012. Geological Survey of Namibia. Diamonds. http://www.mme.gov.na/gsn/diamond.htm [accessed 11 April 2012].

NASA. 2011. *National Aeronautics and Space Administration*. Visible Earth. http://visibleearth.nasa.gov/ [accessed 6 February 2011].

NASRI, V., CARRANZA-TORRES, C. & PETTERSSON, N. 2008. Design of large and shallow caverns of New York Second Avenue Subway. *In*: ROACH, M. F., KRITZER, M. R., OFIARA, D. & TOWNSEND, B. F. (eds) *Proceedings of the 9th North American Tunneling Conference (NAT2008)*. San Francisco. Society for Mining, Metallurgy & Exploration Ltd. (SME), Littleton, CO, 294–301.

NEN. 2011. *Netherlandse Norm*. http://www.nen.nl [accessed 6 February 2011].

NCHRP. 2007. Cone Penetration Testing. National Cooperative Highway Research Program (NCHRP); Synthesis Study 368. Transportation Research Board, Washington.

NEPEC. 2012. National Earthquake Prediction Evaluation Council (NEPEC). http://earthquake.usgs.gov/aboutus/nepec/[accessed 17 February 2012].

NICKSON, R. T., MCARTHUR, J. M., RAVENSCROFT, P., BURGESS, W. G. & AHMED, K. M. 2000. Mechanism of arsenic release to groundwater, Bangladesh and West Bengal. *Applied Geochemistry*, **15**, 403–413.

NISHIDA, Y., FABILLAH, H., ICHIHARA, S., NISHI, J. & CHO, K. D. 2007. The underground images in Japan, Korea and Indonesia. *In*: KALIAMPAKOS, D. & BENARDOS, A. (eds) *Underground Space: Expanding the Frontiers*. Proceedings of the 11th ACUUS International Conference, NTUA Press, Athens, 169–174.

NOFERINI, L., PIERACCINI, M. ET AL. 2007. Using GB-SAR technique to monitor slow moving landslide. *Engineering Geology*, **95**, 88–98.

NORBURY, D. 2010. *Soil and Rock Description in Engineering Practice*. Whittles, Caithness.

NORDIC. 2010. *E6 Trondheim–Stjørdal, Mid-Norway: A Pilot Project Using Competitive Dialogue*. http://www.nordicroads.com/website/index.asp?pageID=387 [accessed 16 October 2010].

NORDMARK, A. 2002. Overview on survey of water installations underground: underground conveyance and storage facilities. *Tunneling and Underground Space Technology*, **17**, 163–178.

NRIAGU, J. O. 1983. Did lead poisoning contribute to the fall of the empire? *New England Journal of Medicine*, **308**, 660–663.

NSW. 2012. *Geological Survey of New South Wales, New South Wales, Australia*. http://www.dpi.nsw.gov.au/__data/assets/pdf_file/0020/350372/Dick_GS20100470.pdf [accessed 8 March 2012].

NUCLEAR POWER. 2011. Nuclear Power in the World Today. http://www.world-nuclear.org/info/inf01.html [accessed 27 March 2012].

NYER, E. K., PALMER, P. L. ET AL. 2001. *In Situ Treatment Technology*. Environmental Science and Engineering series, Lewis Publishers, CRC Press, Boca Raton, FL.

OGC. 2012. Open Geospatial Consortium (OGC). http://www.opengeospatial.org/ [accessed 7 March 2012].

OLESEN, O., DEHLS, J. F., EBBING, J., HENRIKSEN, H., KIHLE, O. & LUNDIN, E. 2007. Aeromagnetic mapping of deep-weathered fracture zones in the Oslo Region – a new tool for improved planning of tunnels. *Norwegian Journal of Geology*, **87**, 253–267.

ONEGEOLOGY. 2012. OneGeology. http://www.onegeology.org/ [accessed 7 March 2012].

ÖNORM B2203-1/2. 2001/2005. *Underground works – Part 1: Cyclic driving (conventional tunnelling). Part 2: Continuous driving (TBM tunnelling)*. INFO-TECHNO Baudatenbank GmbH, Austria. http://www.bdb.at.

OSANLOO, M. & GHOLAMNEJAD, J. 2005. *Environmental Impact of pit lake formation on near by land*. 20th World Mining Congress, Tehran, **1**, 123–128.

OTTO, J. M. 2000. 4. Mineral policy, legislation and regulation. *Mining, Environment and Development*. A series of papers for the UN Conference on Trade and Development (UNCTAD), Geneva.

OVESEN, N. K. 1999. Geotechnical aspects of the Storeboelt Project. *In*: *Proceedings of the 14th International Conference on Soil Mechanics and Foundation Engineering*. 6–12 September

1997, Hamburg. ISSMGE, Taylor & Francis, Balkema, Rotterdam, 2097–2104.

OYAMA, T. & CHIGIRA, M. 2000. Weathering rate of mudstone and tuff on old unlined tunnel walls. *Engineering Geology*, **55**, 15–27, doi: 10.1016/S0013-7952(99)00103-9.

OZMUTLU, S. & HACK, H. R. G. K. 1998. Excavability evaluation and classification with knowledge based GIS. *In*: MOORE, D. P. & HUNGR, O. (eds) *Engineering Geology, A Global View from the Pacific Rim. Proceedings 8th Congress of the International Association for Engineering Geology and the Environment (IAEG)*, 21–25 September, Vancouver, Canada, **I**, 591–598.

OZMUTLU, S. & HACK, H. R. G. K. 2003. 3D modelling system for ground engineering. *In*: ROSENBAUM, M. S. & TURNER, A. K. (eds) *New Paradigms in Subsurface Prediction; Characterization of the Shallow Subsurface Implications for Urban Infrastructure and Environmental Assessment*. Lecture Notes in Earth Sciences **99**, Springer-Verlag, Berlin-Heidelberg, 253–259.

PACHER, F., VON RABCEWICZ, L. & GOLSER, J. 1974. Zum der seitigen Stand der Gebirgsklassifizierung in Stollen- und Tunnelbau. *In*: *Auswirkung geologischer Faktoren auf Bauabwicklung und Vertrag, Die gestaltung von Böschungen in Lockermassen und in Fels. Proceedings of the XXII Geomechanik-Kolloquium (Rabcewicz-Kolloquiums)*. 11–12 October 1973, Salzburg. Österreichische Gesellschaft für Geomechanik, Salzburg & Bundesministerium für Bauten und Technik, Strassenforschung, Vienna, 51–58 [in German].

PAKIANATHAN, L., KWONG, A. K. L., MCLEARIE, D. & CHAN, W. 2002. Pipe jacking: case study on overcoming ground difficulties in Hong Kong SAR harbour area treatment scheme. *In*: *Proceedings Trenchless Asia 2002*. 12–14 November, Hong Kong. China Hong Kong Society for Trenchless Technology (CHKSTT). Available from: http://www.chkstt.org/ [accessed 12 February 2012].

PALMSTROM, A. & BROCH, E. 2006. Use and misuse of rock mass classification systems with particular reference to the Q-system. *Tunnels and Underground Space Technology*, **21**, 575–593.

PARK, C. B., MILLER, R. D., XIA, J. & IVANOV, J. 2007. Multichannel analysis of surface waves (MASW)-active and passive methods. *The Leading Edge*, **26**, 60–64.

PATTERSON, M. C. L. & BRESCIA, A. 2008. Integrated sensor systems for UAS. *In*: *Proceedings UAVs Twenty-Third International Conference/Bristol International Unmanned Aerial Vehicle Systems Conference*. 7–9 April, Bristol. University of Bristol, Cranfield College of Aeronautics, & Royal Aeronautical Society, Department of Aerospace Engineering, Bristol, 19.1–19.13.

PEDERSEN, K. 2000. Exploration of deep intra-terrestrial microbial life: current perspectives. *FEMS Microbiology Letters*, **185**, 9–16.

PELLS, P. J. N. 2004. Substance and mass properties for the design of engineering structures in the Hawkesbury Sandstone. *Australian Geomechanics Journal*, **393**, 1–22.

PÉREZ-ROMERO, J., OTEO, C. S. & DE LA FUENTE, P. 2007. Design and optimisation of the lining of a tunnel in the presence of expansive clay levels. *Tunnelling and Underground Space Technology*, **221**, 10–22.

PESENDORFER, M. & LOEW, S. 2004. Hydrogeologic exploration during excavation of the Lötschberg base tunnel (AlpTransit Switzerland). *In*: HACK, R., AZZAM, R. & CHARLIER, R. (eds) *Engineering Geology for Infrastructure Planning in Europe: A European Perspective*. Lecture Notes in Earth Sciences, **104**, Springer-Verlag, Berlin, Heidelberg, 347–358.

PFAFFHUBER, A. A., GRIMSTAD, E., DOMAAS, U., AUKEN, E., FOGED, N. & HALKJÆR, M. 2010. Airborne EM mapping of rockslides and tunneling hazards. *The Leading Edge*, **29**, 956–959.

PHILPOTTS, A. R. & AGUE, J. J. 2009. *Principles of Igneous and Metamorphic Petrology*. 2nd edn. Cambridge University Press, Cambridge.

PICKLES, A. 2005. *Rock Mass Classification for Pile Foundations. The Characterization of Rock Masses for Engineering Purposes*. The Geological Society, Hong Kong Regional Group. City Univ., Hong Kong. http://www.gsregionalgroup.org.hk/downloads/RockMass Class for Piles [Read-Only].pdf [accessed on 30 July 2010].

PL. 2010. I. Permafrost Laboratory, Geophysical Institute, University of Alaska Fairbanks. http://www.gi.alaska.edu/snowice/Permafrost-lab/projects/projects_completed/proj_transects.html [accessed 14 October 2010].

PLAXIS. 2010. Plaxis bv. http://www.plaxis.nl/ [accessed 15 November 2010].

PORT IN ACTION. 2010. *Port of Rotterdam; Port in Action*. November 2010. http://www.portofrotterdam.com/en/News/newsletters/Port-in-action/Newsletters/Port in Action November 2010.pdf [accessed 9 September 2011].

POULOS, H. G. & BUNCE, G. 2008. Foundation design for the Burj Dubai – the world's tallest building. *In*: *Proceedings of the 6th International Conference on Case Histories in Geotechnical Engineering & Symposium in Honor of Professor James K. Mitchell*. 11–16 August, Arlington, VA. Missouri University of Science and Technology, Rolla, MO, USA, Paper: 1.47.

PREETHI, C. 2001. *The Advisability of Constructing Underground Spaces for Living/Working in Today's World*. Arizona State University, Tempe, AZ.

PRICE, D. G., DE FREITAS, M. H., HACK, H. R. G. K., HIGGINBOTTOM, I. E., KNILL, J. L. & MAURENBRECHER, M. 2009. *In*: DE FREITAS, M. H. (ed.) *Engineering Geology – Principles and Practice*. Springer-Verlag, Berlin, Heidelberg.

PRING, A. 2000. 2. International law and mineral resources. *Mining, Environment and Development. A series of papers for the UN Conference on Trade and Development (UNCTAD)*. UNCTAD, Geneva, Switzerland.

QIAN, Q. & CHEN, X. 2007. Evaluation of the status quo and outlook of the urban underground space development and utilization in China. *In*: KALIAMPAKOS, D. & BENARDOS, A. (eds) *Underground Space: Expanding the Frontiers*. Proceedings of the 11th ACUUS International Conference, NTUA, Athens, Greece. 15–21.

QIAO, Z. G. 2006. *Large Jacked Tunnel in Shanghai Using Steel Pipe Screen Support*. Shanghai No. 2 Municipal Engineering Co., Ltd. No-Dig Award 2006. The International Society for Trenchless Technology (ISTT), London. Available from: http://www.istt.com [accessed 10 February 2012].

RAGLAND, D., HAWLEY, J. & CASSON, E. 2003. NATM tunneling in soft rock in San Diego: Integrating design and construction. *In*: *Proceedings of the 2003 Rail Transit Conference*. San Jose, California, 8–12 June, American Public Transportation Association, Washington.

RAILWAY TUNNELS. 2011. The Worlds's Longest Railway Tunnels. http://www.lotsberg.net/data/rail.html [accessed 29 June 2011].

RAINEYA, T. P. & ROSENBAUM, M. S. 1989. The adverse influence of geology and groundwater on the behaviour of London Underground railway tunnels near Old Street Station. *Proceedings of the Geologists' Association*, **100**, 123–134.

RAJU, V. R. & YEE, Y. W. 2006. Grouting in limestone for SMART tunnel project in Kuala Lumpur. *In*: COWLING, K. G. & AUN, O. T. (eds) *Tunnelling and Trenchless Technology in the 21st century. Proceedings of the International Conference and Ex. on Tunnelling and Trenchless Technology.* 7–9 March, Subang, Selangor, ITA-AITES/The Institution of Engineers, Malaysia, 45–69. Available from ITA-AITES: http://www.ita-aites.org/fileadmin/filemounts/general/pdf/ItaAssociation/Organisation/Members/MemberNations/Malaysia/S6-10.pdf [accessed 12 February 2012].

RANDALL, F. A. & RANDALL, J. D. 1999. *History of the Development of Building Construction in Chicago.* 2nd edn. University of Illinois, Chicago.

READ, R. S. 2004. 20 years of excavation response studies at AECL's Underground Research Laboratory. *International Journal of Rock Mechanics and Mining Sciences*, **41**, 1251–1275.

RED DOG. 2011. World's Largest Lead and Zinc Mine Agrees to Pay $4.7 Million, http://yosemite1.epa.gov/opa/admpress.nsf/6427a6b7538955c585257359003f0230/5168e7a2f973663e852570cb0075e126!OpenDocument [accessed 27 March 2012].

REISSMÜLLER, M. 1997. *In*: EBERHARDT, E., THURO, K. & LUGINBUEHL, M. 2005. Slope instability mechanisms in dipping interbedded conglomerates and weathered marls: the 1999 Rufi landslide, Switzerland. *Engineering Geology*, **77**, 35–56.

REMMELTS, G. 1997. Gebruiksmogelijkheden van de diepe ondergrond van Nederland. Rapport 97-223-B. Delft. Nederlands Instituut voor Toegepaste Geowetenschappen TNO, Utrecht, The Netherlands [in Dutch].

RENGERS, N., HACK, H. R. G. K., HUISMAN, M., SLOB, S. & ZIGTERMAN, W. (2002) Information technology applied to engineering geology. *In*: VAN ROOY, J. L. & JERMY, C. A. (eds) *Engineering Geology for Developing Countries – Proceedings of the 9th Congress of the International Association for Engineering Geology and the Environment (IAEG).* 16–20 September, Durban, South Africa. IAEG & South African Institute for Engineering and Environmental Geologists (SAIEG), Houghton, South Africa, 121–143.

ROAD TUNNELS. 2011. The World's Longest Road Tunnels. http://www.lotsberg.net/data/tun10.html [accessed 29 June 2011].

ROBERTS, D. V. 1996. Sustainable development and the use of underground space. *Tunneling and Underground Space Technology*, **114**, 383–390.

ROCKWARE. 2012. Rockware Inc. Earth science and GIS software. http://www.rockware.com/ [accessed 9 June 2012].

ROCSCIENCE. 2011. Rocscience Inc. – software tools for rock and soil. http://www.rocscience.com [accessed 6 February 2012].

ROGERS, S. & HORSEMAN, S. 1999. *Underground space – The final frontier?* Earthwise, 13/11. British Geological Survey.

ROHRBOUGH, M. J. 1997. *Days of Gold: The California Gold Rush and the American Nation.* University of California Press, Berkeley.

ROME. 2011. Roman Catacombs. http://www.newadvent.org/cathen/03417b.htm [accessed 27 March 2012].

ROMERO, V. 2002. NATM in soft-ground: a contradiction of terms? *World Tunnelling*, **15** (7), 338–343.

RÖNKA, K., RITOLA, J. & RAUHALA, K. 1998. Underground space in land-use planning. *Tunnelling and Underground Space Technology*, **13/1**, 39–49.

ROSE, E. P. F. 2001. Military engineering in the Rock of Gibraltar and its geoenvironmental legacy. *In*: EHLEN, J. & HARMON, R. S. (eds) *The Environmental Legacy of Military Operations.* Geological Society of America, Reviews in Engineering Geology, Boulder, CO, **14**, 95–121.

ROYSE, K. R., RUTTER, H. K. & ENTWISLE, D. C. 2009. Property attribution of 3D geological models in the Thames Gateway, London: new ways of visualising geoscientific information. *Bulletin of Engineering Geology and the Environment*, **68**, 1–16.

RPD. 1999. Ruimtelijke perspectieven in Europa. *In*: *Ruimtelijke verkenningen.* Ministerie van VROM, The Hague, The Netherlands [in Dutch].

RPD. 2000. Het belang van een goede ondergrond. *In*: *Ruimtelijke verkenningen.* Ministerie van VROM, The Hague, The Netherlands, [in Dutch].

RUPKE, J. & CAMMERAAT, E. 2001. *Geomorphologisch-Geotechnische Kartierung des Erosionsgebietes Widentobel.* Alpine Geomorphology Research Group. University of Amsterdam. Berichte Amt fur Umweltschutz des Kantons St. Gallen, Austria. [in German].

RUSSELL, M., COLGLAZIER, E. W. & ENGLISH, M. R. 1991. *Hazardous Waste Remediation: The Task Ahead.* University of Tennessee, Waste Management Research and Education Institute, Knoxville.

RYBAK, M. & BROWN, T. 2008. Battery Park truck sewer emergency tunnel project. *In*: ROACH, M. F., KRITZER, M. R., OFIARA, D. & TOWNSEND, B. F. (eds) *Proceedings of the North American Tunneling 2008*, San Francisco, Society for Mining, Metallurgy & Exploration, Litttleton, 677–686.

SAINSBURY, D. P. 2008. Analysis of river bed cracking above longwall extraction panels in the Southern Coalfield of New South Wales, Australia. In SHIRMS 2008. *In*: POTVIN, Y., CARTER, J., DYSKIN, A. & JEFFREY, R. (eds) *Proceedings of the 1st Southern Hemisphere International Rock Mechanics Symposium*, Perth. 1, Australian Centre for Geomechanics, 325–338.

SALMINEN, R. (ed.), BATISTA, M. J. ET AL. 2005. *Geochemical Atlas of Europe.* Geological Survey of Finland.

SALVUCCI, F. P. 2003. The 'Big Dig' of Boston, Massachusetts: lessons to learn. *In*: SAVEUR, J. (ed.) *(Re)claiming the Underground Space.* Swets and Zeitlinger, B. V., Lisse, The Netherlands, 37–42.

SANDBERG, H. 2003. Three dimensional partition and registration of subsurface space. *Israel Law Review*, **1**, 119–167.

SANDVIK MINING AND CONSTRUCTION. 2011. Sandvik Mining and Construction. http://www.miningandconstruction.sandvik.com/ [accessed 15 April 2011].

SANKARAN NAIR, V. 2004. *Etymological Conduit to the Land of Qanat.* http://www.boloji.com/environment/24.htm [accessed 5 May 2011].

SCHALL, G. & SCHMALSTIEG, D. 2008. Interactive urban models generated from context-preserving transcoding of real-wold data. *In*: *Proceedings of the 5th International Conference on GIScience (GISCIENCE 2008)*, abstracts volume, Park City, Utah, USA, 23–26.

SCHALL, G., SCHÖNING, J., PAELKE, V. & GARTNER, G. 2011. A survey on augmented maps and environments: approaches, interactions and applications. *In*: LI, S., DRAGICEVIC, S. & VEENENDAAL, B. (eds) *Advances in Web-based GIS, Mapping Services and Applications.* CRC Press, Taylor & Francis Group, London, 207–225.

SCHAMINÉE, P. E. L. & KLAPWIJK, A. A. 2010. The STREAM's testdefinition facilities type of test independent database storage. *In*: TOLL, D. G., ZHU, H. & LI, X. (eds) *Information Technology in Geo-Engineering. Proceedings of the 1st International Conference (ICITG).* 16–17 September, Shanghai. Joint Technical Committee 2 (JTC2) & Tongji University, Shanghai, IOS Press, Amsterdam, 274–281.

REFERENCES

SCHMITZ, R. & SCHROEDER, C. 2009. Urban site investigation in the Belgian karst belt. *In*: CULSHAW, M. G., REEVES, H. J., JEFFERSON, I. & SPINK, T. W. (eds) *Engineering Geology for Tomorrow's Cities. Proceedings 10th Congress International Association for Engineering Geology and the Environment*. 6–10 September 2006, Nottingham, UK. Engineering Geology Special Publication, **22**, Geological Society, London, paper no. 801 (on CD-Rom).

SCHNAID, F. 2009. *In Situ Testing in Geomechanics*. The main tests. Taylor & Francis, London, New York.

SCHOFIELD, A. N. 2005. *Disturbed Soil Properties and Geotechnical Design*. Thomas Telford, London.

SCHULSON, E. M. 1999. The structure and mechanical behavior of ice. *JOM Journal of the Minerals, Metals and Materials Society*, **51**, 21–27.

SCHWARZ, L., REICHL, I., KIRSCHNER, H. & ROBL, K. 2003. Risks and hazards caused by groundwater during tunneling: geotechnical solutions used as demonstrated by recent examples from Tyrol, Austria. *RMZ – Materials and Geoenvironment*, **50** (1), 333–336.

SCOSS. 2004. *The collapse of NATM tunnels at Heathrow airport*. SC06.101/scoss5011. Standing Committee on Structural Safety, Structural-Safety, Incorporating CROSS and SCOSS, UK. http://www.scoss.org.uk/publications/rtf/Collapse of the NATM Tunnels at Heathrow Airport.pdf [accessed 18 February 2012].

SEG. 2010. *Society of Exploration Geophysicists*. SEG Digital Library. http://www.segdl.org/ [accessed 19 December 2010].

ŠEJNOHA, J., JARUŠKOVÁ, D., ŠPAČKOVÁ, O. & NOVOTNÁ, E. 2009. Risk quantification for tunnel excavation process. *In: Proceedings WCSET 2009*. 28–30 October, Venice, Italy. *International Journal of Engineering and Physical Sciences, World Academy of Science, Engineering and Technology (WASET)*, **3** (34), 393–401. http://www.waset.org/ [accessed 17 February 2012].

SELINUS, O., ALLOWAY, B., CENTENO, J. A., FINKELMAN, R. B., FUGE, R., LINDH, U. & SMEDLEY, P. 2005. *Essentials of Medical Geology; Impacts of the Natural Environment on Public Health*. Elsevier Academic Press, Amsterdam, The Netherlands.

SERRANO, I. R. (ed.) 2005. *Large-Scale Mining: Its Environmental, Social, Economic and Cultural Impacts in the Philippines*. CBIS Monograph 02. PRRM Conrado Benitez Institute for Sustainability, Quezon City, Philippines.

SEXTON. 2011. *Tar sands in Canada*. http://ffden-2.phys.uaf.edu/102spring2002_web_projects/m.sexton

SHAH, T., MOLDEN, D., SAKTHIVADIVEL, R. & SECKLER, D. 2000. *The Global Groundwater Situation: Overview of Opportunities and Challenges*. Monograph for the World Water Vision of the World Water Commission, International Water Management Institute, Colombo, Sri Lanka.

SHAMIR, U. 1998. Water agreements between Israel and its neighbors. *In*: ALBERT, J., BERNHARDSON, M. & KENNA, R. (eds) *Transformations of Middle Eastern Natural Environments: Legacies and Lessons*. Bulletin Series, **103**, Yale School of Forestry and Environmental Studies, Yale, 274–296.

SHANG, Y., WANG, S., YANG, Z. & WU, F. 2009. Lessons from one tunnel boring machine project in Kunming city, China. *In*: CULSHAW, M. G., REEVES, H. J., JEFFERSON, I. & SPINK, T. W. (eds) *Engineering Geology for Tomorrow's Cities. Proceedings 10th Congress International Association for Engineering Geology and the Environment*. 6–10 September 2006, Nottingham, UK. Engineering Geology Special Publication, **22**, Geological Society, London, paper no. 717 (on CD-Rom).

SHEN, Y., GAO, B., WANG, Z. & WANG, Y. 2009. Dynamic behavior of portal part of traffic tunnel in high-intensity earthquake area. *In*: PENG, Q., WANG, K. C. P., QIU, Y., PU, Y., LUO, X. & SHUAI, B. (eds) *Proceedings of the 2nd International Conference on Transportation Engineering (ICTE 2009)*, American Society of Civil Engineers (ASCE), Chengdu, **4**, 3417–3422.

SHENGWEN, Q., ZHONG QI, Y., FAQUAN, W. & ZHONGHUA, C. 2009. Deep weathering of a group of thick argillaceous limestone rocks near Three Gorges Reservoir, Central China. *International Journal of Rock Mechanics and Mining Sciences*, **46**, 929–939.

SHI, X. 2009. Development and utilization of underground space based on large-scale infrastructure construction of Beijing. *In*: QIHU, Q., PENGLIN, Z. & SI, Y. (eds) *Proceedings of the 12th International Conference ACUUS*, 18–19 November 2009, Shenzhen, China. Associated Research Centers for Urban Underground Space (ACUUS), Dollard-des-Ormeaux (Quebec), Canada & China Publishing Group, Beijing, China, 31–35.

SHIELDS, D. J. 1998. Non-renewable resources in economic, social, and environmental sustainability. *Non-renewable Resources*, **7**, 253–263.

SHIN, H. 2000. *Going Underground in Japanese Perspective*, http://www.lookjapan.com/LBopinion/00OctJP.html [accessed October 2000]

SHIN, H.-S. & PARK, E. S. 2007. The current status of Underground Space Utilization in Korea. *In*: KALIAMPAKOS, D. & BENARDOS, A. (eds) *Underground Space: Expanding the Frontiers, Proceedings of the 11th ACUUS International Conference*, NTUA Press, Athens, Greece, 181–186.

SIEGENTHALER, U., STOCKER, T. F. ET AL. 2005. Stable carbon cycle–climate relationship during the Late Pleistocene. *Science*, **310**, 1313–1317.

SIEMS, M. M. 2006. *Legal origins: reconciling law & finance and comparative law*. Centre for Business Research, University Of Cambridge. Working Paper No. 321.

SIMPSON, S. R. 1976. *Land Law and Registration*. Cambridge University Press, London.

SINDING-LARSEN, R., HOVLAND, M., SHIELDS, D. & GLEDITSCH, N. P. 2006. *Resource issues – towards sustainable use*. International Year of Planet Earth Corp., **6**, Leiden, The Netherlands.

SINGH, A. 2004. FRHI-a system to evaluate and mitigate rock fall hazard in stable rock excavations. *Journal Division of Civil Engineering Institute of Engineering (India)*, **85**, 62–75.

SINGHAL, B. B. S. & GUPTA, R. P. 2010. *Applied Hydrogeology of Fractured Rocks*, 2nd edn. Springer, Dordrecht, New York.

SkyTEM Surveys ApS. 2010. SkyTEM Surveys. http://www.skytem.dk/ [accessed 23 April 2011].

SLOB, S., VAN KNAPEN, B., HACK, R., TURNER, K. & KEMENY, J. 2005. Method for automated discontinuity analysis of rock slopes with three-dimensional laser scanning. Geology and Properties of Earth Materials 2005. Transportation Research Record, *Journal of the Transportation Research Board*, **1913**, 187–194.

SMARTEC SA. 2007. SmarTec, Rocktest Group. http://www.smartec.ch [accessed 7 March 2012].

SMITH, B. D., WALVOORD, M. A., CANNIA, J. C. & VOSS, C. I. 2010. *Airborne electromagnetic surveys for baseline permafrost mapping and potential long-term monitoring*. American Geophysical Union, Fall Meeting 2010, abstract #NS31A-1387.

SNEE, C. 2008. Engineering Geology and cavern design for New York City. *In*: ROACH, M. F., KRITZER, M. R., OFIARA,

D. & TOWNSEND, B. F. (eds) *Proceedings of the North American Tunnelling. NAT 2008*, San Francisco, Society for Mining, Metallurgy & Exploration, Littleton, 364–372.

SNIJDERS, C. 2009. *Modelling moves to 5D*. Build, December 2008/January 2009, 41-41. Available from: http://www.graphisoft.co.nz/news_items/B109Pg41Model5D.pdf [accessed 22 December 2010].

SOGA, K., CHAIYASARN, K., VIOLA, F., YAN, J., SESHIA, A. & CIPOLLA, R. 2010. Innovation in monitoring technologies for underground structures. *In*: TOLL, D. G., ZHU, H. & LI, X. (eds) *Information Technology in Geo-Engineering. Proceedings of the 1st International Conference (ICITG)*. 16–17 September, Shanghai. Joint Technical Committee 2 (JTC2) & Tongji University, Shanghai, IOS Press, Amsterdam.

STADT OPPENHEIM. 1993. *Die unterirdische Stadt Oppenheim/Rhein*. Stadtgeschichte und Stadtentwicklung auf historischen Fundamenten.

STAMATIOU, E. 2002. *Land Policy: Overview of Greek Land Property and Relations*. Dept. of Planning and Regional Development, University of Thessaly, Volos, Greece. Discussion Paper Series, **8/7**, 145–176.

STATE OF QUEENSLAND. 2011. *State of Queensland*, Queensland Government, Australia. http://www.epa.qld.gov.au/chims/placeDetail.html?siteId=16279 [accessed 15 April 2011].

STERLING, R. L. & GODARD, J. P. 2001. *Geo-engineering considerations on the optimum use of underground space*. ITA-AITES, 1–18. http://www.ita-aites.org/fileadmin/filemounts/association/publications/CD30th/ITA_01_1-18.pdf [accessed 9 February 2012].

STEUER, A., SIEMON, B. & AUKEN, E. 2009. A comparison of helicopter-borne electromagnetics in frequency- and time-domain at the Cuxhaven valley in Northern Germany. *Journal of Applied Geophysics*, **67**, 194–205.

STOTER, J. E. 2004. *3D Cadastre*. PhD-thesis, Technical University Delft, The Netherlands.

STOTER, J. E., VAN OOSTEROM, P. J. M., PLOEGER, H. D. & AALDERS, H. J. 2004. *Conceptual 3D Cadastral Model Applied in Several Countries*. FIG Working Week, The Netherlands.

STRACHER, G. B. 2007. *Geology of Coal Fires: Case Studies from Around the World*. Geological Society of America. Reviews in Engineering Geology, Boulder, CO.

STRUCKMEIER, W., RUBIN, Y. & JONES, J. A. A. 2005. *Groundwater – reservoir for a thirsty planet?* International Year of Planet Earth Corp., **2**, Leiden.

SWART, P. D. 1987. *An Engineering Geological Classification of Limestone Material*. Memoirs Center Engineering Geology, **49**, Technical University Delft, The Netherlands.

SWITZER, J. 2001. *Armed Conflict and Natural Resources: the Case of the Minerals Sector*. Discussion paper for Expert's workshop Mining, Minerals and Sustainable Development project.

TARBUCK, E. J., LUTGENS, F. K. & TASA, D. 2010. *Earth: An Introduction to Physical Geology*. 10th edn., Prentice Hall, Upper Saddle River, NJ.

TASELAAR, F. 2002. *Geokoepels, de ruimte in nieuw daglicht*. Ingenieursbureau Amsterdam, Amsterdam, The Netherlands [in Dutch].

TAUTONA. 2011. TauTona Mine. http://en.wikipedia.org/wiki/TauTona_Mine [accessed 9 August 2011].

TAYLOR, L. E., HILLIER, J. A. & BENHAM, A. J. 2005. *World Mineral Production: 1999–2003*. British Geological Survey, Nottingham, UK.

TCB. 1996. *Rapport Diepe ondergrond en bodembescherming*, **TCB R06**, The Hague, The Netherlands [in Dutch].

TCB. 1997. Advies Ondergronds beluchten, **TCB A24**, The Hague, The Netherlands [in Dutch].

TCB. 2004. *Advies inzake concept beleidsaanbevelingen project 'Bodem als Energiebron en –Buffer'*, **pnTCB S13**, The Hague, The Netherlands [in Dutch].

TCFE. 2012. Technical Council on Forensic Engineering (TCFE). American Society of Civil Engineers (ASCE). http://www.asce.org/Content.aspx?id=2147488650 [accessed 12 March 2012].

TEGTMEIER, W., ZLATANOVA, S., VAN OOSTEROM, P. J. M. & HACK, H. R. G. K. 2009. Information management in civil engineering infrastructural development: with focus on geological and geotechnical information. *In*: KOLBE, T. H., ZHANG, H. & ZLATANOVA, S. (eds) *GeoWeb 2009 Academic Track – Cityscapes, Proceedings of the ISPRS Workshop*. 27–31 July, Vancouver. ISPRS Archives, Vol. XXXVIII-3-4/C3 Comm. III/4, IV/8 and IV/5, 68–73. Available from: http://www.isprs.org/ [accessed 7 March 2012].

TELFORD, W. M., GELDART, L. P. & SHERIFF, R. E. 1990. *Applied Geophysics*. 2nd edn. Cambridge University Press, UK.

TENNESSEE. 2011. Abandoned Mines Land Reclamation. http://www.state.tn.us/environment/wpc/programs/abandmine [accessed 11 April 2012].

TERRAFIRMA. 2011. Terrafirma. A Pan-European Ground Motion Hazard Information Service. http://www.terrafirma.eu.com [accessed 22 April 2011].

THAPA, B. B., MARCHER, T., MCRAE, M. T., JOHN, M., SKOVAJSOVA, Z. & MOMENZADEH, M. 2009. NATM strategies in the U.S. – Lessons learned from the initial support design for the Caldecott 4th Bore. *In*: ALMERARIS, G. & MARIUCCI, B. (eds) *Proceedings of the Rapid Excavation and Tunneling Conference*, Las Vegas, Society for Mining, Metallurgy & Exploration Ltd., (SME), Littleton, CO, 96–107.

THE ECONOMIST. 2005, 2009. The Economist Commodity-Price Index. http://www.economist.com/ [accessed 20 July 2011].

THOMAS, A. H. 2008. *Sprayed Concrete Lined Tunnels*. Taylor & Francis, Abingdon.

THOMAS, A. H., LEGGE, N. B. & POWELL, D. B. 2004. The development of sprayed concrete lined (SCL) tunnelling in the UK. *In*: SCHUBERT, W. (ed.) *Rock Engineering. Theory and Practice. Proceedings of the ISRM Regional Symposium EUROCK 2004 & 53rd Geomechanics Colloquy*. 7–9 October, Salzburg. Austrain Society for Geomechanics. VGE Verlag Glückauf, Essen, Germany, 25–30.

THOMS, R. L. & GEHLE, R. M. 2000. *A brief history of salt cavern use*. Proceedings of the 8th World Salt Symposium, Elsevier, The Hague.

THURO, K. 2003. Predicting roadheader advance rates: geological challenges and geotechnical answers. *In*: VARDAR, M., GÜNEY, A. & DEMIRBAG, E. (eds) *50th Years Symposium of the Faculty of Mines, Istanbul Technical University – The Underground Resources of Turkey Today and Future*. 5–8 June, Istanbul. Istanbul Technical University, Istanbul, 1241–1247. Available from: http://www.geo.tum.de/people/thuro/pubs/2003_itu_roadheader.pdf.

THURO, K. & PLINNINGER, R. J. 2003. Hard rock tunnel boring, cutting, drilling and blasting rock parameters for excavatability. *In*: HANDLEY, M. & STACEY, D. (eds) *Proceedings 10th ISRM congress; Technology roadmap for rock mechanics (ISRM 2003)*. 8–12 September, Sandton. South African Inst. of Mining and Metallurgy, Johannesburg, 1227–1233.

THURO, K. & SCHOLZ, M. 2004. Deep weathering and alteration in granites-a product of coupled processes. *In*: STEPHANSON, O.,

HUDSON, J. A. & JING, L. (eds) *GeoProceedings of the 2003 International Conference Coupled T-H-M-C Processes in Geosystems*, Royal Inst. of Technology, Stockholm, Sweden. 13–15 October. Elsevier Geo-Engineering Book Series, Amsterdam, **2**, 785–790.

TIMILSINA, B. P. 2004. *Country Report on Management of Urban Water Environment in Nepal*. JICA Executives' Seminar on Public Works and Management, 17–31 October, 2004, Tsukuba, Japan. Japan International Cooperation Agency (JICA), Tokyo, Japan, 403–436.

TOLL, D. G. 2007. Geo-engineering data: representation and standardisation. *Electronic Journal of Geotechnical Engineering*, **12**, special issue, http://www.ejge.com/Index_ejge.htm.

TONGJI UNIVERSITY. 2006. *1st China–Japan–the Netherlands Tunnel Seminar*. Proceedings, Tongji University, China.

TONINI, A., GUASTALDI, E., MASSA, G. & CONTI, P. 2008. 3D geo-mapping based on surface data for preliminary study of underground works: A case study in Val Topina (Central Italy). *Engineering Geology*, **99**, 61–69.

TONON, F. 2010. Sequential excavation, NATM and ADECO: What they have in common and how they differ. *Tunnelling and Underground Space Technology*, **25**, 245–265.

TORP, T. A. 2003. CO_2 Subsurface storage – Experience and Expectations. Presentation at First Int. Conf. On Subsurface Development & Management of the Subsurface, Utrecht (unpublished).

TORRES ACOSTA, C. A. 2008. *Geometric characterization of rock mass discontinuities using terrestrial laser scanner and ground penetrating radar*. MSc thesis. University Twente-ITC, The Netherlands.

TØRUM, E., SANDVEN, R., HOVEM, S. G. & RØNNING, S. 2010. Design of cut-and-cover tunnel in quick clay based on CPTU and laboratory tests. *In*: *2nd International Symposium on Cone Penetration Testing, CPT'10*. 9–11 May, Huntington Beach, California. Technical Committee TC-16 of the ISSMGE/California State Polytechnic University, San Luis Obispo, CA, USA, paper 3-42. Available from: http://www.cpt10.com/ [accessed 7 February 2012].

TUGRUL, A. 2004. The effect of weathering on pore geometry and compressive strength of selected rock types from Turkey. *Engineering Geology*, **75**, 215–227.

TUNCONSTRUCT. 2012. Tunconstruct; advancing the European underground construction industry, through technology innovation. http://www.ifb.tugraz.at/tunconstruct/.

TUNNEL CANADA. 2011. Tunnelling Association of Canada. http://www.tunnelcanada.ca [accessed 27 March 2012].

TURNER, A. K. & GABLE, C. W. 2007. A review of geological modeling. *In*: *Three-Dimensional Geological Mapping for Groundwater Applications Workshops*. 2007 Annual Meeting, Geological Society of America, 27 October, Denver, CO. Illinois State Geological Survey (ISGS), Geological Survey of Canada & Minnesota Geological Survey. ISGS, Champaign, IL, USA (extended abstract). http://www.isgs.illinois.edu/research/3DWorkshop [accessed 7 March 2012].

TURNER, A. K. & D'AGNESE, F. A. 2009. The role of geological modeling in a web-based collaborative environment. *In*: *Three-Dimensional Geological Mapping for Groundwater Applications Workshops*. 2009 Annual Meeting, Geological Society of America, 17 October, Portland, OR. Illinois State Geological Survey (ISGS), Geological Survey of Canada & Minnesota Geological Survey. ISGS, Champaign, IL, USA, 58–62 (extended abstract). http://www.isgs.illinois.edu/research/3DWorkshop [accessed 7 March 2012].

TYRRELL, H. G. 1911. *History of Bridge Engineering*. The G. B. Williams Co. printers, Chicago, Illinois (Googlebooks, 2007).

UIC. 2011. Uranium Information Centre Website. http://www.uic.com.au/nip41.htm [accessed 11 September 2006].

UN. 2000. *United Nations Millennium Declaration*. Resolution A/RES/55/2. New York.

UNEP. 1972. *Declaration of the United Nations Conference on the Human Environment*. http://www.unep.org/Documents.Multilingual/Default.asp?documentid=97&articleid=1503 [accessed 10 February 2012].

UNEP. 1992. Agenda 21 – Earth Summit: The United Nations Programme of Action from Rio.

UN ESCAP. 1985. *Geology for Urban Planning; selected papers on the Asian and Pacific Region*. ESCAP, Bangkok, ST/ESCAP/394.

UN GENERAL ASSEMBLY. 2005. *The International Year of Planet Earth*, 2008 (A/RES/ 60/192), New York.

UNIVERSITY OF OTTAWA. 2011. World Legal Systems. http://www.juriglobe.ca [accessed 11 August 2011].

UN POPULATION DIVISION. 2004. *World Population in 2300*. United Nations, New York.

UN POPULATION DIVISION. 2011. *World Population Prospects: The 2010 Revision, Highlights and Advance Tables*. Working Paper No. ESA/P/WP.220. Population Division, Department of Economic and Social Affairs, United Nations, New York.

UN POPULATION DIVISION. 2012. *World Urbanization Prospects: The 2011 Revision: Highlights*. Working Paper No. ESA/P/WP/224. Population Division, Department of Economic and Social Affairs, United Nations, New York.

UPTUN. 2009. *Cost-effective, Sustainable and Innovative Upgrading Methods for Fire Safety in Existing Tunnels*. European RTD-project funded by the European Commission in FP5. http://www.uptun.net/ [accessed 30 January 2011].

USACE. 2010. *US Army Corp of Engineers*. http://www.usace.army.mil/Pages/default.aspx [accessed 23 December 2010].

USARMY. 2009. *U.S. Army images*. http://search.ahp.us.army.mil/search/images/ [accessed 10 December 2010].

USDOF. 2009. *US Department of Transportation*. Federal Highway Administration. http://international.fhwa.dot.gov/ [accessed 30 January 2011].

USGS. 2001. *Database. 93 Minerals*. Reston, VA.

USGS. 2006. *USGS Mineral Resources Program – Supporting Stewardship of America's Natural Resources*. Circular 1289. Reston, VA.

USGS. 2011. *Earth Resources Observation and Science (EROS) Center*. http://eros.usgs.gov/#/Home [accessed 6 February 2011].

USGS. 2012. United States Geological Survey (USGS). http://www.usgs.gov/ [accessed 1 March 2012].

USGS EARTHQUAKE. 2011. *Earthquake Topics* http://earthquake.usgs.gov/learn/topics/USGS [accessed 14 January 2011].

VAGT, G. O. & IRVINE, R. D. 1998. Construction aggregates in Canada – an overview. *In*: BOBROWSKY, P. (ed.) *Aggregate Resources, a Global Perspective*. Balkema, Rotterdam, 9–26.

VAN DER MAAREL, E. & DAUVELLIER, P. L. 1978. *Naar een Globaal Ecologisch Model voor de ruimtelijke ontwikkeling van Nederland*. Ministerie van Volkshuisvesting en Ruimtelijke Ontwikkeling, Den Haag [in Dutch].

VAN DER MEER, F. D. & DE JONG, S. 2001. *Imaging Spectrometry: Basic Principles and Prospective Applications*. Kluwer Academic Publishers, Dordrecht, The Netherlands.

VAN DER MOOLEN, B., RICHARDSON, A. F. & VOOGD, H. 1998. *Mineral Planning in a European Context*. Geo Press, Groningen.

VAN MEURS, G., VAN DER ZON, W., LAMBERT, J., VAN REE, D., WHIFFIN, V. & MOLENDIJK, W. 2006. The challenge to adapt soil properties. *In*: THOMAS, H. R. (ed.) *Proc. Opportunities, Challenges and Responsibilites for Environmental Geotechnics, 5th ICEG Environmental Geotechnics Congress*. 26–30 June, Cardiff. Thomas Telford Publication, London, 1192–1199.

VAN OOSTEROM, P. & MEIJERS, M. 2011. Towards a true varioscale structure supporting smooth-zoom. *In: Proceeding of the 14th Workshop of the International Cartographic Association (ICA) Commission on Generalisation and Multiple Representation & the ISPRS Commission II/2 Working Group on Multiscale Representation of Spatial Data*. 30 June–1 July, Paris. ICA Commission on Generalisation and Multiple Representation. Available from: http://aci.ign.fr [Accessed 8 March 2012].

VAN PAASSEN, L. A., DAZA, C. M., STAAL, M., SOROKIN, D. Y., VAN DER ZON, W. & VAN LOOSDRECHT, M. C. M. 2009. Potential soil reinforcement by biological denitrification. *Ecological Engineering*, **36** (2), 168–175.

VAN REE, D. & CARLON, C. 2003. New technologies and future developments – Is there a truth in site characterisation and monitoring? *Land Contamination & Reclamation*, **11**, 37–47.

VAN STAVEREN, M. Th. 2006. *Uncertainty and Ground Conditions: A Risk Management Approach*. Butterworth-Heinemann, Elsevier, Oxford.

VERHOEF, P. N. W. 1997. *Wear of Rock Cutting Tools: Implications for the Site Investigation of Rock Dredging Projects*. Taylor & Francis, Rotterdam.

VERMEULEN, H. 2002. *De bodem onder Europa, Bodembescherming in internationaal en Europeesrechtelijk perspectief*. SKB, Gouda, The Netherlands [in Dutch].

VERVOORT, A. & DE WIT, K. 1997. Correlation between dredge ability and mechanical properties of rock. *Engineering Geology*, **473**, 259–267.

VOGELHUBER, M., ANAGNOSTOU, G. & KOVÁRI, K. 2004. The influence of pore water pressure on the mechanical behaviour of squeezing rock. *In*: OHNISHI, Y. & AOKI, K. (eds) *Contribution of Rock Mechanics to the New Century. Proceedings 3rd Asian Rock Mechanics Symposium (ARMS 2004)*. 30 November–2 December, Kyoto. Japanese Geotechnical Society (JGS), International Society for Rock Mechanics (ISRM). Mill Press, Tokyo, **1**, 659–664.

VON RABCEWICZ, L. 1964a. The new Austrian tunnelling method. Part one. *International Water Power & Dam Construction*, **16**, 453–457.

VON RABCEWICZ, L. 1964b. The new Austrian tunnelling method. Part two. *International Water Power & Dam Construction*, **16**, 511–515.

VON RABCEWICZ, L. 1965. The new Austrian tunnelling method. Part three. *International Water Power & Dam Construction*, **17**, 19–24.

VROM. 2007a. *Conference Report: Echte schatten vind je onder de grond*. Almere, The Netherlands [in Dutch].

VROM. 2007b. *Dossier Ondergrond*. http://www.vrom.nl/pagina.html?id=23773

W3C. 2012. *World Wide Web Consortium (W3C)* – Extensible Markup Language (XML). http://www.w3.org/XML/ [accessed 7 March 2012].

WACKERNAGEL, M., ONISTO, L. ET AL. 1997. *Ecological Footprints of Nations*. Centro de Estudios para la Sustentabilidad, Mexico.

WALLIS, S. 2010. *Symptoms of the collapse syndrome*. Tunnel Talk. July. http://www.tunneltalk.com/Discussion-Forum-Jul10-Collapse-syndrome.php [accessed 15 November 2010].

WANG, S. 2004. *Regulating Death at Coalmines: Changing Mode of Governance in China*. Department of Government & public administration, Chinese Univ. of Hong Kong, Hong Kong.

WANG, A. 2007. *Property Rights in China under the New Property Law*. Angela Wang & Co, Solicitors, Hong Kong.

WANG, Z. Z., GAO, B., JIANG, Y. & YUAN, S. 2009. Investigation and assessment on mountain tunnels and geotechnical damage after the Wenchuan earthquake. *Science in China Series E: Technological Sciences*, **522**, 546–558.

WASSING, B., VELDKAMP, H. & BREMMER, C. 2003. Assessment of uncertainties in volume and yield estimations of classic deposits. *In*: ROSENBAUM, M. S. & TURNER, A. K. (eds) *New Paradigms in Subsurface Prediction. Characterization of the Shallow Subsurface; Implications for Urban Infrastructure and Environmental Assessment*. Lecture Notes in Earth Sciences, **99**, Springer-Verlag, Berlin, Heidelberg, 313–322.

WEIDINGER, F. & LAUFFER, H. 2009. The Tauern tunnel first and second tubes from the contractor's viewpoint. *Geomechanics and Tunnelling*, **2**, 24–32.

WENNER, D. & WANNENMACHER, H. 2009. Alborz Service Tunnel in Iran: TBM tunnelling in difficult ground conditions and its solutions. *In: Proceedings of the 1st Regional and 8th Iranian Tunnelling Conference*. 18–20 May, Tehran, Iran. Iranian Tunnelling Association (IRTA) & Tarbiat Modares University, Tehran, Iran, 342–353.

WEST. 2008. *West's Encyclopedia of American Law*, 2nd edn. The Gale Group, Inc., Ann Arbor, MI.

WICANDER, R. & MONROE, J. S. 2009. *Historical Geology*. 6th edn. Brooks/Cole, Belmont, CA.

WIEDEMEIER,, TODD, H., WILSON, J. T., KAMPBELL, D. H., MILLER,, ROSS, N. & HANSEN, J. E. 1999. *Technical protocol for implementing intrinsic remediation with long-term monitoring for natural attenuation of fuel contamination dissolved in groundwater volume I*. US Airforce guidance. Air Force Center for Environmental Excellence, Technology Transfer Division, Brooks Air Force Base San Antonio, Texas.

WILDENBORG, A. F. B., BOSCH, J. H. A., DE MULDER, E. F. J., HILLEN, R., SCHOKKING, F. & VAN GIJSSEL, K. 1990. A Review: effects of (peri-)glacial processes on the stability of rock salt. *In: Proceedings of the 6th International Congress on International Association Engineering Geology*, **5**, Balkema, Rotterdam, 2763–2770.

WILLIAMSON, I. P. 1983. *A Modern Cadastre for New South Wales*. UNISURV S-23, School of Surveying, The University of New South Wales.

WILLIAMSON, I. P. 2001. *Land Administration Best Practice, Providing the Infrastructure for land Policy Implementation*. Department of Geodesy, Delft University of Technology, Delft, The Netherlands.

WINCHESTER, S. 2002. *The Map that Changed the World*. Penguin Books, London.

WOODWARD, J. 2005. *An Introduction to Geotechnical Processes*. Spon Press, Taylor & Francis Group, Abingdon.

WORLD BANK. 2006. *Water Resources Management in Japan*. Policy, Institutional and Legal Issues. World Bank Analytical and Advisory Assistance (AAA) Program. China: Addressing Water Scarcity. Background Paper 1.

WORLDCOAL. 2011. World Coal Association. http://www.worldcoal.org [accessed 27 March 2012].

WOSKOV, P. & COHN, D. 2009. *Millimeter wave deep drilling for geothermal energy, natural gas and oil*. Annual report 2009. PSFC/RR-09-11. Plasma Science and Fusion Center, Massachusetts Institute of Technology (MIT), Cambridge, USA.

REFERENCES

WRB. 2010. *World Reference Base (WRB)*. International Standard Soil Classification. International Soil Reference and Information Centre (ISRIC), International Union of Soil Science (IUSS) & FAO. http://www.fao.org/nr/land/soils/soil/en/ [accessed 24 December 2010].

WSSD. 2003. Plan of Implementation of the World Summit on Sustainable Development. http:www.un.orgesasustdevdocumentsWSSD_POI_PDEnglishWSSD_PlanImpl.pdf [accessed 3 April 2012].

WU, C., XIA, C. & LI, Z. 2006. Safety assessment system for evaluating spontaneous combustion of sulfide ores in mining stope. *In*: HUANG, P., WANG, Y., LI, S., ZHENG, C. & MAO, Z. (eds) *Progress in Safety Science and Technology, Proceedings International Symposium on Safety Science and Technology (2006 ISSST)*. 24–27 October, Changsha, China. China Occupational Safety and Health Association, Beijing Institute of Technology, Science Press, Beijing, 1599–1603.

WYCISK, P., HUBERT, T., GOSSEL, W. & NEUMANN, Ch. 2009. High-resolution 3D spatial modelling of complex geological structures for an environmental risk assessment of abundant mining and industrial megasites. *Computers and Geosciences*, **351**, 165–182.

WYLLIE, D. C. & MAH, C. W. 2004. *Rock Slope Engineering: Civil and Mining*, 4th edn. Taylor & Francis e-Library, London.

XMML. 2012. *eXploration and Mining Markup Language*. http://www.seegrid.csiro.au/twiki/bin/view/Xmml/WebHome [accessed 7 March 2012].

XU, Y.-S., SHEN, S.-L. & DU, Y.-J. 2009. Geological and hydrogeological environment in Shanghai with geohazards to construction and maintenance of infrastructures. *Engineering Geology*, **109**, 241–254.

YANBING, W., LIXIN, W., WENZHONG, S. & LIU XIAOMENG, L. 2007. On 3d GIS spatial modeling. *In*: JIANG, J. & ZHAO, R. (eds) *Proceedings ISPRS Workshop on Updating Geo-spatial Databases with Imagery & 5th ISPRS Workshop on DMGISs*. 28–29 August, Urumchi, Xingjiang, China. ISPRS, Beijing, China. ISPRS Archives, Volume XXXVI-4/W54, 2007, 237–240. Available from: http://www.isprs.org/ [accessed 7 March 2012].

YANG, C. & RASKIN, R. 2009. Introduction to distributed geographic information processing research. *Int. Journal of Geographical Information Science – Distributed Geographic Information Processing Research*, **23**, 553–560.

YEFIM CAVALIER. 2003. Depth of the Deepest Mine. http://hypertextbook.com/facts/2003/YefimCavalier.shtml [accessed 9 August 2011].

YU, C.-W. 1998. *Creep characteristics of soft rock and modelling of creep in tunnel*. PhD thesis. Department of Civil and Environmental Engineering, University of Bradford, UK.

YU, M., WANG, H. & NI, D. 2009. Underground space development of Expo Axis of Shanghai Expo Park. *In*: *Proceedings of the 12th International Conference ACUUS*, Beijing, China, 110–114.

ZHANG, J. & PENG, S. 2005. Water inrush and environmental impact of shallow seam mining. *Environmental Geology*, **48**, 1068–1076, doi: 10.1007/s00254-005-0045-8.

ZHANG, X., LAI, Y., YU, W. & WU, Y. 2004. Forecast analysis for the re-frozen of Feng Huoshan permafrost tunnel on Qing-Zang railway. *Tunnelling and Underground Space Technology*, **19**, 45–56.

ZHANG, S., LAI, Y., SUN, Z. & GAO, Z. 2007. Volumetric strain and strength behavior of frozen soils under continement. *Cold Regions Science and Technology*. **47**, 263–270.

ZHOU, G. 2006. *Water resources management in an arid environment. The case of Israel*. World Bank Analytical and Advisory (AAA) Program. China: Addressing Water Scarcity. Background Paper No. 3.

ZIGTERMAN, W. 2009. Design interactions of underground and surface structures. *In*: CULSHAW, M. G., REEVES, H. J., JEFFERSON, I. & SPINK, T. W. (eds) *Engineering Geology for Tomorrow's Cities. Proceedings 10th Congress International Association for Engineering Geology and the Environment*. 6–10 September 2006, Nottingham, UK. Engineering Geology Special Publication, **22**, Geological Society, London, paper no. 445 (CD-Rom).

ZOBL, F. & MARSCHALLINGER, R. 2008. Subsurface GeoBuilding Information Modelling GeoBIM. *GEOinformatics*, **811**, 40–43.

Index

Page numbers in *italic* denote figures. Page numbers in **bold** denote tables and text boxes.

Aare Granite *80*
Århus Convention 149
abandoned mines 14, 171
Aboriginal land, mining **12**, 135, 140
abrasive, abrasiveness 48, 65, *66*
accuracy 128
acid mine drainage 14
acid water 70, *73*
acidification, allowed values 148
Aeschertunnel glacial till **55**
air bubble and TBM 77
air raid shelter 21
alluvial exploitation 134, 138–140, **141–142**
Al Najaf, underground city **16**
Amsteg *82*
Amsterdam 55, 84, **85**, **86**, 98
Analysis of Controlled Deformation methodology 93
analysis, 5D 170
angle of shearing resistance *see* φ
anhydrite 62, 63, 98
Antarctic Treaty 141
anthrax 98
Apennines *117*
aquiclude 57
aquifer 23, *57*
 deep-seated 29
aquitard 55, *57*
archaeological data source 126
archaeological heritage 12, **17**, 41
archaeological heritage sites 147, 154
 at risk 163, 165
 preservation 147
 water withdrawal 145
archaeological sites 115
arche *56*, *88*
arching 62, 63, *64*
 collapse 54
archive, subsurface 41
atmosphere 15, 23, 32, 41, 62
arsenic in groundwater 58, *156*, 157
arsenic, allowed values 148
artesian *56*, *57*
as built/as designed/as is 119
asbestos mining 141
Äspö slightly fractured diorite and granite **55**
asset control 133–135
ASTM (American standards) 101
atmosphere auditing framework 161
auger *107*
augmented visualization *131*

Australian subsurface homes 17, *19*
Aznacollar Zinc mine 30, *32*

backfill 43, 77
 procedures 77
bacteria in subsurface 64, 101
Bath, abandoned mines 80
Bath, Roman infrastructures *10*
bathymetric maps 102
bearing capacity 20, 40–41, 57
Beijing underground development 173, 175, 177
bell-pit 6, 76, *79*
bentonite 67
Bhatwari *56*
Bieniawski's Rock Mass Rating 86–88
Big Dig, Boston **37–38**
biochemical weathering 58
biodiversity 11, 41, 175
biogeochemical processes 175
biohazard 98–99
biological activity in subsurface 62
bioremediation 175
biosphere 1, 14, 34
black coal 172
blast 73, 88
 hole *66*
 induced damage 71, *73*, **91**
Bodio *92*
bolt 72
Boolboonda Railway Tunnel 74
borehole, super-deep, Kola *177*
boreholes 105–106, *107*, *108*, *116*
Beskyd Tunnel *105*
Boston 36, **37–38**
Boston Central Artery **37–38**
boulder 48, **55**, 80
breccia 60
brine 17, **33**, 115
brittle 48, 53–55, 81
bronze 8, 9
Brundtland Commission 1, 160
BS (British Standard) 99
burial sites and infectious disease 98–99
Burj Khalifa Tower 84

Cu **52**
CaCO₃ *see* calcium carbonate
cadastre for subsurface registration 161
calcium carbonate 48
calcrete 61, 79
caprock 60

carbon capture 154
Carbon Capture and Sequestration (CCS) 150, 154
carbon dioxide *see* CO_2
    underground storage **32–34**
carbon sequestration 150
$CaSO_4$ *see* anhydrite
$CaSO_4 \cdot 2H_2O$ *see* gypsum
Castaic dam *56*
cave 55
cave homes 16
caves and archaeological sites 115–116
caves, voids, geophysical detection 110
CCS *see* Carbon Capture and Sequestration
cement, cementation (ground) 46–48, 50, 61, 68, 77, 80
cement grout *see* grout
cement mix 68
cementing (particles in PFC) 101
CEN (European Committee for Standardization) 101
central artery **37**
chalk 12, 35, 131
chemical materials 6, 12
chemical waste 29
China, transport tunnels 37
China, underground cities 39
    see also Beijing
chipping 106
Chuquicamata copper mine 25, *26*
city, underground **16**, *18*, 19–21, 37, 39–40
    Coober Pedy 16, *19*, 34, 39
    see also urban development
CityGML 125, *128*
civil engineering and underground structures 43–44
civil engineering information system 128
civil law 133, 152
clay 43, **47**, 48, 49, 53, 55, *56*, 68, 79, 81, 94, *94*
    cement 50
    deformation **55**
    quick 57
    squeezing 63
    strength **52, 53**
    swelling 48, 58, 62, 63, 80
claystone 48
clean-up technology 30, 150, 152
cleavage 48
    (in ground or rock mass) *see* discontinuity
climate (*see also* weather) 58, 59, 61
climate and change 104, 154
    and Earth history 177
    groundwater 177
    permafrost 76
Club of Rome 24
CO_2 31, 33–34, 136, 150, 175
    emissions 32
    reduction 175
    storage **32–34**
Coal Bed Methane 172

coal fire 30, 98
coal mine *30*
    reclamation 29–30
coal production 28
coal waste 98
coal, historic mining 14
coal, mining law 140
cobble 48
cockroaches 62
cohesion **50, 52–53**
collapse and mining *14*
collapsible soil 48, 57
commodities, demand 171
commodity prices 24–25
common law 133–134, 152
condominium law 136
conductivity 108
Cone Penetration Test (CPT) 106, *116*
confined/unconfined (water) *56*
confining 50, 52–54
    pressure **52**
conglomerate 60, 93
continuum (numerical model) 94
construction materials 12
    current use 25–27, *28*
    legislation **140–141**
    sterilization of 175
construction stability 17, 20
contaminants, active removal 165
contaminants, sustainability issues 163
contaminated sites 174
conveyor 31, 77
    belt *77*
    screw *77*
Coober Pedy, cave homes 16, *19*, 39
cooperative model, ownership 137
copper mine 25, *26*
copper, historic mining 6
coral reef *see* reef
CPT *see* Cone Penetration Test
crack (in construction) 44, 76, 98
crack (in ground, rock or soil mass) *see* discontinuity
cracking sound 64
creep 53–54, 59, 75, 78
crusts 59
culvert *78*
    box *78*
customary land holding 134, **139–140**
customary law 134
cut and cover support 72, 83, *83*, **85–86**
cutter head *67, 77*

Daejon *115*
data collection and intergration 162
data collection methods 101
data density in digital modelling 119–120, *121*
data format 118–119, *120*

data processing 99, 125
    CityGML 125, *126*
    data sources 125–126
    standardization 125
database 127–128
database management systems **127**, 128
decision support system (DSS), 175
decomposed *see* weathering
decontamination 174
deformation 51, **54**, *56*, 63
    measurement 110
deformation modulus **55**
deformation *v.* time, tunnel wall *92*
deformation phase **81**
Delft 45
Deriner Granodiorite **53**
dense **52**
desertification 3
design of underground structures 83–84, 86–96
    classification systems 86
    monitoring 90, *92*
    numerical modelling 95–96, *96*, *97*
dewatering 68, *72*
diamond mine 25
diaphragm (wall) 66, **85–86**
digital data 118–119, *120*, 122–128
dilatometer (DMT) 106
dimension (model) 117
dimensionality 118
DIN (German standards) 101
DINO databank **127**, 128
discontinuities 50–51, **52**, **55**, 56–58, 62–65, 68, 70, 73, 76,
    78, 88, *89*, *91*, 94–95, 98–99
discontinuous (numerical model) 96
discoloured *see* weathering
disease 99
disintegrated *see* weathering
dissolution mining 27
DMT *see* dilatometer
Dobříš Report 154
Doha marine loose sand **55**
Doomsday Book 134
dormant spores 99
dozing 92
drained **52**
DPSIR [Driving forces Pressures State Impact Responses]
    framework *155*
dredging 66
drill/drilling 63, 64, 102, *105*
    auger 106
    bit *66*
    casing 106
    chisel *107*
    core barrel *107*
    fluid 106
    percussion 106
    rotary 106
    shell 106
    string 106
    tools *70*, 105, *107*
drill rate *66*
DSS *see* decision support system
Dubai 83
duricrust 60
dust control 98, 99
Dutch Base Registration Subsurface 159
Dutch Cadastre 138, 146
Dutch Directive on Soil Protection 149

Eagle Ford Shale **52**
early warning systems 127
earth observation techniques 23
earth materials 3, 5–6, 12, 21, 23–25, 48, 103, 171
Earth Mother 3, *4*
Earth Summit 1, 31
earthquake 97–98
earth pressure balance shield (EPB) *77*
earth processes 1, 2, 14, 21, 42, 178
easement (for access) 137
ecological disaster 31
ecological footprint 41
ecosystem 144, 160–163
effective 23, 44, 50, 53, 84, 97, 149, 151, 170
    cohesion' **52**
    φ' **52**
    stress **52**
Eispalast *see* Jungfrau Glacier
elastic *44*, 49, 53–54
elasto-plastic *see* plastic
electro-magnetic methods 103, 112, *114*
energy materials 27
energy prices 172
energy saving, undergound structures 177
environment and mining 29–31
environment and politics 23, 41
environmental awareness 169, *170*
environmental compartments 23
environmental concern 23, 29, 31, 35, 42
environmental constraints 25
environmental damage 3, 30, 149, 152
Environmental Impact Assessment 149
environmental policy 149, 152
Environmental Protection Agency 165
Environmental Protection Law 149
environmental protection legislation 148–153
environmental protection strategies 157
EPB *see* earth pressure balance shield
equivalent quartz content 65, *66*
erosion features and excavations 61
ESR *see* Excavation Support Ratio
European Environment Agency 154
European Union, legislation and directives 138, 144–155
evaluation of legislation 152–153
excavation damage 89, *91*

excavation in
   man-made material 81, 83
   mining areas 76–78, *79*
   permafrost 74–76
   snow and ice 74–75, *79*
   soft and weak ground 74, *76–78*
   strong and weak ground 78–80
   swelling and sqeezing material 80, **81–82**
   volcanic terrain 80–81, *83*
excavation methods **68**
   economic viability 71–72
   groundmass damage 71, *73*
excavation support 62–72
   selection *63*
   stress *64, 65*
   techniques 65
Excavation Support Ratio (ESR) 88, *89*
excavation support systems 73–83
   cut and cover 72, 83, *83*, **85–86**
   flexible/rigid 72–74, *75*
   unsupported 72, *74, 80, 83*
exogenic processes 177
expert knowledge and risk 97–98, 130
exploration rights 140
extensometer 107
extractability 173
extraction, historic 5–15
extraction methods, future 171

failure 51–52
failure and time effects 63–64, *92*
failure mechanism 62–63
fatalities 31, 52, *94*
Falset **52, 55**, *61*
Fanagalo 97
fault (in ground, rock or soil mass) *see* discontinuity
faulted 50
feldspar 48
fence diagrams 118
fertility rate 169
fibre-optic sensor 93, *92*
field survey 104, *105*
file formats 119
fire underground 99
fissure *see* discontinuity
flat jack 106
flexible support 72–74, *75*
flint, historic mining 12, *13*
flowing failure 63
fluid pressure, permeability and flow 108–109
fluoride pollution 161, *163*
folded 50, 95
foraminifera 49
forensic site investigation 116
fossil fuels 6, 24, 31–32
foundations 20, 40–41
   in ice 75–76

interaction in subsurface 83
stress distribution 64–65, *65*
fracture (in ground, rock or soil mass) *see* discontinuity
freezing 68–70, 74, 99
   and thawing 58, 75, 99
fresh *see* weathering
fungi 62
future developments and trends 169–171
   infrastructure and public space 171
   mining and storage 171
   subsurface management 172–175

Gaia theory 3
gamma-ray 115, *116*
gas bubble 41
gas extraction, law 140
gas field, subsidence *30*
gas flaring 172
gas intrusion 98
gas (methane) hydrate 29, 172
gas production 27–29
gas reservoir 32, 150
gas resources 155, 172
Gas To Liquid 173
gas, underground storage 17
GeoBIM *126*
geo-data subsurface information system **127**
geo-dome 174, **175**, 176
geo-electrical methods 114, *116*
geographical information system (GIS) 120
geohazard 97–98
   and operational risks 95–99
geohazard, legal aspects 151
geology 44
geological constraints 173–176
geological data source 127
geological heritage 2
Geological Strength Index (GSI) 88, *90*
geophysics 4, 103, 110, 123
   downhole logging 115, 116
   electro-magnetic methods 103, 112, 113
   gamma–gamma density 115
   gamma-ray 115, 116
   geo-electrical methods 114, 116
   geophysical techniques 110–116
   GPR, Ground Penetrating Radar 111
   gravity 115
   magnetic 115
   neutron density 115
   resistivity 114
   seismic 110–112, 116
   seismic resonance (TISAR) *112*
   suitability of methods 115
   surface waves (MASW) *112*
   tomography *112*
GeoScience Markup Language (GeoSciML) 122, 129

geosphere 1
geotechnical data source 127–128
geotechnical engineering 44–51
geotechnical model 129–130
geotechnical properties and weathering 58–61
geotechnical site investigation 101–116
    boreholes 105–106, *107*
    caves and archaeological sites 115–116
    field survey 104, *105*
    fluid pressure, permeability and flow 108–109
    geophysics *see* geophysics
    laboratory and simple field tests 107–108, *110*
    maps, special purpose 101–103
    maps, subsurface 103
    pits and trenches 105, *106*, 108
    remote sensing 103–104, *114*
    standards and codes 101
    strength and deformation tests 106–107, *109, 110*
    stress, in-situ test 108
    unit 129
German Civil Code 133
GIS *see* geographical information system
glacier 48, 75, 76, *79*, 104
    and climate change 76
    deposits **55**
    erosion 61
glaciation 177
Glavaniev zone *82*
global warming, permafrost 76
GML 122
gneiss 48, 53, **81**, 95
gold mine 25, *27*
gold, historic mining 11, 14
Gotthard Base Tunnel *67*, 81–82, 173
    squeezing rock mass **81–82**
gouge *87*
GPR, Ground Penetrating Radar 112
granite **47**, 49, 51, **52**, **55**, *61*, *66*, 72, 74
granodiorite 50, **53**, *61*
gravel 48, **52**, *76*, 84, 93
gravity **33, 47**, 63, 87, 94, 113
gravity failure 63
green revolution 170
greenhouse gases 23, 32
grinding **68, 71, 105**
ground conditions 67, 84, 96, 99
ground modification, chemical and biological 177–178
ground penetrating radar survey 112–113
ground properties 43, 49, 95, 174
groundmass and discontinuities 51
groundmass improvement 68–69
groundwater
    mining 6, 8–14
    recharge 6–7, 23, 57, 68, 162
    resource management *74*, **142**, 162
        international 155, 175
    rights to 135–136

groundwater and
    crust development 60–61
    excavation 66
    mechanical property 55, *57*
    shear strength **50–51**, *56*
    subsidence 151
groundwater flow velocity 109–110
groundwater flow, redirected 174, 175
groundwater-related directives 23, 60
grout, grouting 68, 74, 78, 82, **85**, 98, *158*, 177
gunite 80
gypsum 12, 25, 48, 58, 62, 64, 80

hardpan 60–61
Hawkesbury sandstone **52**, **55**
hazardous waste 150, **153**
health and safety provision 157
heat insulation 62, 99
Heathrow Express Tunnel, collapse 93, 95, 97
Helsinki, coordination of subsurface plans **159**
Helsinki, utility tunnels 146
hepatitis 98
historical use of subsurface 3–21
Hoek-Brown failure criterion 88, 94
host rock 32, 51, 173–174
HTML 125
hydraulic/pneumatic hammer **68**
hydrocarbon reserves 171
hydrocarbon resources 29
hydrocarbon underground storage 17
hydrological engineering **8**, 175, 177
hydrological isolation capacity, 173
hydrosphere 1, 46

ice 41, **51–53**, **55**, 74–75, 98–99
    Age *15*
    erosion 61
    deformation **48–51**, **54**
    freezing 68–69, 74, 177
    excavation 61–72, 74–89
    hazard in 48
    strength **50**
    unsupported tunnel *74, 79*
ICMM *see* International Council for Minerals and Metals
igneous rock properties 48–49
illite 48
industrial minerals 6, 140, 171
Industrial Revolution 12
infrastructure 15–17
    data source 125–126
    planning policy 157–159
    public spaces **17**, 31–32, 171–173
    registration 135–137, 146–148, **148**
    spatial planning 163–165
    small 125
InSAR *see* Interferometric Synthetic Aperture Radar
insert *92*, **127**

Institut Cartogràfic de Catalunya (ICC) *102*
intact ground strength 50
intact rock 48
intact rock strength 71, 76
Interferometric Synthetic Aperture Radar (InSAR) 23, 93, 103
International Council for Minerals and Metals (ICMM) 31
International Geophysical Year 3
International Union for the Conservation of Nature 160
International Year of Planet Earth 127
insulation 62
INTI (Argentinean standards) 101
iron, historic mining 10
Islamic law 134
ISO (International Organization for Standardization) 101
isolation, 15–16, 32, 167, 173
isolation capacity Israel, groundwater 144

jack hammer **68**, 72
Japan, Special Measures Act **137**
Japan, transport tunnels 36, *39*
Japan, water law 152
joint (between diaphragm pannels) **85**
joint (in ground, rock or soil mass) *see* discontinuity
Joint Set Number **88**
Jungfrau Glacier *79*

kakiritic **81**
Kallidromo Tunnel *92*
kaolinite 48
karst 55, 58–59, 75, 78, 111
    geophysical detection **111**
Kimberlite pipe 25
Königshain granite **52**, **55**
Kyoto Treaty 150

laboratory and simple field tests 107–109
Lake Isabella *106*
land degradation 170
land registration system 161
landscape and mining 29–30
landscaping and reclamation 170
land subsidence 13–14, 23, *27*, *30*, 59, 152, 157
landslide 40, 104, 112
land-use planning policy 155–158
LandXML 125
Lascaux caves *15*
lava tubes 80
Lavrio silver mines **11**
lead, historic mining 10
leasing of subsurface 137
legal systems of the World *135*
legislation 133–154
    environment 148–152
    natural resources, ownership 138, 140–148
    ownership and spatial planning 135–138, **139–140**, 153–154

life expectancy 169
life support system 2, 41
Light Detection And Ranging (LiDAR) 103, *104*, 126
lignite mine 29, *31,* 155
    open cast 155
    historic mining 6, 115
    reclamation 29–30
limestone 12–13, 16–17, 47–48, **52–53**, 55, 76, **111**, *115*, 173
    construction material 12, 25
    excavation in *11*, **17**
    properties **53**
limits to growth 23
linear-elastic *see* elastic 62
lining 62
Liquefied Natural Gas (LNG), 29
lithification 48
LNG *see* Liquefied Natural Gas 29
LIDAR *see* Light Detection And Ranging
loading 43–44, 53, 55, 68, 75, 107–108
    clay and peat 68
    ice 75
London Clay **55**
loose 46
    material 74
Lötschberg Tunnel 61
Luguon (packer) test 109

Madeira *75*, 80
Madrid Antarctica Treaty 141
Maginot Line 16, *20*
magnetic survey 114
Malta Convention 41, 147
management strategies 154, 160, 162, 164
Manila 80, *83*
man-made material, excavation in 44, 80
map 101
    geological 103
    special purpose 101–103
    topographical 101
marble, strength 48
markup language 125
mass 49
megacity *178*
mercury 14
metals 11–12
metamorphism 46
methane hydrates 172
micro-tunnelling 67
military data source 126
mineral resources 25, 136, 140, **141–144**, 149, 155, 159, 162, 165
    management 159–165
    non-renewable 3, 5–6, 24
        current use 23–28
    ownership 136

policy 154–155
renewable 5–6, *7*, 23
minerals, metallic, historic mining 6, 115, 138
minerals, non-metallic 6, 138
mines, deep 25, *26*
mining 43
   historical review 6, 115
   impact 14, 16–17
   license 23
   longwall 78
   rights 139–140, **139**
   room and pillar **11**, 75
   safety 43
   stoop and retreat 7
mining and reclamation 29–30
mining and storage 171
mining areas, excavation in 76–78, *79*
Mining Rock Mass Rating 87, *89*
mix shield 77
modelling
   2.5D 118
   3D 101, 116, *117*, 118, *118*, 119, **122**, 131
      infrastructure *126*
      programmes 120–121, **122**
   4D 121, 128, 132
   5D 170
   computer aided 120–121
   data format 118–119
   dimensionality *118*
   file format 119
   modelling, mechanical 93, *96*
   scale, detail, resolution, data density 119
   temporal data 119–120
   uncertainty 119, 128–129
   see also numerical
monitoring subsurface structures 91, 93
montmorillonite 48, 62
Montréal, underground city **39–40**
Mu-Cha Tunnel Fault 53, **55**
mudstone 62, *105*
Munich *94*
multiple use of space 135
Mycenae 62, *64*

Napoleonic Code 133
Nathpa Jhakri Hydro Power Project *81*
national database 128
National Geoscience Data Centre 128
National Parks, mining in 140
NATM *see* New Austrian Tunnelling Method 72, 93–95
natural attenuation concept 165
natural gas 6, 14, 17, 27–30, 35, 98, 134–136, 138, 140–141, **143**, 150–152, 155, 172–174
natural hazards 151, 174
natural resources, ownership 138, 140–148
nature reserves 138, 170

NEN (Dutch standards) 101
New Austrian Tunnelling Method 72, 93–94
New Delhi *76*
New York 94, *96*
nickel, allowed values 148
nickel mine 25
nitrate pollution 144
nodule 79
non-renewable resources, 5–6, 24
Norway, transport tunnels 35, *36*
Nubian aquifer 6
nuclear industry and underground storage 17
nuclear waste 23, 29, 32, 34, 64
   damage to model canister *97*
   disposal 150
   storage 160
null measurement 91
numerical modelling, stability 93–94, *95*

OECD countries 25, 29, 32
offshore seismic investigation 111
oil extraction, legal issues 140
oil production 28
oil, historic mining 14
oil reserves 28
oil, underground storage 17
Old Church Delft *45*
OneGeology 125, 127
opal mine, abandoned 16
open cast *see* open pit 13, 25
open pit (open cast) 167, 171–172
operational risks 95–99
Oppenheim, underground city **17**, *18*
ore grades 25
Oude Kerk Delft *45*
Ouvrage Schoenenbourg, Alsace *20*
overbreak *56*, 63
over exploitation 23
ownership and spatial planning 135–138
ozone depletion 154

φ **52–53**
packer test *see* Luguon test
packing **52**
Particle Flow Code 94, *97*
Peace Treaty, groundwater 144
peat 43, *108*
   historic mining 14
penetrometer *110*
penetration test 106
penstock **87**
percussion drilling 105, *107*
permafrost 74–75, 112–113
   excavation in 74–76
   mapping 112
permeability 55, 57, 70, 106, 108
permits, use in subsurface management 161

pH 108
phyllite 81
piezometer 108
piezometric level *56*
pile foundations 20
pillar and stall 7, **11**, 75, 76
pingo 99
pipe installation 66, 68
pipe roof *89*
Pipe Roof Box Culvert Jacking 74, *78*
pits and trenches 105, *106*, 108
plague 98
plastic (deformation) 48, 49, 53–54, 75
plastic (material) 84
plate bearing test 107, *110*
platinum, historic mining 11
PMT *see* pressuremeter
pneumatic hammer *see* hydraulic hammer
Poisson's ratio **55**, 111
polar ice 41, 75
policies for subsurface development 133, 157, 160, 163, 165, 175–177
    groundwater 157
    mineral resources 155, 162
    transport and infrastructure 157
political issues 165, 170–172
polluter-pays principle 154
pollution and groundwater 157
pollution, remote sensing 103
population 169, *170*
    growth 1, 3, 24, 40
pore space, ownership 136
pore water pressure *see* water
porosity 54, 55
Porto 80
post-mining operation 29
precipitation 23, 46, 58, 60
pre-splitting (blasting) **68**
pressuremeter (PMT) 106
private land 134, 137, **139**
profit, sustainable development 160
programmes for 3D modelling **122–124**
property variation 129–130
psychological issues 171, 176, 177, 178
public land 134, **139**
public space underground 172
public works data source 127
pyrite 58, 62, 64, 98
pyroclastic 62
pyrometallurgical processing 34

qanat, water collection **8**, *9*
Q-system 88, *88*
quartz 48, 50, 66
quartzite 48
quartz-sandstone 48
quick clay 57

radar *see* ground penetrating radar survey
radioactive waste 29, 34, 172
radon 98
raise borer **68**
Rammelsberg 73
raster format 118
ravelling 88
recharge area *56*
reclamation 29–30, 171
    legal obligations 133, 146
    mining areas 151, 171
recycling 34, 163, 171
reef 49
registered reserves 24
registration system for subsurface development 157–160
remediation, legal obligations 151–153
remote sensing 103–104, *126*
renewable resources 5, 6, 23, 160
reserves-to-production ratio *29*
residual soil **52–53**, **55**, **59**, 63
resistance phase 81
resolution in digital modelling 119
resources and subsurface development 171, *172*
rib *88*
rigid support 72–74, *75*
Rijswijk *70*, 83
riparian rights 141
ripping 92
risk management **33**, 96, 97
    data limitation 96
    routine 95–96
risks and liability in subsurface development 152–153
road header **68**
rock and soil cycle 48–49
rock bolt *see* bolt
rock burst 25
rock definition 45
rock mass *see* mass
Rock Mass Rating 86, *87*, *89*
rock properties 44–62
    and potential use 172–175
Roman
    construction 62
    law 133, 136, 142
    mine waste 14
    ship *5*
    water supply 6, *10*
rooftop 154
room and pillar *see* pillar and stall
rotary drilling 105, *107*
routine 96

S_u **52**
safety standards 43
salinization 13, 149, 162, 176
salt mine 12
    storage in **32**

salt, historic mining 12, *27*
salt, strength 48, 49
sampler 106
sand 47–48, 55, *58*, 60, 62–63, 76, 83, *85*, 93
    abrasive 48
    deformation 57
    quick 59
    strength **52**, 48
San Diego 80, 93, 97
sandstone 47, 52–53
    abrasive 48
    deformation 50
    strength **52**, 48
Sankt Gotthard *see* Gotthard 67
satellite imagery 104
saw/sawing **68**
scale in geological models 119–120, *121*
schist 48
schistosity 48
    (in ground or rock mass) *see* discontinuity
scouring 177
screw conveyor *see* conveyor 77
Sedrun **81**
seed banks 41
seepage 60, 72–73, *73*, **85**, *86*, 99
seismic survey 110–112
settlement effects *45*, **86**
settlement, human 15–16, **17**, *18–19*
shafts, concealed 76
shale 47, 48, **52–53**, **55**, 62–63, 80, *129*
shale gas 172
Shanghai 23, 39, *78*, 84
Sharia law 134
shear
    discontinuity 49, **50**, **51**, 54, **59**, 86, *104*, 108
    displacement **50**, **54**, 63, 65, 68, 70
    failure **50**
    forces on TBM 80
    strength **50–51**, 51, *56*, 57–58, *58,* 62–63, 65, 70, 76, 86, 108
    stress **50**
sheared
    (flysch) **53**, **55**
    (rock mass) **81**
    (sandstone) **53**, **55**
shearing resistance **50**, **53**
shell 48, 106
shield 68–69, 77
shotcrete 72, 80, 87–88, 93
Sibbe limestone **52**, **55**
silent spring 23
silt 48, **55**, 76, 111
silver, historic mining **11**
simulation models, **33**
sinking cities 23
site investigation *see* geotechnical site investigation
    forensic 115

size
    block **50**, 53, 57–58, 65, 70–71, 86–88
    grain 48
slate **17**, 72, **81**
slaty cleavage 48
Sleipner field, CO_2 storage **33–34**
slickensided *87*
slurry shield 77
smooth wall (blasting) **68**
snow and ice, excavation in 74–75, *79*
snow, excavation hazard 99
soft and weak ground, excavation in 74, *76–78*
soil **52**
    residual *see* residual soil
soil definition 45
soil mechanics 20
soil pollution 29
soil properties 44–62
Soil Protection Law 149
soil remediation 151
soil sampling 106, 107, *110*
soil testing *109*
solution cavern 17
spalling 62
span (excavation/tunnel) 88
spatial planning legislation 138, **139–140**, 153–154
spatial planning policies (Dutch) 144–147, 157
spilling *see* pipe roof
spontaneous combustion 98
spot bolting 88
spring *57*
SPT *see* Standard Penetration Test
squeezing 62
squeezing failure 62, 80, **81–82**
stabilizing techniques 68–69
Standard Penetration Test (SPT) 106
standardization of data 122
standards and codes 101
stand-up time 63–64, 74
steel rib *see* rib
storage underground 23, 31–34
    future use 171–172
    heat and cold 150, 157, 159
    historic use 15, 16–17
storage, rock properties 173
Storebœlt Project 79
strength **52–53**, 48–50, 52–53, 57, 58
    and deformation tests 106–107, *109, 110*
stress and excavations 62, *64,* 65
stress
    distribution under foundations 64, *65*
    field *44*, 64
    loading *44*
    radial *64*
    tangential *64*
    test, in-situ 108

stress (*Continued*)
    time effects 53–54
    virgin field *64*
strong and weak ground, excavation in 79–80
submarine slides 172
submarine tunnels 173
subsidence
    aquifer 23
    gas field *30*
    groundwater extraction 146, 152, 153
    salt mining 25, *27*
subsidiarity principle 148
subsurface
    archives 77, 115, 127
    biota 175
    construction legislation 146–148, **148–149, 153–154**
    definitions 1
    development policy 162–165
    Dutch authorities and regulations **149, 151**
    management 172–173
    planning **159**
    role in human society 1–2, 23
subsurface/underground space 4, **16**, 39, 70, 76–77, **137**, **159**
subway development 173
sulphate, strength 49, 50
sulphide 58, 80–81
support structures 63, *65*
surface engineering 43
sustainable development 1–2
    management 169–178
    strategies 160–163
swelling 48, 58, 61–63, 68, 80
    and sqeezing material in excavation 80, **81–82**
Sydney-Gunnedah Basin coal **52**
system Earth 1

tailings 8, 14, 29–31, 70, 138
tar sands 27, 29, 172
TauTona gold mine 25
Tavetsch Intermediate Massif 81–82
taxation and groundwater management 163
TBM *see* tunnel boring machine
temperature 108
tensile strength 48, **50**, **52–53**, 54
tension piles 83
test
    laboratory 107
    field 107
Thames Gateway, confidence data *131*
thermal isolation capacity 172
tilt *45*
timber (support) 62
time and geological processes 131–132
time effects and
    deformation *92*

    and failure 63–64
    and stress field 53–54
time records 41
tin, historic mining 9
tomography *112, 115*
topographic data source 126–127
Torrens System, title registration 161
toxic waste 31
transparent decision making 161, 175
transport and infrastructure 156–159
trench cutters 66, *69*
triaxial **50–51**
troglodite dwellings 16, *19*
TS *see* tensile strength
tunnel
    boring machine *67*, 68, *71*, *77*, *80*, 177
    break through *99*
    collapse 61, *94*
    ownership 136–137
    safety directives 146
    support requirement *88*
tunnel stability 56
    numerical model *95*
tunnels, transport 20–21, *24*, 35–37
tunnelling boom 170
Tweede Heinenoord Tunnel *117*

uncertainty 128–130
unconfined compressive strength 50, **53**, **60**
unconfined *see* confined (water)
UCS *see* unconfined compressive strength
underbreak 104
underground
    cities 16–18, 37–38
    engineering 43
    infrastructure 17, 21, 23, 34, 66, *131*, 137, 173
    storage 14, 15, 17, 23, **33–34**, 172–173
undrained 44
Union Canal Tunnel 72
UN Millennium Development Goals 154
uranium production 29
uranium, historic mining 12, **12**
urban development underground 175–178
    artist's impression *178*
    see also city, underground
urban planning 175
urbanization 1, 3, 40–41
urbanization trends 169, *170*
urban sprawl 27, 40, 137, 153, 175
USA Bureau of Land Management 138
USA Environmental Impact Assessment 149
USA Environmental Protection Agency 165
USA Federal Land Policy and Management Act 138
USA legal cost of remediation 152
USA mining law 138, 140
USA National Historic Preservation Act 147
USA Safe Drinking Water Act 145

USGS Organic Act 162
utility tunnels, multi-purpose 146

Valletta Treaty 41, 147, 154
vane test *107*
vector format 118–119
Vinalmont limestone **52**, **55**
Vindhyan sandstone 52
virus and disease 98
visualization *117*, 131–132
volcanic eruption, forecasting 98
volcanic rocks, excavation in 80–82, *83*
voxel 118

waste disposal and storage, legislation 148–149
waste disposal, geological constraints 173
waste dumps, excavations in 82, 98
waste repositories *51*
water *see also* groundwater
    collection channels (qanat) **7**, *8*
    cycle 5, 6, 23
    hazard in excavation *99*
    management **8**, 23, 29, 162, 164, 172, 175
    policies 154
    polluted/pollution 70
    pore pressure *57*
    pressure 69
    purification 10
    resource 5–6, *7*
    table *55*
    testing 101, 112
    wear *61*
weather (*see also* climate) 97
weathering 44, 46–48, 58, *59*, 63, 66, 84, 87, 104
    classification *59*, **85**, 86–88
    and failure 63
    and geotechnical properties 52, 58–59
    and swelling 62
well point dewatering *72*
wetland vulnerability 162
World Finance Centre (Shanghai) 84
World Heritage Areas, mining 140
World Summit 31
world urbanization prospects 169
worldwide legislation, subsurface issues 148, 164, *166*
worship and ritual 3, *4*

XML 125
XMML 125

Yemen, groundwater depletion 162
yielding *81*
yielding steel arch **81**
Young's modulus **54**
Yucca Mountain *97*
Yucca Mountain-Topopah Spring Tuff, 52

zinc mine, reclamation 30, *32*